Order, Chaos, Order

Order, Chaos, Order

Order

*The Transition from Classical
to Quantum Physics*

PHILIP STEHLE

Department of Physics and Astronomy
University of Pittsburgh

New York   Oxford
OXFORD UNIVERSITY PRESS
1994

Oxford University Press

Oxford   New York   Toronto
Delhi   Bombay   Calcutta   Madras   Karachi
Kuala Lumpur   Singapore   Hong Kong   Tokyo
Nairobi   Dar es Salaam   Cape Town
Melbourne   Auckland   Madrid

and associated companies in
Berlin   Ibadan

Published by Oxford University Press, Inc.,
200 Madison Avenue, New York, New York 10016

Oxford is a registered trademark of Oxford University Press

Library of Congress Cataloging-in-Publication Data
Stehle, Philip.
Order, chaos, order: the transition from classical to
quantum physics / Philip Stehle.
p. cm. Includes bibliographical references and index.
ISBN 0-19-507513-7
ISBN 0-19-508473-X (pbk)
1. Physics — History.
2. Quantum theory — History.
I. Title.
QC7.S78 1994
530.1'2'09 — dc20   92-41506

2 4 6 8 9 7 5 3 1
Printed in the United States of America
on acid-free paper

Contents

Introduction

Since the time the Greeks dominated intellectual activity, including science, there have been two "revolutions" in the physical sciences that I would classify as major. One of these occurred in the seventeenth century, the age of Galileo and Newton, when it became possible to regard the universe as a machine operating according to the laws of mechanics. The second started just before the beginning of the twentieth century, in the age of J. J. Thomson, Max Planck, Albert Einstein, Ernest Rutherford, and Niels Bohr, when the classical model of the mechanical universe became untenable and began to be modified by a patchwork of rules involving the energy quanta introduced by Planck in 1900. There were comparable upheavals in the life sciences when Charles Darwin announced his theory of evolution and when molecular biology took off with the discovery of the double helix by Francis Crick, James Watson, and others.

Revolutions are rare and complicated things, and there can be no definitive account of one that makes all others redundant. A revolution is seen from the context provided by the present, an ever-changing context quite different from that in which the revolution took place. Understanding the second revolution in the physical sciences requires reconstruction of the situation in which it started, and this reconstruction itself depends on both the present and the past. We must not project our present knowledge back onto the past so as to make the development that eventually emerged seem to have been inevitable. We must, however, be guided by our present knowledge in connecting simultaneous developments that seemed unconnected at the time. To the people involved in the process, it was anything but clear that a radical change in principles was required. Later, when that need became more credible, it was not known what the new principles should be. There was a solidly founded basis on which classical physics had been built, and those parts of the classical edifice that had been confirmed had to be retained. The new principles would have to yield the same consequences as the old ones did when applied to the old problems and yet yield new and confirmed results where the classical ones failed. The starting point had to be what was then known, but how to proceed from there had to be found by trial and error based on experimental findings, on those parts of the classical theories that seemed untouched by the flood of new experimental results, and on inspired guesses by those faced with intractable problems. Both the classical theories and these exploratory efforts toward a new and more widely valid theory are the subject of this book.

My interest lies more in the evolution and interplay of ideas than in the personalities of the participants. The changes during this period of development in physics led to a new set of principles lying at the base of our conception of nature. These new principles were much more abstract than the old ones. According to the old principles the quantities that could be measured in an experiment were represented in the theory by numerical quantities, and the connection between theory and experiment was direct. The means of making an observation of a system did not enter into the theoretical description of that system in any essential way, and there was no inevitable disturbance of a system by the observing apparatus. According to the new principles the means of making a measurement have to be included in the theory predicting the results of an experiment. This need arose because of the discovery that there are irreducible effects on a system produced by making observations. Such a fundamental change in the relations between theory and experiment did not come easily, and how a reluctant community was driven to accept it is one of our main concerns.

Many aspects of nature were well described by the classical picture of particles moving according to Newton's laws of motion under the influence of forces such as those arising from gravity, stretched springs, or the presence of an electromagnetic field. The electromagnetic field itself was perfectly described by Maxwell's equations. Newton and Maxwell had provided a framework into which specific theories, such as those of the solar system and radio communication, fitted extraordinarily well. In the few places that difficulties appeared they were ascribed to ignorance of exactly how the force between two particles should be described or exactly how radiation interacts with matter, rather than to fundamental deficiencies in the framework. No one doubted that the position and velocity of a particle could be observed with a degree of accuracy limited only by the ingenuity of the observer or that electromagnetic waves traveled through space as described by Maxwell's equations. This framework of ideas was a very comfortable one. Within it there was no lack of challenges for scientists; there was always the possibility of making more precise measurements of quantities, such as the charge on the electron or the speed of light. More interesting, there was always the discovery of what this set of principles implied in the new situations that were provided by new materials and new experimental techniques. It was universally expected that the results of experiments probing these new situations would fit into the prevailing mold.

There are many ways that theory and experiment can disagree. The theoretical treatment of a system must idealize that system to a certain extent in order to make the theory at all tractable so that the quantities of interest can actually be evaluated. The experiment must be designed so that all influences that affect the result are controlled and no extraneous influences are allowed to act. Instruments must be well calibrated and must measure what is meant to be measured. When theory and experiment disagree, it is not clear where to place the blame. The theory may have been incorrectly applied; some vital factor may have been overlooked or underestimated; or an outright error may have been made in the calculation. The experiment may have allowed some unexpected effect to influence the result; some piece of apparatus may have malfunctioned; the experimentalist may not have interpreted the results correctly. Did the theorist err or leave something out? Did the experimentalist allow something unknown or unrecognized to influence the system? Or is there a fundamental problem with the conceptual framework within which the experiment was supposed to be described? Experimentalists prefer the first explanation, theorists the second, and generally no serious scientist prefers the third until driven to believe that the other two explanations have been excluded.

When theory and experiment do disagree and both the theory and the experiment seem to be reliable, it still does not follow that the conceptual framework is in danger. There may be relatively minor changes in the theory or in the conduct of the experiment that will bring them into agreement. When there are disagreements in a variety of cases, then if one is able to see some relation among these cases, one may begin to sense a fundamental problem. Starting in 1895 in physics, difficulties were arising in such varied fields as the specific heats of gases and of crystals, the ionization of gases by X-rays, the spectrum of blackbody radiation, the radioactive decay of certain elements, and the emission of spectral lines by substances raised to high temperatures. Who was in a position to recognize these discrepancies as coming from a single source?

Things do not always fit into such a simple pattern as that just outlined. Sometimes there are alternative theories that are equally successful in describing what is known. Then the choice among them can be made on aesthetic grounds until experimental results force a decision. Simplicity is often the criterion used to prefer one theory to another. Einstein's special theory of relativity at first did not predict phenomena that were not already predicted by the theory of Maxwell and Lorentz. What it did was to make explicit a profound symmetry that was hidden in the older theory. The expression of this symmetry required space and time to be related in a way different from that taken for granted since antiquity and stated explicitly by Newton and Kant.

There were effects that had to be analyzed in distinct terms in the old theory that could be analyzed in a unified way in the new one. What had previously appeared to be pure coincidence now emerged as a necessary consequence from the new viewpoint. This led to greatly simplified ways of attacking some problems. Only after the application of relativity to the mechanics of particles did such world-shaking ideas as that embodied in the equation $E = mc^2$ emerge. Even before this happened, however, the beautiful symmetry of the theory of special relativity persuaded many of its superiority to the old picture while leaving many others completely unconvinced of its bearing on the real world.

To study the course of any intellectual revolution, we need to know not only what was done that later would be regarded as right but also much of what was attempted that turned out to be irrelevant or even wrong. Seldom do new frameworks emerge whole and perfect in one step. Although this has happened, the theory of special relativity is the only example that comes to my mind. In the case of quantum physics there was a long trail of incomplete and occasionally erroneous attempts at understanding. The persistence and controlled originality of those who worked on these difficult problems, their ability to accept and profit from criticisms of their efforts, and the beauty of the edifice that emerged are wonderful to behold.

The indispensable source of information about these labors is the scientific literature in which the original accounts of the work are given. This is a difficult source, being huge in extent, hard to follow for the earlier years, and largely free of clues to the importance of the work being reported. Easier sources are those biographies of the principal actors that include some description of the subject's scientific work, detailed scientific histories of experiments and of theories, lectures given by the participants, and broad-scale reviews written for a variety of audiences. These secondary sources necessarily involve selection and evaluation of material by their authors and so cannot be used as sole sources, but they are valuable guides and make the work of later writers much easier.

This book has developed out of my own fascination with the subject. I organized a course in the history of physics designed for undergraduate students with some knowledge of physics. The preparation to teach it made me go back to the original papers of those people whose names are associated with spectacular ideas—Maxwell, Boltzmann, Einstein, Planck, Bohr—and then to those of many others whose names are less widely known. Reading them has been a relevation. What we do not appreciate from our present situation is the state of knowledge and scientific opinion at the time the work was done. We know the victories, but we do not necessarily know the battles in which they were won. We do not know what the opposition was to ideas that turned out to be successful and on what it was based. We tend not to recognize the great successes of theories that had to be abandoned in the light of later information but that were valuable syntheses of what was then known. The textbooks do not dwell on the false trails and false starts along what turned out to be the road to the future. In short, by exploring the recent past I acquired a new appreciation for science as a whole, including what is being done today.

People who do research have few illusions about what doing scientific work entails. Most of their time is spent applying well-established ideas to well-defined problems to see what should be expected in particular cases. This is the way science is used. The particular cases may be ones with useful applications to technology; they may be ones for which the validity of the theory is uncertain and has to be tested; they may be purely exploratory, to see whether interesting and unexpected consequences arise. The application of current science to new problems is not always a routine matter. All real systems are too complicated to be treated exactly, so there has to be an estimation of what is important and what can safely be simplified. The work is done by humans, and humans have lapses. In breaking new ground it is hard always to be right about everything. If the work is to be published, a serious author will be as careful as possible that what is said is correct, but he or she will not always succeed. Boltzmann and Planck, to name only two great scientists, both published major papers that turned out to be wrong. A measure of their greatness is their reaction to the criticism they received; they improved their work until it turned into the new basis of the physical sciences, statistical mechanics and quantum mechanics.

My purpose in writing this book is to describe the sequence of developments that constituted the second revolution in the physical sciences in a way that will convey the excitement and the suspense they produced in the scientific community to an audience much larger than this community. In the course of this revolution people took steps whose implications could not be imagined when they were taken. The perception gradually spread that the old, well-established, and cherished view of the nature of the physical world had flaws that might well prove fatal. There was, however, no foreknowledge of what new view was to replace the old one, and the path from the old to the new was unmarked and booby-trapped. At the end of the revolution, four constants dominated physics. They were e, the electron charge; h, the quantum of action known as Planck's constant; k, the gas constant per molecule known as Boltzmann's constant; and c, the speed of light. At the beginning of our story the electron was unknown; the concept of a quantum of action did not exist; the gas constant per molecule was a minor concept that did not even get a symbol until 1900; and the speed of light, while known and regarded as amazingly large, was not regarded as anything especially important or profound. I shall point out that the steps taken in exploring the path from the old to the new, some leading forward and some leading astray, some being large and some small, but all being important to going from the start to the finish. My main sources are the published works of people working at research in physics, supplemented in important ways by biographies and histories of various special topics. I have not tried to explore the psychology of these people; that is beyond my knowledge and my serious interest. I have not tried to unearth a universal pattern of evolution of such revolutions. This revolution provides a good story, and it may or may not have a moral.

Writing about a subject as technical and abstract as physics for an audience not familiar with the jargon and mathematical language of the professionals presents a challenge. Jargon saves time and avoids confusion among specialists. When only a limited field of ideas is under discussion a word or phrase can carry a greater load of contextual baggage than when the discussion is more general. I have tried to keep the jargon to a minimum by supplying the necessary context. The use of mathematics in physics is more central; it is not just an economy but is often indispensable in communicating general principles clearly and unambiguously among working scientists. A principle is a general one only if it applies to many situations, and it must therefore be stated in general or abstract terms. Mathematics is a natural language for making such statements, but it does present a barrier to those who do not use it regularly. I have tried to tell the story using only basic mathematics explicitly and even sparingly. In many cases, expressing ideas in words without using mathematics has forced me to think more clearly about them. I have provided "amplifications" containing some explicit mathematics and other more technical matters when they might contribute substantially to the appreciation of the story by those able and willing to examine them. I have tried, however, to make the story comprehensible to those to whom the amplifications are not interesting.

Order, Chaos, Order

CHAPTER 1

The Old Order:
Classical Physics

Natural philosophy, or physics, had reached an advanced and largely satisfactory status by the last decade of the nineteenth century. Classical mechanics was triumphant, especially in the realm of celestial mechanics in which the behavior of the solar system seemed to be completely accounted for by Newtonian gravitation. The few discrepancies were so small, though definite, that they could not be imagined to require fundamental modification of the philosophical basis of physics for their resolution. Mechanism was accepted by the great majority as the ultimate explanation for everything.

Electricity and magnetism was almost as successful as mechanics. It had provided a physical basis for optics and the beginnings of electrical engineering. The prevalent opinion was that eventually it too would become a branch of mechanics as soon as a satisfactory model of the ether was achieved. Fluid mechanics was a branch of mechanics based on Newton's laws of motion, even though the description of the motion of a continuous fluid required mathematics different from that for the description of the motion of discrete systems, as in particle mechanics, and most problems turned out to be mathematically intractable.

Heat and thermodynamics were subjects apart. They gave rise to a comprehensive and sophisticated theory based on the general principles of the conservation of energy and the nondecrease of entropy. The theory was apparently independent of particular models of whatever it was used to characterize and therefore appealed to some as a good candidate for the "theory of everything." The attempt to make thermodynamics a branch of mechanics through the atomic–molecular model of matter had more success than did the corresponding attempt in electricity and magnetism but seemed to generate less widespread enthusiasm. To many, atoms were less desirable than the ether.

It was into this challenging but comfortable world that a stream of new results was about to flow, a stream that cut new channels as it went. In this chapter I shall describe the landscape that was so dramatically altered by this stream.

From Galileo and Newton: Classical Mechanics

From the time of Copernicus, Galileo, Kepler, and Newton to the very recent past, the goal of physicists was to describe all natural phenomena in terms of the motion of material bodies. There was, however, no unanimity on what constituted a material body. Some scientists and philosophers followed René Descartes in conceiving the universe as a *plenum* completely filled with a continuous, infinitely divisible form of matter. Others were of the atomistic school and believed that matter was composed of discrete *atoms*. An atom was conceived of as the smallest unit of a material exhibiting the characteristics of that material. Atoms could not be divided, at least not without destroying their identity as units of that kind of matter. An atom of a particular material had a mass characteristic of that material and could be assigned a definite position in space. Much later, in the nineteenth century, the idea of a *field* was introduced by Michael Faraday in the form of the electric and magnetic fields. A field has a value at each point of space and so forms a kind of plenum. Atoms and fields were considered to be distinct entities, described in distinct ways and having little in common with each other. In modern times this dichotomy has become blurred. As we shall see later in this book, these alternatives no longer appear to be mutually exclusive; they are present simultaneously in the quantum picture of the world. It occasionally happens that a distinction, sharp when first made, simply loses its meaning in the light of later developments. Atoms versus plenum is a distinction that has met this fate.

Galileo started the modern chain of development in the study of motion, though this is not to say that motion had not been studied earlier. Names such as Democritus, Zeno, Aristotle, Archimedes, Ptolemy, and many others are associated with the study of various aspects of matter and how it behaves. Aristotle, whose physics dominated the scientific world during the first part of the Renaissance, did not contemplate truly universal features of motion; rather, he ascribed different natural motions to terrestrial and celestial bodies. He supposed the stars and planets to be carried about a fixed earth on rotating celestial spheres. The celestial sphere carrying the moon around the earth divided the outer, celestial realm of the stars and planets from the inner, terrestrial one. The bodies of the sublunar world supposedly tended to move along straight lines, and those in the celestial realm were assumed by their nature to move eternally in perfect circles. This distinction between the two realms persisted into the seventeenth century. What then emerged was the possible existence of universal laws that apply to the motion of any material body without distinction between celestial bodies and terrestrial ones.

Before Galileo the feature of motion that was felt to need explanation was velocity, the rate at which a body changes its position. This was a feeling based on the familiar fact that we see bodies in motion, at least those moving horizontally, slow down and eventually stop. Falling bodies do speed up for a while, but they too eventually come to rest when they land on what is below them . It was taken for granted that heavier bodies fall faster than lighter ones. Galileo took a giant step when he identified acceleration, the rate at which the velocity of a body changes with time, as the quantity characterizing the motion of a body under the influence of gravity.[1] He did experiments on spheres rolling down inclined planes that sufficiently diluted the influence of gravity for him to time their motions with the crude timing devices available, such as water clocks or his finger on his pulse. Galileo thus established that the acceleration of gravity at the surface of the earth is the same for all compact bodies no matter what their size or location or velocity may be. Very light and extended bodies such as feathers and leaves obviously did not follow this rule. His famous, though probably apocryphal, experiment of dropping masses from the Leaning Tower of Pisa has been cited for centuries as the crucial demonstration of this fact. He further showed that the distance covered by a falling body is proportional to the square of the elapsed time, not to the time itself. In double the time, the distance fallen from rest is quadrupled. Galileo did not extend this idea beyond motion under the gravitational attraction of the earth.

It was Isaac Newton who made the grand generalization of Galileo's discovery.[2] He stated that at all places and at all times the surroundings of a body specify that body's acceleration, its rate of change of velocity, but not the velocity itself. Acceleration is much harder to identify than is velocity when simply observing the motion of a body. It can involve a speeding up or a slowing down and also a change in the direction of the motion. The most obvious acceleration is that of a body dropped from rest, the one studied by Galileo. According to Newton the acceleration of a body can depend on many different external influences such as gravity, friction, or the act of being thrown. Note that all these require the presence of bodies other than the one whose motion is being described: Gravity requires the earth; friction requires the presence of something to rub against; and throwing requires the arm and hand of the thrower. This necessity for the presence of something to produce the acceleration of a body led Newton to relate the behavior of bodies that exert forces on other bodies to the behavior of the bodies on which those forces are exerted. In this way he showed how to describe the mutual action of bodies on each other.

One special force that Newton proposed was that of universal gravitation, an attractive force that every body in the universe exerts on every other one. This force depends on the separation of the gravitating bodies, decreasing inversely as the square of their separation. The force studied by Galileo that causes the falling of bodies near the surface of the earth is Newton's gravity between the earth and the falling body; the force that causes the moon to revolve around the earth is Newton's gravity between the moon and the earth; and the force that causes the planets to revolve around the sun is Newton's gravity between the planets and the sun. His success in describing all of these phenomena with a single law of force abolished the distinction between the celestial realm and the terrestrial realm.

Newton's laws of motion apply to a body whatever the nature of the force on that body is. According to them it is force on the body, not the body's velocity, that is determined by the circumstances in which the body finds itself. But this was far from self-evident to Newton's contemporaries.

In 1895 classical mechanics retained all the features that Newton had given to it, although it had undergone an extensive mathematical development in the intervening two centuries. At its base were Newton's laws of motion. We state them here because they furnished the conceptual framework of nearly all of physics at that time and of much of physics even now.[3]

Newton's first law of motion states:

> Every body continues in its state of rest, or of uniform motion in a right line, unless it is compelled to change that state by forces impressed on it.

This statement, anticipated in part by Galileo, was a revolutionary one directly contradicting the Aristotelian idea that continuing motion requires a continuing cause.[4]

Motion involves the change in a body's position with time. Newton was explicit in stating what he meant by the position of a body and by time. He conceived of a unique reference frame or space in which the body's location was to be given, calling it *absolute space*. Exactly how he thought of this space as being defined is unclear. In his *Opticks* he had introduced an "æther," a continuous medium that could support vibrations that traveled with a speed close to that of light.[5] These vibrations were able to affect the behavior of the corpuscles that he considered to constitute light and that enabled him to account for certain optical phenomena. It is possible that he thought of absolute space as that in which this ether was at rest. His definition of it was as follows:

> Absolute space, in its own nature, without relation to anything external, remains always similar and immovable. Relative space is some movable dimension or measure of the absolute space; which our senses determine by its position to bodies; and which is commonly taken for immovable space; such is the dimension of a subterraneous, an aerial, or celestial space, determined by its position in respect to the earth. Absolute and relative space are the same in figure and magnitude; but they do not remain always numerically the same. For if the earth, for instance, moves, a space of our air, which relatively and in respect of the earth remains always the same, will at one time be one part of the absolute space into which the air passes; at

another time it will be another part of the same, and so, absolutely understood, it will be continually changed.

Newton also explicitly defined what time meant to him:

Absolute, true, and mathematical time, of itself, and from its own nature, flows equably without relation to anything external, and by another name is called duration.

It was with respect to this absolute space and absolute time that Newton described the motion of a body.

The carrying along of "subterraneous spaces" by an earth moving through absolute space is easy to picture; what is difficult is to pick out the absolute space in which this motion of the earth is to be described. Is it the space in which the sun is at rest, or the stars, or something else? Any space in which a body not acted on by other bodies moves with constant velocity would seem to be a likely candidate. But the matter remained obscure, and eventually it was accepted that there was no need of absolute space in mechanics. The principle of *Galilean* relativity is that any one of Newton's "relative spaces" to which his first law of motion applies can play the role of absolute space in the science of mechanics. Time was the same in all relative spaces. (Later, starting in the early years of the nineteenth century, the wave theory of light seemed to distinguish between an absolute space, the ether in which light waves propagate, and the relative spaces moving through this ether, observable by measurements on the speed of light. This remained a conceptual possibility until the introducton of *Einsteinian* relativity in which time, too, was relativized.) The form of the laws of mechanics is the same in all relative or inertial spaces. These got the name of inertial spaces or inertial reference frames because inertia connotes resistance to change, and a free body in an inertial space does not change its velocity. Its inertia keeps it moving uniformly, as seen in any one of these spaces.

One can interpret Newton's first law of motion in more modern terms to mean that inertial reference frames exist. In any such frame the velocity of any free body, that is, a body not subject to impressed forces, is constant, and this is the content of the first law of motion.

Newton's second law of motion states:

The change of motion is proportional to the motive force impressed; and is made in the direction of the right line in which that force is impressed.

The environment is the source of the force acting on a body, a force that does not arise from the body itself but from the outside. The effect of the environment is to change a body's velocity; it does not require it to have a particular velocity. A body can be at any position and have any velocity at a given instant. The *law* of force governing its future motion starting from these initial conditions does not depend on these conditions, but the *value* of the force generally does depend on the position and sometimes on the velocity of the body acted on. Once this position and velocity are determined, the environment determines the force on the body and hence its acceleration, so one is not free to specify the initial values of the position, the velocity, *and* the acceleration. The specification of the initial position alone or the initial velocity alone is not sufficient to determine the future motion.

It is useful to formulate this explicitly in a simple case. If a particle of mass *m* moves along a straight line, its position can be specified by giving its distance, *x*, from some fixed point on the line. Its velocity *v* is the ratio of the change in its position, denoted in the notation introduced by Newton's contemporary Gottfried Wilhelm Leibnitz, by the *differential dx* of *x* to the length of time *dt* it takes this change to occur. Similarly, its acceleration *a* is the ratio of the change in its velocity, *dv*, to the length of time *dt* it takes this change to occur. In mathematical form, then $v = dx/dt$ and $a = dv/dt$. If a force *F* is applied to this particle, Newton's second law of motion written mathematically is

$$F = m\frac{dv}{dt} = m\frac{d\left(\dfrac{dx}{dt}\right)}{dt} = ma.$$

Written in this way the law has the form of a *second-order differential equation* for the position as a function of the time. It is differential because it involves differentials of the position and of the time. It is of second order because *a* involves the differential of the differential of the position. One particular motion out of the set of all motions possible under the influence of this force is selected by specifying the particle's position *x* and velocity *dx/dt* at some one time, the initial time. The differential equation then uniquely specifies the motion at other times.

Newton's third law of motion observes:

> To every action there is always opposed an equal reaction: or, the mutual actions of two bodies upon each other are always equal, and directed to contrary parts.

Here the word *action* means force. The feature of the environment of a body that produces a force on it is the presence of other bodies. The third law states that the body influenced by this environment also influences the bodies that give rise to this influence. Defining the momentum of a particle as the product of its mass and its velocity, *mv*, these mutual interactions are such that the change in the momentum of a body due to the influence of other bodies is exactly opposite to the change in momentum of the bodies exerting that influence. In this way the total momentum of an isolated system of particles exerting forces on one another does not change. The change in the momentum of one particle is always compensated by the changes in the momenta of the particles causing the momentum of the first one to change. Stated in this way, it is the law of conservation of momentum. Applying it to an isolated system of two bodies, one can determine the ratio of the masses of the two bodies by letting them interact with each other and measuring their accelerations;

their masses are in inverse proportion to their accelerations. A particular body can be chosen as a standard, and the masses of other bodies can then be expressed as multiples of this standard. This is more than just a definition of the mass ratio because it has consequences that can be checked experimentally. If I have bodies A, B, and C, I can determine the mass ratios m_A/m_C and m_B/m_C by letting these two pairs of bodies interact with each other. I can then determine the mass ratio m_A/m_B either by letting them interact with each other and measuring their accelerations or by simply calculating the ratio $(m_A/m_C)/(m_B/m_C) = m_A/m_B$. If these two procedures gave different results, mass as defined by Newton's third law would be an unimportant and uninteresting quantity. As long as the velocities occurring are much less than that of light, these two procedures do give the same answer.

According to Newton, these laws were to apply everywhere, to the motion of objects on the surface of the earth, to that of the planets around the sun, and to atoms, which Newton suggested to be "solid, massy, hard, impenetrable, moveable particles" in Query 31 of his *Opticks*. These sweeping statements provided a general framework in which the motion of any material system was expected to fit. It was the generality of this framework that permitted his successors to consider the universe as a machine. Newtonian mechanics leads to a deterministic picture of the world. If the initial conditions are known, then so is the motion at all other times, past or future. This was stated dramatically by Pierre Simon Laplace:

> Given for one instant an intelligence which could comprehend all the forces by which nature is animated and the respective positions of the beings which compose it, if moreover this intelligence were vast enough to submit these data to analysis, it would embrace in the same formula both the movements of the largest bodies in the universe and those of the lightest atom; to it nothing would be uncertain, and the future as the past would be present to its eyes.[6]

This view persisted until the introduction of quantum mechanics, when it required some modifications. Even in the classical case, however, doubts of the relevance of this extreme form of determinism have recently arisen in connection with the study of *chaotic* systems. Simple classical systems are now known for which the system's future motion depends so sensitively on the initial conditions that it can be impossible to predict that future motion even qualitatively for large but finite times from initial conditions determined with only finite precision. These chaotic systems do not fit into Laplace's picture. Such reservations about determinism are summed up in the "Butterfly effect": A butterfly stirs its wings in Beijing today and changes the weather in New York next month.[7]

Newtonian mechanics has had a longer history of unbroken elaboration of a few basic ideas than has any other branch of science. Some of these elaborations were primarily technical in nature, making the theory easier to apply to complicated systems. Others introduced new concepts that led to new ways of thinking about the subject. The biggest technical advance was made by Joseph Louis Lagrange.[8] The direct writing down of Newton's second law of motion in equation form as was done earlier in this section is simple only if the position of the particle is referred to a set of straight-line coordinate axes at right angles to each other, a Cartesian coordinate system. The reason is that the motion from which the action of a force causes deviation is uniform straight-line motion. Such coordinates are not always convenient. Take, for example, a particle acted on by a force always directed toward or away from a fixed point and depending only on the distance from that point, a central force such as the force of the earth on the moon. If Cartesian coordinates x, y, z are used, the component of the force along any one axis is generally a complicated function of all three coordinates. Because there are three axes, this makes Newton's second law have the form of three simultaneous differential equations, and so solving them usually poses mathematical difficulties. It is advantageous to introduce the distance r of the particle from the force center as one coordinate, as the strength of the force depends only on it. The other two coordinates needed can be chosen as similar to the latitude and longitude of a point on the surface of the earth. When these spherical coordinates r, θ, ϕ are used with the force center as the origin, only the radial component of the force is nonzero, but the expressions for the components of acceleration are complicated. Motion with constant r, for example, is along a path on the surface of a sphere and is therefore not along a straight line. Uniform motion along a straight line not passing through the force center is not motion with a constant radial velocity. Lagrange found a simple way to write the acceleration associated with coordinates other than Cartesian ones, and this permitted him to write equations of motion in a coordinate system in which the forces are simple, such as spherical coordinates with their origin at the center of force for a central force system. These equations, like Newton's original ones, are differential equations of the second order, known as Lagrange's equations of motion. In order to pick one particular possible motion out of all the possible ones, we must specify the values of the coordinates and their rate of change at some initial time, just as in the case of Newton's equations in their original form.

A more basic development in mechanics was the introduction of the concept of *mechanical energy*. The term *vis viva* had been used for a long time in mechanics with a variety of meanings, all associated with some aspect of motion, but eventually it came to mean *kinetic energy*. The kinetic energy, E_{kin}, of a particle is defined to be one half the product of the particle's mass and the square of its velocity;

$$E_{kin} = \tfrac{1}{2}mv^2.$$

This quantity is intimately related to the quantity *work*. In mechanics the work done by a force on a body is defined to be the product of the magnitude of the force and the distance in the direction of the force through which it acts. A vertical force such as gravity does positive work on a body only to the extent that the body moves down, in the direction of this force. If the body moves up, the work done by gravity is negative, and if it moves horizontally, this work is zero. It is an immediate consequence of Newton's second law of motion that the work done by the total force acting on a particle is equal to the increase in the kinetic energy of that particle. Thus a particle moving up while acted on by gravity loses both kinetic energy and speed, whereas one moving down under this influence gains both kinetic energy and speed.

In his *Celestial Mechanics* of 1799 Laplace developed the idea of potential energy, the energy of position.[9] He showed that to move a particle (planet) from one place to another while it is acted on by the gravitational forces of other bodies (the sun and other planets) without changing its kinetic energy in the process, the amount of work required to be done by the mover depends only on where the particle is in the beginning and where it is at the end, and not on the path it follows from one place to the other. He called this amount of work the change in potential energy of the particle in the presence of the other gravitating bodies. It is denoted by ΔV, the Δ indicating change in the following quantity. Not all forces share this property, however. For example, the work done by friction on a body depends on the length of the path from the starting point to the finish, and not just on those places. If we move a body once around a circle so that it returns to its starting point, no gravitational force on that body does any net work, but a frictional force would do negative work on it. In the absence of other forces, friction always reduces the kinetic energy of a body and so slows it down. There is no potential energy that can be associated with frictional forces.

The result of these considerations is that when friction is absent, it is possible to define the mechanical energy of a system of particles in such a way that it is conserved, so that its value remains unchanged throughout the course of a motion. This energy is the sum of the kinetic energies of the particles constituting the system and the potential energy associated with the forces acting on these particles. The idea of the conservation of energy had been born.

A fundamental change in the way of writing the equations of motion was made by William Rowan Hamilton. Newton and Lagrange had treated the coordinates of the particles in the system as the variables governed by the laws of motion. There was one equation for each coordinate, and so in order to pick out a particular motion from all those satisfying the equations, it was necessary to give the values of both these coordinates and their rates of change, the corresponding velocities, at some initial time. Hamilton introduced an additional set of variables that he called the *momenta* conjugate to the coordinates. These were equivalent to, but not identical with, the particle velocities. He then could write a set of equations for this doubled set of variables that were equivalent to Newton's or Lagrange's equations but that had a very different and very symmetrical mathematical form.[10] It was this symmetry that gave them importance. Half of them defined the velocities in terms of the momenta, and the other half gave the rate of change of the momenta.

One great advantage of Hamilton's way of writing the equations of motion was that it led to the introduction of a *phase space* with very useful properties. Phase space is an abstract mathematical space. It can be thought of as a space defined by a set of mutually perpendicular axes, there being one axis for each coordinate and one for each momentum, so there are $2f$ axes all together if f coordinates are needed to specify the configuration of the system. The state of motion of the system at any one time can then be represented by the position of a single *phase point* in this space, the coordinates of this point in phase space being the values of the coordinates and the momenta needed to specify the configuration and motion of the system at that time. Phase space is distinct from *configuration space*, another abstract mathematical space having an axis for each coordinate of each particle in the system but none for the momenta. The configuration of the system at a certain time is given by the location of a point in this space at that time, but no information about how the system is moving is provided by this location alone. Hamilton's equations of motion specify how the phase point representing the state of motion of the system moves through phase space. We shall see how useful this way of describing the motion of a mechanical system can be in areas such as the kinetic theory of gases and statistical physics.

The second volume of Newton's *Principia* is entitled "The System of the World (in Mathematical Treatment)." Here the theorems of the first volume were applied to the motion of the satellites of Jupiter and Saturn around those planets, to the motion of the five planets Mercury, Venus, Mars, Jupiter, and Saturn around the Sun, and to the motion of the Moon around the Earth. All these motions were shown to be produced by a force of gravity between bodies proportional to their masses and inversely proportional to the square of their distance from the central body around which they move. Account was given of the motion of the planets and comets, of tides produced on Earth by the Moon, of the effect produced by the Sun on the Moon's orbit around the Earth, and of a host of other phenomena that had previously been considered unrelated to one another.

In the hands of later workers, among them Laplace, Lagrange, Hamilton, C. G. J. Jacobi and Henri Poincaré, celestial mechanics became the model scientific theory. It led to the prediction of the presence of the outer planet Neptune. Neptune was discovered by its influence on the motion of Uranus. William Herschel had discovered Uranus in 1781. By 1820 it had a position that deviated more than a minute of arc from the one predicted by celestial mechanics. This discrepancy was resolved in 1846 when Neptune was observed at the location predicted by both U. J. J. Leverrier and J. C. Adams where it would provide the necessary perturbation of the orbit of Uranus. Neptune had been seen before but had been mistaken for a star. Successive observations of it to detect its motion had not been made. (Later it was found that Neptune did not quite account for all of the observed perturbation of Uranus, and in 1930 Pluto was observed at the place where it could account for most but not all of the remainder. There may be an even more distant "tenth planet" accounting for the rest. It has not yet been seen.) The only astronomical fly in the ointment was the motion of the planet closest to the Sun, Mercury. There can be no undiscovered planet close enough to the Sun to account for any anomaly in Mercury's motion. According to the best celestial mechanics calculations, the perturbations of Mercury's orbit by the other planets should cause the major axis of Mercury's Keplerian orbit to change direction at a rate of 529.2 seconds of arc per century, as against an observed precession of 572.7 seconds per century. This discrepancy of 43.5 seconds of arc per century remained unaccounted for until the formulation of the general theory of relativity in 1916.[11]

The motions of bodies large on the scale of atoms and molecules but small on the scale of the cosmos are, with few exceptions, accurately described by Newtonian mechanics. At the small end of the scale of size, quantum effects are important. At the large end of the scale, gravity is the dominating principle, and here Einstein's theory of gravitation, general relativity, plays a dominant role. In between Newton still reigns, and in 1895 he reigned everywhere.

The idea that matter consisted of bodies subject to Newton's laws of motion was consistent with everything that was then known about nature in both the large and the small, and it formed the basis of the intuition that generations of people used to interpret everyday experiences. It took overwhelming evidence and much time and soul-searching to alter this picture. The necessary evidence began to accumulate at the end of the nineteenth century.

Notes

1. Galileo Galilei, *Discourse on Two New Sciences* (Leyden: Elsevier, 1638).

2. I. Newton, *Philosophiæ Naturalis Principia Mathematica* (London, 1686).

3. I. Newton, *Mathematical Principles of Natural Philosophy*, trans. Andrew Motte (Berkeley and Los Angeles: University of California Press, 1934).

4. Aristotle, *Physics*, trans. H. G. Apostle (Bloomington: Indiana University Press, 1969), bk. H5: "If A is the mover, B the thing in motion, S the length over which the motion occurred, and T the time taken, (1) in time T a force equal to that of A will cause a thing which is half of B to move over length 2S, and (2) it will cause it to move over length S in half the time of T." The first of these says that the average speed resulting from a given force is inversely proportional to the mass, not inconsistent with the Newtonian result. The second, however, makes the distance covered by a given body linear in the time, not quadratic, as is the motion under constant acceleration, and so implies a direct relation between force and speed.

5. I. Newton, *Opticks* (New York: Dover, 1952), E. T. Whittaker's "Introduction," p. lxx, and bk. 2, pt 2.

6. P. S. Laplace, *Oeuvres complètes*, vol. 7: *Théorie analytique de probabilités*, "Introduction," quoted in Bartlett, *Familiar Quotations* (Boston: Little Brown, 1968), p. 479.

7. James Gleick, *Chaos* (New York: Viking Press, 1987).

8. L. de Lagrange, *Mecanique analitique* (1788).

9. P. S. Laplace, *Théorie des attractions des spheroids et la figure des planètes* (Paris Academy, 1785).

10. W. R. Hamilton, "On a General Method in Dynamics," *Philosophical Transactions of the Royal Society*, pt. II, 1834, p. 247, pt. I, 1835, p. 95.

11. A. Pais, *Subtle Is the Lord* (New York: Oxford University Press, 1982), chap. 14.

Amplification

Newton formulated laws of classical mechanics that remained unchanged until the early twentieth century when the theory of special relativity required some modifications of them without changing their essential character. In this Amplification I shall describe some of the mathematical developments associated with these laws and indicate their importance in providing new ways to think of the motion of mechanical systems.

The simplest situation is one in which the force acting on a particle is always along the direction in which the particle is moving. Then there is no sideways acceleration, and there is a single equation of motion in the original Newtonian form for the single coordinate of the particle, measuring its distance along the line from its starting point. No further development is called for here aside from that of the mathematical theory needed to solve the particular equation that arises from the particular force acting on the system being considered. This may be very difficult, but most systems of interest are more complicated than this.

If the force acting on a particle is not always parallel to its velocity, Newton's second law must be applied separately to the motion in each of the possible directions; the law takes the form of a vector equation involving directions as well as distances. On referring the motion to an inertial reference frame furnished with a Cartesian coordinate system, three equations of Newtonian form are obtained, one for motion parallel to the x-axis, one for motion parallel to the y-axis, and a third for motion parallel to the z-axis. Here a complication often arises because the force in, say, the x-direction depends not only on the x-coordinate of the particle but also on the y- and z-coordinates. The three equations of motion then become coupled to one another and must be solved simultaneously. These coupled equations can be difficult to solve, and so approaches that make them more tractable are of great value.

Lagrange introduced a way of writing the equations that can yield relatively simple equations of motion in situations in which the method of resolving forces into the Cartesian components does not. He introduced as *generalized* coordinates q_i, a set of quantities whose values completely specify the positions of the particles of a system. He also introduced as *generalized* velocities $\dot{q}_i \equiv dq_i/dt$, the rates of change of the generalized coordinates. The generalized velocities are linear combinations of the Cartesian velocities with coefficients that may depend on the coordinates and so are easy to calculate. He did *not* introduce the second time derivatives of these generalized coordinates as generalized accelerations because they are not simply related to the Cartesian accelerations. Instead he found another way of proceeding. He introduced the kinetic energy of the system, the sum of the kinetic energies of all the particles comprising the system. This depended on the squares of the velocities of the particles and was therefore simply a quadratic expression in the generalized velocities with coefficients that might depend on the generalized coordinates, denoted by $T(\dot{q}_i, q_i)$. He then also introduced the potential energy $V(q_i)$ of the system of particles depending on the generalized coordinates but not on the velocities. On writing $L = T - V$ he showed that the equations of motion could be written in the form

$$\frac{d}{dt}\left(\frac{\partial L}{\partial \dot{q}_i}\right) - \frac{\partial L}{\partial q_i} = 0, \qquad i = 1, \cdots, f.$$

These are the celebrated *Lagrangian equations of motion*, L being called the *Lagrangian* function of the system. One of their great advantages is the freedom to choose the generalized coordinates in such a way that a particular system's equations are especially simple. If, for example, the Lagrangian L does not depend on a certain coordinate, say q_1, then in the equation with $i = 1$ the term $\partial L/\partial q_1$ on the left in the equation for that coordinate vanishes, and the equation reads

$$\frac{d}{dt}\left(\frac{\partial L}{\partial \dot{q}_1}\right) = 0, \qquad \frac{\partial L}{\partial \dot{q}_1} = \text{constant}.$$

The second-order equation is thus reduced to a first-order one containing an arbitrary constant.

For a central force problem such as that of the moon moving around the earth, the potential depends only on the earth–moon distance. If this distance is introduced as a generalized coordinate r, then the potential energy does not depend on the other two coordinates, the longitude ϕ and the colatitude θ (like the latitude of the geographers except that it is measured from the pole rather than from the equator) of the moon relative to the earth, and the kinetic energy does not depend on ϕ, even though it does depend on θ. Then the ϕ equation is in the above form, and $\partial L/\partial \dot{\phi}$ is constant. This is the angular momentum of the moon around the polar axis from which the colatitude is measured. Using this result, the coordinate θ can be eliminated from the Lagrangian, and the θ-equation then acquires this form. The corresponding constant is the magnitude of the angular momentum of the moon in its motion around the earth. This simplification was vital to the systematic development of celestial mechanics. By suitably extending the idea of a coordinate, even Maxwell's equations can be written in Lagrangian form.

Half a century later Hamilton introduced still another way of writing the equations of motion of a system. Instead of a single differential equation for each coordinate, Hamilton found a set of twice as many equations but of only first order, that is, containing only first derivatives with respect to the time. To do this is easy, but to do it in a useful way is not. To be useful Hamilton defined a set of *canonical momenta* by

$$p_i = \frac{\partial L}{\partial \dot{q}_i}$$

which is usually a set of equations linear in the generalized velocities that can be solved to give the velocities as functions of the momenta and the coordinates. He then introduced the function H, now called the *Hamiltonian*, defining it by

$$H = \sum_{i=1}^{f} (p_i \dot{q}_i - L).$$

In many cases the Hamiltonian function H is nothing but the energy of the system, $T + V$, expressed as a function of the coordinates and momenta. Then the equations of motion of the system are of the remarkably symmetrical form

$$\dot{q}_i = \frac{\partial H}{\partial p_i}, \qquad \dot{p}_i = -\frac{\partial H}{\partial q_i}.$$

These are *Hamilton's canonical equations of motion.*

The phase space of the system is a set of $2f$ dimensions in which the coordinates q_i and the momenta p_i serve as Cartesian coordinates. The state of motion of the system at any instant is completely specified by giving the values of these $2f$ quantities and, therefore, by specifying the point in the phase space with these values as coordinates. In time this phase point traces out a curve in the phase space, with this phase curve specifying the way the system moves as time progresses.

As I mentioned earlier, it is easy to write down a doubled set of first-order equations that are equivalent to a set of second-order equations, but in general this gains little. The reason that Hamilton's equations are uniquely interesting is that they are so symmetrical. Only the algebraic sign distinguishes one set from the other. This symmetry opens up opportunities for further simplifications and other developments, especially in connection with complex systems such as gases.

If there are many systems constituting an *ensemble*, all described by the same Hamiltonian, the state of motion of each system will specify the phase point of that system, and because all the systems share the same Hamiltonian, their phase spaces will be alike. The points specifying the state of motion of the individual systems of the ensemble can then be thought of as all lying in a common phase space, and these points will move through this common space in accordance with Hamilton's canonical equations of motion. The left-hand side of these equations gives the components of the velocity with which the point at any given location in this common phase space moves.

We will find it useful in connection with the study of gases to think of these phase points as molecules of an abstract mathematical fluid that flows through the phase space. The form of Hamilton's equations guarantees that this fluid is incompressible, so that the number of phase points per unit volume of phase space does not change as the points move in accordance with the equations of motion. Mathematically this follows from the equation

$$\sum_{i=1}^{f} \left(\frac{\partial \dot{q}_i}{\partial q_i} + \frac{\partial \dot{p}_i}{\partial p_i} \right) = \sum_{i=1}^{f} \left(\frac{\partial^2 H}{\partial p_i \partial q_i} - \frac{\partial^2 H}{\partial q_i \partial p_i} \right)$$

$$= 0$$

which is a direct consequence of Hamilton's equations. This is one of several theorems bearing the name Liouville's theorem. This theorem does not, however, guarantee a simple behavior of the phase fluid. Points that start out close to one another may be widely separated at later times. J. Willard Gibbs, who used this idea heavily in his groundbreaking work in statistical mechanics, gave an example. A drop of ink is placed in a vessel of water, and the water is then stirred. Both the ink and water are incompressible, and so their density does not change in this process, but what was a drop of ink is spread out into a long filament wandering around the entire volume of the containing vessel. On looking at the fluid without a sufficiently powerful magnifying glass, it appears to be of a uniform pale color. But if appropriate liquids are used and if they are not stirred for too long a time, a reversal of the stirrer's motion will actually reverse the apparent diffusion of the "ink" and restore the original drop, giving a vivid illustration of the time reversibility of the mechanical equations of motion. If the stirring is continued too long, effects of random or chaotic motions in the fluid not directly caused by the stirrer and therefore not reversed along with that of the stirrer, will accumulate to such an extent that the drop will not be reconstituted by reversing the stirrer's motion.

From Oersted and Faraday: Electromagnetism

By the last decade of the nineteenth century the theory of electromagnetism was firmly established in the form given to it by James Clerk Maxwell.[1] This theory developed out of two discoveries connecting electric with magnetic phenomena. The first was made by Hans Christian Oersted and the second by Michael Faraday.

In 1819 during a lecture demonstration Oersted noticed that an electric current affects a magnet in a completely unanticipated way. When a magnet was at right angles to the current-carrying wire, it was thought that the poles might be either attracted or repelled by the wire, but no such effect had ever been found. In his fateful lecture Oersted finally placed the magnet parallel to the wire. To his extreme surprise the magnet tended to set itself at right angles to the wire. This did not conform with the idea that the forces on the poles of magnets are always attractive or repulsive. This force was not toward or away from the wire but was sideways, one direction on the north pole of the magnet and the other direction on the south pole. The magnet tended to align itself in a direction tangent to a circle centered on the wire and not along a radius. A mathematical description of this force was promptly developed by André-Marie Ampère, but no clear physical picture emerged.

Faraday made a key discovery in 1831, that if a magnet aligned along the axis of a loop of wire is moved either toward or away from it, a current is induced in the loop.[2] As with Oersted's experiment, the outcome did not fit into the attraction–repulsion picture. Here the induced current flowed at right angles to the direction of motion of the magnet. Both these effects involved motion: Oersted's required moving charges in the form of an electric current, and Faraday's required a moving magnet.

These two effects, influencing a magnet by a motion of electric charge and producing an electric current by the motion of a magnet, demonstrated dramatically that electricity and magnetism are not separate disciplines. Instead, they are intimately related to each other and are just two aspects of a larger subject, electromagnetism. The unification of these two relations of experience was comparable in importance to the unification of celestial and terrestrial physics by Newton.

As a result of many wide-ranging experiments in electric and magnetic phenomena, Faraday was led to conceive of electric and magnetic forces being transmitted by stresses in a medium, later called the *ether*, filling the space between electrified bodies and magnets, much as mechanical stresses are transmitted by rigid bodies. This contrasted with the prevailing view that the force between two charged bodies or two magnets was an *action-at-a-distance* attraction or repulsion like the force of gravity between two massive bodies; in the action-at-a-distance view no time was required for the influence of one body to be transmitted to another, all effects being transmitted instantaneously. Faraday represented his stresses by *lines of force* beginning on positively electrified bodies and ending on negatively electrified ones.

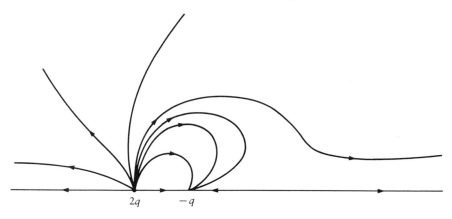

Fig. 1.1 When two charges were present, Faraday's lines of force were curved. The lines of force arising from two charges of unequal size and opposite sign are shown.

In the presence of more than one charged body, these lines were curved, as seen in Fig. 1.1. He thought of each of these lines as elastic, tending to get as short as possible and hence causing an attraction between bodies with opposite kinds of charge on which the lines begin and end. He considered the lines to repel each other sideways so that they tended to spread out and fill all space. No two lines could cross, because if they did it would mean that at the point of crossing the force had two directions. The repulsion between two bodies with the same kind of charge Faraday thought to be due to this repulsion between lines, the lines starting on two positively charged bodies crowding each other. At the time this view—being expressed more in words and pictures than in mathematical language—was not taken seriously. The attitude of his contemporaries was neatly summed up in Lord Kelvin's (William Thomson's) preface to the English edition of Heinrich Hertz's papers on electric waves:

> During the fifty-six years which have passed since Faraday first offended physical mathematicians with his curved lines of force, . . .

In a major work published in 1856 entitled "On Faraday's Lines of Force," Maxwell gave a mathematical formulation of Faraday's ideas, which was the beginning of Maxwell's development of a theory of the electromagnetic field:

> When a body is electrified in any manner, a small body charged with positive electricity, and placed in any given position, will experience a force urging it in a certain direction. If the small body be now negatively electrified, it will be urged in the opposite direction.
> The same relations hold between a magnetic body and the north or south pole of a small magnet. If the north pole is urged in one direction, the south pole is urged in the opposite direction. In this way we might find a line passing through any point of space, such that it represents the direction of the force

acting on a positively electrified particle, or an elementary north pole, and the reverse direction of the force on a negatively electrified particle or an elementary south pole. Since at every point of space such a direction may be found, if we commence at any point and draw a line so that, as we go along it, its direction at any point shall always coincide with that of the resultant force at that point, this curve will indicate the direction of that force for every point through which it passes, and might be called on that account a *line of force*. We might in the same way draw other lines of force, till we had filled all space with curves indicating by their direction that of the force at any assigned point.

We should thus obtain a geometrical model of the physical phenomena, which would tell us the *direction* of the force, but we should still require some method of indicating the *intensity* of the force at any point. If we consider these curves not to be mere lines, but as fine tubes of variable section carrying an incompressible fluid, then, since the velocity of the fluid is inversely as the section of the tube, we may make the velocity vary according to any given law, by regulating the section of the tube, and in this way we might represent the intensity of the force as well as its direction by the motion of the fluid in these tubes. This method of representing the intensity of a force by the velocity of an imaginary fluid in a tube is applicable to any conceivable system of forces, but it is capable of great simplification in the case in which the forces are such as can be explained by the hypothesis of attractions varying inversely as the square of the distance, such as those observed in electrical and magnetic phenomena. In the case of a perfectly arbitrary system of forces, there will generally be interstices between the tubes; but in the case of electric and magnetic forces it is possible to arrange the tubes so as to leave no interstices. The tubes will then be mere surfaces, directing the motion of a fluid filling up the whole space. It has been usual to commence the investigation of the laws of these forces by at once assuming that the phenomena are due to attractive or repulsive forces acting between certain points. We may however obtain a different view of the subject, and one more suited to our more difficult inquiries, by adopting for the definition of the forces of which we treat, that they may be represented in magnitude and direction by the uniform motion of an incompressible fluid.[3]

During his development of the theory, Maxwell used a sequence of mechanical models of Faraday's ether, starting with the incompressible fluid just introduced. No mechanical model behaved in all ways like the system he was studying, so in the final work published in 1873 this scaffolding had vanished.[4] What remained was a set of equations, Maxwell's equations, governing the behavior of Faraday's lines of force. These lines had been transformed into an electric field $E(x, y, z, t)$ and a magnetic field $B(x, y, z, t)$ describing the forces on charged particles and magnetic poles, respectively, at any position defined by the coordinates x, y, z at time t. Here E and B are vector quantities giving both the direction and the magnitude of the respective fields. The direction of the vector E at any point is the direction of the Faraday line going through that point, and the magnitude of E is the strength of the force on a unit positive charge at that point. The problem of the ether had

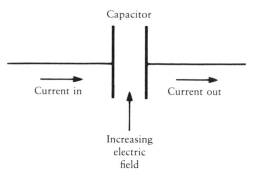

Fig. 1.2   When a current enters one side of a capacitor, giving that plate a positive charge, and leaves the other, giving the second plate a negative charge, a *displacement current* is present in the space between the plates where the electric field is increasing.

been inverted; instead of finding the theoretical consequences of a particular model, the task undertaken by many was to find a model that led to Maxwell's theory as embodied in his equations.

In addition to giving Faraday's ideas of fields a sharpened and mathematical expression, Maxwell made a crucial contribution going beyond the ideas of his predecessors. He showed that a changing electric field produced a magnetic field and so was in a sense equivalent to an electric current. For example, in the charging of the parallel plate capacitor shown in Fig. 1.2, current flows into one plate increasing the charge on it, and current flows out of the other plate increasing the negative charge on it. No charge flows in the space between the plates, but the changing charges on the plates result in an increasing electric field between the plates. Maxwell treated this increasing field as a continuation of the current flowing in the rest of the circuit, calling it a *displacement current*. The magnetic field around the wire, proportional to the current in the wire, is also present around the space between the capacitor plates, being produced by the displacement current present there.

Electromagnetism as formulated by Maxwell accounts for the electric and magnetic phenomena known to the ancients, for the production of magnetic fields by electric currents and changing electric fields, and for the production of electric fields by changing magnetic fields. Maxwell's theory describes how stresses are transmitted between charged bodies at speeds never exceeding the speed *c* appearing in his equations, thus eliminating the concept of instantaneous action at a distance embodied in the law of Coulomb for the force between charges and the law of Ampère for the force between current elements. The stresses were thought to be transmitted through a medium, the ether. The ether played a dual role in electromagnetism, transmitting forces between charges and currents and providing a preferred reference frame in which Maxwell's equations governed the behavior of the electromagnetic field just as absolute space provided Newton with a preferred reference frame in which Newton's equations governed the motion of particles.

Faraday's law of induction states that a changing magnetic field generates an electric field. The changing electric field so generated constitutes a Maxwellian displacement current that in turn generates a magnetic field. The cycle then starts over. Maxwell showed that this combination of effects leads to the existence of electromagnetic waves. Each field, electric or magnetic, in the wave changes with time, and the change of one generates the other. This self-renewing process continues and constitutes the electromagnetic wave. The electric and magnetic fields in these waves are perpendicular to the direction in which the wave moves, so that the waves are transversely polarized. They travel through the ether free of matter with the speed c. The quantity c occurs in Maxwell's equations wherever electric and magnetic quantities appear together. The force between the wires carrying electric currents is the result of the electric current in one wire moving through the magnetic field produced by the electric current in the other. The expression for this force therefore involves the constant c. This means that the constant c can be evaluated by measuring the force between two wires carrying currents and equating it with the value of the theoretical expression for it in terms of the currents. The value so obtained was close to the known speed of light. Also, the polarization properties of these waves were just those needed to account for the polarization of light, and so light waves could be considered as electromagnetic in nature. This was soon confirmed by the work of Heinrich Hertz.

Hertz demonstrated the finiteness of the speed with which magnetic effects travel in the laboratory. Even more important, he demonstrated the existence of electromagnetic waves having all the properties ascribed to them by Maxwell.[5] He generated them by electric circuits in which rapidly oscillating currents flowed, and he detected them by the currents they produced in other circuits placed at a distance from their source. The fact that Hertz's waves shared the polarization properties and the speed of light waves permitted the identification of light as electromagnetic in nature. This in turn made Faraday's stress-supporting medium and the luminiferous ether of optics the same.[6] Hertz's experimental work, together with his clear presentation of the theory, established the theory of the electromagnetic field among European scientists, who had generally not appreciated it as early as the British had. In this presentation Hertz wrote:

> From the outset the conception was insisted upon, that the electric and magnetic forces at any point owe their action to the particular condition of the medium which fills the space at that point; and that the causes which determine the existence and variations of these conditions are to be wholly sought in the conditions in the immediate neighborhood, excluding all actions-at-a-distance.[7]

This is the specification of what we now call a *local field theory*.

By 1895, then, electromagnetism as defined by Maxwell's equations was in the form in which we know it now. The physics behind it was, however, based on the idea of a material elastic medium, the ether of elusive properties,

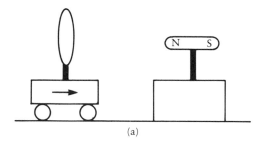

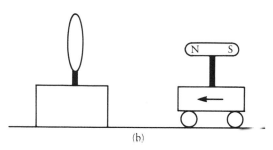

Fig. 1.3 The same current flows in a wire loop when (a) the loop is moved toward a magnet and (b) when a magnet is moved toward the loop, as long as the *relative* motion of the two is the same.

and the theory was not recognized as having the relativistic character that we now see in it. While others concerned themselves with constructing mechanical models of the ether, Hertz, who was not interested in models of the ether, noted that Maxwell's theory consisted of his equations and nothing more. These equations involved observable physical quantities, electric and magnetic fields, charges and currents, and correctly described the relations among these quantities. The description of observable quantities was enough for Hertz, however, as it was for Heisenberg four decades later in another connection. This view had served him and physics well, as it had led him to do the work that put electrodynamics on an unassailable foundation.

The medium, the ether, in which the stresses corresponding to the electromagnetic fields resided, defined a unique reference frame—the rest frame—in which Maxwell's equations held. It was assumed that motion with respect to this ether could be detected so that the electrodynamics of moving bodies would be different from that of bodies at rest. This difference showed up in Faraday's induction. Consider the system consisting of a loop of wire and a bar magnet illustrated in Fig. 1.3. If the loop is held fixed relative to the ether and the magnet is brought near, the magnetic field at the loop will change with time, and this changing magnetic field, according to Maxwell's equations, will produce an electric field that causes the electric charges free to move in the conducting material to flow around the loop. If instead the magnet is held fixed and the loop is brought near it, the magnetic field will not change and so there will be no induced electric field. The charges in the wire that are free to carry a current are now moving through a magnetic field and therefore experience a force that causes them to move around the loop. If the relative motion of the bar magnet and loop is the same in the two cases, the current is the same, but the dynamics leading to its presence is different. The electrodynamics of moving bodies became an important area of research. Hertz's reformulation of Maxwell's theory appeared in two parts, "On the Fundamental Equations of Electrodynamics for Bodies at Rest" and "On the Fundamental Equations of Electrodynamics for Bodies in Motion." H. A. Lorentz devoted much time and effort to understanding why these two situations resulted in the same current, but not until Albert Einstein introduced the special theory of relativity was the matter cleared up. This will be discussed in more detail later.

When light became an electromagnetic phenomenon, the luminiferous ether became the medium in which Faraday's stresses occurred, and the problem of the earth's motion through the ether became a problem of electrodynamics. As early as 1887 W. Voigt noted the following remarkable fact: If he took Maxwell's equations for free space, that is, when no charges or currents are present, and rewrote them in terms of coordinates referring to a reference frame moving uniformly through the ether, in general he obtained a set of equations differing in mathematical form from those he started with.[8] He found, however, that he could define coordinates and time in the moving reference frame in terms of those in the rest frame in such a way that the transformed Maxwell equations he obtained had exactly the *same* mathematical form as the original ones. In a sense, then, uniform motion through the ether did not necessarily affect the mathematical form of Maxwell's equations describing waves in free space. Voigt used this invariance of form to derive the first-order Doppler effect for light, but he drew no general conclusions from it. These transformations of Maxwell's equations were rediscovered by H. A. Lorentz and extended to the Maxwell equations including charges and currents.[9] Lorentz's transformation of the charge and current did not quite retain the form of the original equations, but Poincaré showed how the transformation could be modified to keep the form completely unchanged. None of these physicists interpreted the invariance of the equations under these transformations to mean that no distinction could be made among coordinate systems moving uniformly with respect to one another, rendering the ether as superfluous as Newton's absolute space had become in mechanics. The required transformations meant introducing a new time variable that depended on both the original time and the original space coordinates. This mixing of space and time made impossible for them the acceptance of the new "local time" variable as equivalent to the "true" time originally used. The ether survived but became infinitely elusive.

Notes

1. J. C. Maxwell, *A Treatise on Electricity and Magnetism* (Oxford: Clarendon Press, 1873; 3rd ed., New York: Dover, 1954).

2. M. Faraday, *Experimental Researches in Electricity* (London: B. Quaritch, 1839).

3. *Transactions of the Cambridge Philosophical Society* 10 (1856): pt. 1; *The Scientific Papers of James Clerk Maxwell*, ed. W. D. Niven (New York: Dover, 1965), vol. 1, p. 155.

4. Maxwell, *A Treatise on Electricity and Magnetism.*

5. H. Hertz, *Electric Waves* (London: Macmillan, 1900).

6. A discussion of various views of the ether is given by Tetu Hirosige, "The Ether Problem, the Mechanistic World View, and the Origins of the Theory of Relativity," in R. McCormmach, ed., *Historical Studies in the Physical Sciences*, vol. 7 (Princeton, NJ: Princeton University Press, 1976).

7. H. Hertz, *Electric Waves*, p. 241.

8. W. Voigt, *Göttingen Nachrichten* (1887), p. 41.

9. H. A. Lorentz, *The Theory of Electrons* (New York: Dover, 1952).

Amplification

Maxwell's equations for the electromagnetic field in free space can be written in deceptively simple form through the use of a compact vector notation. At the time of Maxwell and Hertz this notation had not been invented, and so they wrote the equations using Cartesian components of the electric and magnetic fields. As written by Jackson[1] they are

$$\nabla \cdot \mathbf{E} = 4\pi\rho \tag{1}$$

$$\nabla \times \mathbf{B} - \frac{1}{c}\frac{\partial \mathbf{E}}{\partial t} = \frac{4\pi}{c}\mathbf{j} \tag{2}$$

$$\nabla \cdot \mathbf{B} = 0 \tag{3}$$

$$\nabla \times \mathbf{E} + \frac{1}{c}\frac{\partial \mathbf{B}}{\partial t} = 0. \tag{4}$$

In component notation the first equation reads

$$\frac{\partial E_x}{\partial x} + \frac{\partial E_y}{\partial y} + \frac{\partial E_z}{\partial z} = 4\pi\rho.$$

It says that the rate of change of the x-component of the electric field in the x-direction plus the rate of change of the y-component in the y-direction plus the rate of change of the z-component in the z-direction at any point is proportional to the charge density at that point. If the electric field is represented by lines of force in the manner of Faraday, starting on positive charges and ending on negative ones, this equation requires the net number of lines of force leaving a volume to be proportional to the charge included in that volume. The third equation then can be read to require that there is no magnetic charge anywhere. The other equations are more complicated to interpret in words. The second one describes the production of a magnetic field around a wire carrying a current or around a changing electric field. The fourth one is a statement of Faraday's law of induction, that an electric field is produced around a changing magnetic field.

The quantity c appearing wherever electric and magnetic fields are related to each other in these equations turns out to be the speed with which electromagnetic waves travel in free space. It also gives a measures of the magnetic effect of one electric current on another, which can be measured directly, so that the Maxwell equations provide a way to predict the speed of light on the basis of experimental measurement of the magnetic interaction between two currents. It was the near equality of c as determined in this way with the measured speed of light that permitted the identification of light as an electromagnetic wave.

The question of the units in which electromagnetic quantities are expressed is a bothersome one. At the time we are considering, there were two systems in common use, the electrostatic (esu) and the electromagnetic (emu). The electrostatic unit of charge was defined so that Coulomb's law had the simple form

$$f = \frac{q_1 q_2}{r^2},$$

with the force measured in dynes and the distance in centimeters. The electrostatic unit of current was then the flow of one esu of charge per second. The electromagnetic unit of current was defined so that the expression for the force per unit length between two long parallel currents was simply

$$f/L = \frac{J_1 J_2}{r}.$$

The electromagnetic unit of charge was then the amount of charge transported per second by one emu of current. These two units of charge differ in dimension as well as in magnitude, with the conversion between the two systems being the speed of light c. In the second Maxwell equation, J_m replaces j_s/c, so that the numerical value of a charge measured in emu is much smaller than that of the same charge measured in esu, the electromagnetic unit being so much larger than the electrostatic one:

$$1 \text{ emu} = 2.99792458 \times 10^{10} \text{ esu} \approx 3 \times 10^{10} \text{ esu}.$$

(The SI unit of charge, the coulomb, is 1/10 emu or very nearly 3×10^9 esu.)

Note

1. J. D. Jackson, *Classical Electrodynamics*, 2nd ed. (New York: Wiley, 1975).

From Rumford and Carnot: Heat and Thermodynamics

Classical mechanics and electrodynamics are completely deterministic in the sense that given the initial state of a system, the future values of any observable quantities associated with that system can be predicted. They are also reversible in time. That is, if a state *A* of the system develops into a later state *B* and if at this later time one then reverses the velocities of all particles in the system and the direction of all electric currents and magnetic fields, the system will retrace its development and arrive at the reversed form of state *A*. It is assumed that all the variables that affect the development of the system in time are included in the equations of motion and that no *hidden variables* of the system influence this development. This is often a good assumption; celestial mechanics as described earlier is an example. Clearly, a planet is not completely described by giving its mass, the position of its center, and the velocity of its center. Otherwise there would be no new information about Jupiter to be gained from making a flyby. But these data are the only ones that have any observable effect on its orbit under nearly all circumstances, so that if the motions of all the planets were reversed without reversing the motions of the winds on the planets, for example, the solar system would retrace its development with great, if not perfect, accuracy.

The presence of friction changes things. Friction causes mechanical energy to be lost and the equivalent amount of heat to appear, and this does affect the mechanical behavior of a system. The variables describing the internal motions of the particles composing a body that are associated with quantities such as the temperature of the body or the amount of heat gained or lost by the body do not enter into the mechanical or electromagnetic equations of motion for the body as a whole, so that now there are "hidden variables" that affect its mechanical and electromagnetic behavior. Because of their hiddenness, it is generally impossible to reverse all the velocities and currents among these variables, and so the time development of the system containing the body cannot be reversed. Dissipative systems, ones in which mechanical energy is dissipated in the form of heat, are irreversible.

The theory used to describe heat in the eighteenth century was a causal, reversible theory. Heat was thought to be a weightless fluid called *caloric* that could not be created or destroyed; caloric was conserved. It could flow from one place to another, and the amount of it present at a given place determined the temperature at that place. It tended to flow from places of high temperature to places of lower temperature, just as water tends to flow from higher places to lower ones. Many phenomena involving heat were well described by this theory.

Around the beginning of the nineteenth century, however, this view became less widely accepted. Count Rumford (Benjamin Thompson) bored cannon for the Elector of Bavaria, and in 1798 he reported on his observations of the generation of heat in the process. The duller the boring tool was, the more heat was generated. This production of heat by friction continued as long as the boring went on, so that friction as a source of heat "appeared evidently to be inexhaustible." Friction therefore provided a continuous source of caloric, which did not accord with the idea of its conservation. According to Rumford:

> Anything which any insulated body, or system of bodies, can continue to furnish without limitation, cannot possibly be a material substance; and it appears to me to be extremely difficult, if not quite impossible, to form any distinct idea of anything capable of being excited and communicated in the manner in which heat was excited and communicated in these experiments, except it be motion.[1]

This doubt thrown on the concept of heat as a conserved "imponderable" fluid subject to laws like those governing material fluids such as water and air did not, however, immediately lead to a substitute theory or explain what might be in motion.

Sadi Carnot was the first to study in a quantitative way the connection between heat and work or energy. Lamenting the lack of interest in the principle of operation of steam engines, Carnot wrote "*Memoir on the Motive Power of Heat*" examining the amount of mechanical work that could be obtained when a certain amount of heat (caloric) was transferred from a body at a higher temperature to one at a lower temperature, as from the boiler to the condenser of a steam engine.[2] In this memoir Carnot applied the caloric theory. He considered an engine consisting of a cylinder fitted with a piston and containing a "working substance" such as steam that operates between a high temperature T_1 and a low temperature T_2. Initially the cylinder is placed in contact with the source of heat at the higher temperature, the boiler of the engine. The piston moves out for a time under the influence of the pressure force of the working substance, being kept from gaining much speed by an almost equal external force coming from the machinery that the engine is to drive. The cylinder is then removed from the heat source, and the piston continues to move out, being stopped when the working substance cools down to the temperature T_2 of the condenser. The cylinder is then put in contact with the condenser. The external force is then increased just enough to make the piston move in slowly, compressing the working substance at the lower temperature T_2. The cylinder is then removed from the condenser, but the piston is still pushed in slowly, further compressing the working substance and raising its temperature until it again reaches the temperature T_1 of the boiler. The expansions and compressions are arranged so that the final volume is the same as the starting one, so the cylinder and the working substance end up in their starting condition. They have gone through a *Carnot cycle*. In this process the working substance does work on the external machine while it is expanding and receives work from the external machine while it is being compressed. The work it does is greater than the work it receives because the working substance expands when the pressure is higher and is compressed when it is lower. In Carnot's analysis of this cycle, caloric flowed from the boiler into the cylinder while it was in contact with the boiler and expanding, and caloric flowed out of the cylinder into the condenser while it was in contact with the condenser and contracting, so that the cycle transferred caloric from a high to a low temperature.

This Carnot cycle can be reversed in direction by adjusting the external force very slightly so that the working substance undergoes a slow compression rather than an expansion while the cylinder is in contact with the boiler and so that it undergoes a slow expansion rather than a compression while in contact with the condenser. Now the work received by the working substance from the external machine during the compression is greater than the work done during the expansion. In the forward cycle, caloric passes from the higher temperature T_1 to the lower temperature T_2 and work is received by the external machine from the engine, whereas in the reversed cycle, work is done on the engine by the machine and caloric is pumped from the lower temperature T_2 to the higher temperature T_1. From this reversibility of the cycle Carnot arrived at the following conclusion:

> Now if there were any method of using heat preferable to that which we have employed, that is to say, if it were possible that the caloric should produce, by any process whatever, a larger quantity of motive power than that produced in our series of operations, it should be possible, by diverting a portion of this power to effect a return of caloric, by the method just indicated, from the body B to the body A—that is from the refrigerator to the source—and thus to re-establish things in their original state, and to put them in position to recommence an operation exactly similar to the first one, and so on: there would result not only the perpetual motion, but an indefinite creation of motive power without consumption of caloric or of any other agent whatsoever. Such a creation is entirely contrary to the ideas now accepted, to the laws of mechanics and of sound physics; it is inadmissible. We may hence conclude that the maximum motive power resulting from the use of steam is also the maximum motive power which can be obtained by any other means.

This result was arrived at using the caloric theory. Although Carnot was not convinced of the validity of this theory, it provided a means of carrying out the needed calculations. In a footnote he wrote:

> But can we conceive of the phenomena of heat and of electricity as due to any other cause than some motion of bodies, and, as such, should they not be subject to the general laws of mechanics?

He then had a premonition of the identification of heat as a form of mechanical energy. How he fitted electricity into this is obscure. He seems not to have been aware of Rumford's experiments.

The identification of heat as a form of energy interconvertible with other forms was not clearly established until the work of Robert Mayer, a German physician, in 1842 and of James Prescott Joule, a Manchester brewer and scientist, in 1843. Mayer came to the conclusion that the energy of the world is constant, and he made the first rough measurement of the mechanical equivalent of heat, the amount of mechanical work required to generate a unit amount of heat, by measuring the warming of water by shaking. Joule made more accurate measurements of the mechanical equivalent of heat by measuring the warming of water stirred by a paddle driven by a falling weight. Others, including Liebig and Helmholtz, investigated the generation of animal heat from chemical sources. All these activities led to the formulation of the law of the conservation of energy, including heat as a form of energy along with mechanical and electrical energy.

Carnot had introduced fundamental ideas into the study of heat, the reversible cycle by which a heat engine performs mechanical work on an external machine and transfers heat between reservoirs at different temperatures. Above all he had elevated to the level of a fundamental physical principle the statement that it is impossible for any process to extract heat from a source at any temperature and to convert it all into mechanical energy. This impossibility statement is the essence of what became the second law of thermodynamics. The mathematical formulation of this law was first given by Rudolf Clausius much later. It is surprising that the denial of the possibility of the unrestricted conversation of heat into work preceded the recognition of the conservation of energy, the content of the first law of thermodynamics. Much else was contained in Carnot's *Memoir*, such as theorems on the specific heats of gases, but it was the two contributions just outlined that made it memorable and that profoundly altered the approach to the physics of heat.

Rudolf Clausius gave the first and second laws of thermodynamics their current mathematical form. In 1850 Clausius found that Carnot's denial of the possibility of perpetual motion from heat was in large part independent of the assumed conservation of caloric.[3] He restated Carnot's inadmissibility of obtaining unlimited work from heat as the inability by any mechanism to transfer heat—by then regarded as a form of energy—from a cooler body to a warmer one *without making any other changes*. (If an external source of energy is available, heat can be transferred from a low to a high temperature, as is done by an air conditioner.) This inability was obvious if heat conduction was the mechanism of transfer, but it was not obvious if, for example, the mechanism involved radiation. Sunlight focused by a lens produces temperatures at its focus that are much higher than any in the surroundings, even hot enough to ignite tinder. The temperature at the focus is, however, not higher than that at the surface of the sun from which the radiation comes, and Clausius showed that no system of lenses can make it so. His statement applied in this case, too, as it does in all cases yet encountered.

Caloric had been thought of as a conserved quantity. A system in a given state was considered to contain a definite amount of caloric. Heat by itself is not conserved; rather, it is one form of energy and work is another. The energy content U of a body can be specified in terms of the variables that determine the state of the body, but it is impossible to make a unique division of this energy into so much heat and so much work. One can specify the amount Q of heat added to a body during a given process and the amount W of work done by the body on the outside world during that process. The conservation of energy requires that the change ΔU of the internal energy during that process be given by

$$\Delta U - Q \quad W.$$

If forms of energy other than heat and work can be transferred into or out of the body during this process, they too must be included on the right-hand side of this equation. This way of stating the law of conservation of energy is known as the *first law of thermodynamics*.

Something seemed to have been lost in making the transition from a theory in which caloric was conserved to a new theory in which heat, the replacement of caloric, was not conserved. The caloric content of a body had been used in specifying the state of that body, but the heat content of a body was a meaningless concept, even though the heat Q added to a body in a particular process could be specified. Clausius introduced a new quantity, *entropy*, derived from the heat added to a body and the temperature at which that addition took place, in such a way that the change in the entropy of a body during any process could be specified. Then the entropy could be used in defining the state of that body.

Clausius noted that in a Carnot engine using an ideal gas (a gas that obeyed the ideal gas law $pV = nRT$) as its working substance, the ratio Q_1/T_1 of the heat Q_1 added to the working substance at the boiler temperature T_1 to that temperature T_1 is the same as the ratio Q_2/T_2 of the heat Q_2 rejected from the working substance at the condenser temperature T_2 to that lower temperature T_2. Clausius defined the ratio of the amount Q of heat transferred to a body at a temperature T to that temperature to be the change in the entropy of that body and denoted it by ΔS:

$$\Delta S = \frac{Q}{T}.$$

Clausius's observation was that the size of ΔS for the expansion part of the

Carnot cycle is the same as that for the compression part. At the end of a complete Carnot cycle, the two values of ΔS just cancel each other. The quantity S, the entropy content of the working substance, had a value at the end of the cycle just equal to that at the beginning. Only changes in entropy were defined; the value of the entropy was determined only to within an additive constant, a quantity that did not change during the course of any thermodynamic process.

In none of the processes in a Carnot engine is there a net change in entropy. When the engine is in contact with the boiler, the boiler loses the entropy Q_1/T_1, and the engine gains the same amount. During the expansion when not in contact with either the boiler or the condenser, there is no transfer of entropy anywhere. When the engine is in contact with the condenser, it loses the entropy Q_2/T_2 and the condenser gains the same amount. During the recompression to the temperature of the boiler, there is again no transfer of entropy anywhere. Instead of the engine's action transferring a definite amount of caloric from the boiler to the condenser, it transfers a definite amount of entropy. In any *reversible* process the total entropy is conserved. The conservation of entropy in reversible processes replaces the earlier conservation of caloric. In an irreversible process such as the direct transfer of heat from a warm to a cold body, the warm body loses less entropy than the cold body gains, so here there is a net increase of entropy.

With this definition of entropy, the statement of the impossibility of transferring heat from a cooler body to a warmer one without making some other change in the system can be rephrased to produce the second law of thermodynamics in its familiar form: No process, reversible or irreversible, occurring in an isolated system can decrease the entropy of that system. This conclusion does not depend on any model of the structure of matter, but only on the ability to measure heat in units of energy and to identify various other forms of energy such as the kinetic energy, potential energy, chemical energy, and so forth making up the internal energy of bodies. The temperature in Clausius's definition of the entropy is the temperature as measured with an ideal gas thermometer.

Clausius had shown that the ratio Q/T of the heat transferred to the temperature is the same at both temperatures in a Carnot cycle using an ideal gas as the working substance. W. Thomson (later Lord Kelvin) introduced a scale of temperature, now called the thermodynamic or kelvin scale, defined by requiring this ratio to be the same for Carnot engines using any working substances, not just ideal gases. If Q_1 and Q_2 are the amounts of heat transferred at the two temperatures by a Carnot engine using any working substance, the ratio of the temperatures of the hot and cold reservoirs is defined by

$$\frac{T_1}{T_2} = \frac{Q_1}{Q_2}.$$

Today this scale is universally used for scientific purposes.

Thermodynamics underwent a great mathematical development during the latter half of the nineteenth century based on the assumption that the laws of thermodynamics are exactly fulfilled under all conditions. It was applied to an enormous variety of problems arising in physics, chemistry, and engineering and gave accurate results when used with accurate data.

The second law of thermodynamics makes the subject of heat very different from those of mechanics and electromagnetism because it introduces a distinction between past and future. The past is the time when the total entropy of an isolated system was smaller than it is now. Neither of the other subjects does this; to every mechanical or electromagnetic process there corresponds a process going in the other direction in time, although this was not recognized for electromagnetism until Boltzmann pointed it out much later. Gaining an understanding of this difference has been a continuing effort of physicists and philosophers ever since.

Thermodynamics is a theory independent of any model of the structure of matter. It requires the ability to distinguish various forms of energy such as internal energy, heat transferred, radiation, and mechanical and electrical work done from one another, but it does not require describing these forms in more basic language. This led some scientists to consider it the ultimate theory beyond which there was no need and no means to go, but obviously this view did not prevail.

Notes

1. *The Complete Works of Count Rumford*, vol. 1 (Boston: American Academy of Arts and Sciences, 1870; S. C. Brown, ed., *Collected Works of Count Rumford* (Cambridge, MA: Harvard University Press, 1968), p. 22.

2. S. Carnot, *Réflexions sur la pouissance motrice de feu* (Paris, 1824, in W. F. Magie, trans. and ed., *The Second Law of Thermodynamics* (New York: Harper Bros., 1899). The quotations are from the English translation.

3. R. Clausius, Berlin Academy of Sciences, February 1850.

From Daniel Bernoulli and Joule: Kinetic Theory and Statistical Physics

It was not until the seventeenth century that air was recognized as having weight. Evangelista Torricelli was the first to show that the pressure of the atmosphere required a mercury column of some thirty inches in height to balance it. This was described in 1644 but not published until some years later. The experiment was repeated a little later in France by Blaise Pascal, once using red wine in place of mercury. The variation in pressure with altitude confirmed the interpretation of these experiments: The higher the altitude was, the less air there was above to hold up and the shorter the mercury column that was needed to balance its weight. Gases had joined the company of material things under the name of *elastic fluids*.

The model of a gas as consisting of freely moving particles had its first quantitative application at the hands of Daniel Bernoulli in his "Treatise on Hydrodynamics," published in 1738.[1] These particles had nothing to do with the chemical atoms or molecules we know today; chemical elements had not yet been identified. Rather, Bernoulli considered the pressure in air to be due to the bombardment of the walls of the container by "air corpuscles," all moving with the same speed v. He showed that according to his model, the pressure is proportional to v^2/V, where V is the volume of the gas. He also stated—contrary to the prevailing caloric theory—that "heat may be considered as an increasing internal motion of the particles," so that increasing the temperature of the air increases the elasticity, that is, the pressure, of the air in a constant volume. Nothing followed from this work for many years.

In 1808 John Dalton (1766–1844) published his "New System of Chemical Philosophy" in which he advocated an atomic theory of chemical reactions. This theory explained two regularities in the results of the chemical analysis of compounds. The ratio of the masses of two elements entering a compound was a fixed number independent of the sample size or its mode of preparation. Also, if one element forms more than one compound with a second, the ratio of the masses of the first combining with a fixed mass of the second can be expressed as the ratio of two small integers. These regularities are summarized in the laws of definite proportions and the law of multiple proportions. They are easily understood if a compound consists of molecules containing a definite number of atoms of each element in the compound and if different compounds of the same elements differ in the number of atoms of each element present in molecules of those compounds.

In 1811 Amedeo Avogadro (1766–1856) proposed that equal volumes of gases under the same pressure and at the same temperature contain equal numbers of molecules. The grounds for this proposal lay in the behavior of gases that combine with each other. The volumes of gases under the same conditions of temperature and pressure that combine completely with each other to form a chemical compound are in the ratio of two small integers. His proposal provided a direct connection of this fact concerning the chemical combination of gases with the law of definite proportions. According to Avogadro's hypothesis the weights of equal volumes of gases under the same conditions are proportional to the molecular weights of the gases, and it is the molecular weights that determine the combining weights in the law of definite proportions. Today an amount of gas whose mass in grams is equal to its molecular weight is called one *gram molecular weight* or one *mole* of the gas. The number of molecules in one mole of a gas is called *Avogadro's number*, denoted by N_A. Avogadro had no estimate of how large this number was.

Some who found the atomic idea persuasive, including Dalton himself, still rejected Avogadro's hypothesis because of the confusion between atoms and molecules that was widespread at the time. After all, equal volumes of different gases containing equal numbers of molecules may not contain equal numbers of atoms because their molecules may contain different numbers of atoms, so until the distinction between atoms and molecules was clearly understood, there was room for confusion. Avogadro's hypothesis was widely accepted among chemists only after the first international meeting of chemists held in Karlsruhe in 1860.[2] S. Cannizzaro gave a paper there in which he reviewed a vast number of chemical reactions and specific heat measurements and showed how only the hypothesis of the existence of molecules composed of definite numbers of atoms could bring order into their description. Also attending the meeting was Dmitri Mendeléev, the developer of the periodic table of the elements. In his Faraday lecture to the Chemical Society in 1889 Mendeléev described the effect produced by Cannizzaro:

> I vividly remember the impression produced by his speeches, which admitted of no compromise, and seemed to advocate truth itself, based on the conceptions of Avogadro, Gerhardt, and Regnault, which at that time were far from being generally recognized.[3]

Cannizzaro thus removed the doubts of those already disposed to accept discrete particles but did nothing for those fundamentally opposed to a discrete picture of matter composed of countable atoms. The counting process was still undeveloped.

Building on Avogadro's proposal, James Prescott Joule made "Some Remarks on Heat and the Constitution of Elastic Fluids" published in the *Memoirs of the Manchester Literary and Philosophical Society* in 1851 and in the *Philosophical Magazine* in 1857.[4] These "Remarks" contain the first derivation of the mean square of the speed of the molecules in hydrogen:

> Let us suppose an envelope of the size and shape of a cubic foot to be filled with hydrogen gas, which, at 60° temperature and 30 inches barometrical pressure, will weigh 36.927 grs. Further, let us suppose the above quantity to be divided into three equal and indefinitely small elastic particles, each weighing 12.309 grs.; and, further, that each of these particles vibrates between opposite sides of the cube and maintains a uniform velocity except at the instant of impact; it is required to find the velocity at which each particle must move so as to produce the atmospherical pressure of 14,831,712 grs. on each side of the cube.

He obtained a speed of 6225 feet per second, or 1.90×10^5 cm/second and pointed out that

> the above velocity will be found to produce the atmospheric pressure, whether the particles strike each other before they arrive at the sides of the cubical vessel, whether they strike the sides obliquely, and thirdly, into whatever number of particles the 36.927 grs. of hydrogen are divided.

He further noted that the pressure is proportional to the kinetic energy of the molecules, so that

> The specific heats of other gases will be easily deduced from that of hydrogen; for the whole *vis viva* and capacity of equal bulks of the various gases will be equal to one another; and the velocity of the particles will be inversely as the square root of the specific gravity. Hence the specific heat will be inversely proportional to the specific gravity, a law which has been arrived at experimentally by De la Rive and Marcet.

James Clerk Maxwell and Rudolf Clausius were among the first to develop in great depth the kinetic theory of gases. Joule had mentioned that intermolecular collisions would not affect his value for the speed of hydrogen molecules, but he gave no explanation for his assertion. In 1858 Clausius introduced the concept of *mean free path*, the average distance that a molecule travels before colliding with another one. The possibility of molecular collisions had been ignored until it was pointed out that if molecules travel as fast as Joule said, gases should diffuse into one another very rapidly. Ammonia molecules should move at $6225 \times \sqrt{(2/17)} = 2135$ feet per second, according to Joule, 2 being the molecular weight of hydrogen and 17 that of ammonia. The smell from a freshly opened bottle of ammonia should be detectable almost instantly at large distances, instead of after the passage of minutes or more, if ammonia molecules at room temperature move at such speeds. If, however, there are frequent collisions between moving molecules that change the direction of motion without, on average, slowing them down, the molecules will move fast but will not keep going in the same direction long enough to get very far in a short time. In current language, molecules make *random walks* instead of directed marches. This explanation of the slowness of diffusion was proposed by Clausius and immediately exploited by Maxwell in other connections.

Clausius also was the first to consider the effect of molecular rotation on the specific heat of gases. Previously the only form of energy assigned to molecular motion was the kinetic energy of the linear motion of the molecules, and this was not enough to account for measured values of specific heat, the energy required to raise the temperature of the gas. Later this became an acute problem. Forms of internal motion of molecules other than rotation became necessary to account for molecular behavior, but no consistent assignment of these motions accorded with measured values of the specific heat.

In 1859 Maxwell read a remarkable paper at a meeting of the British Association for the Advancement of Science at Aberdeen, Scotland, entitled "Illustrations of the Dynamical Theory of Gases."[5] Here, following Clausius in including the effect of collisions, he treated molecules as perfectly elastic hard spheres of finite size, and he analyzed the effect of collisions on the motions of these molecules. Among other things, he derived the distribution of molecular velocities; he proved the equipartition of energy, the equality of the mean kinetic energy among molecules; and he found an expression for the coefficient of viscosity of a gas. I shall discuss these three results in some detail because of the important role they played in future developments.

The Distribution of Molecular Velocities

Because gravity is too weak to change appreciably the motion of a molecule in the time between its collisions with other molecules or with the walls of the containing vessel, its effect on the gas can be ignored. After many collisions the direction of motion of a molecule is completely random; all directions of motion are equally probable, and the distribution is *isotropic*. This result requires no strong statistical assumption for its derivation.

Maxwell made what he regarded as an obvious statistical assumption to obtain an expression for the probability of a molecule's having a given velocity. He assumed that the probability of a molecule's having a certain component of velocity along one coordinate axis is independent of the components of its velocity along either of the other two coordinate axes. Later others pointed out that it was conceivable that if one component had an improbably large value, the other components might have compensatingly small probabilities of also being large. This would mean that the different components of velocity are correlated with one another. Maxwell excluded this possibility and asserted that the three components were not correlated. On the basis of this assumption he showed that the probability that a molecule has a certain component of velocity along some direction is described by the famous bell-shaped normal curve plotted in Fig. 1.4. The most probable value is zero, and the probabilities of motions with equal components in opposite directions are equal. The distribution in speed without regard to the direction of motion is shown in Fig. 1.5. The most probable speed is not zero; the probability of a very low speed is small because this requires all three components of velocity to be very small simultaneously, and the probability is large that at least one component will have an appreciable value.

Equipartition of Energy

After discussing the distribution of the velocities of molecules Maxwell stated and proved the following theorem:

> Proposition VI. Two systems of particles move in the same vessel; to prove that the mean *vis viva* (kinetic energy) of each particle will become the same in the two systems

This was the famous *equipartition of energy* theorem. It was generalized by Boltzmann in 1868 to include the internal degree of freedom of the molecules. Its implications haunted statistical physics until the introduction of the energy quantum more than three decades later.

The Viscosity of Gases

The viscosity or internal friction of gases depends on the occurrence of collisions between molecules. Maxwell reported:

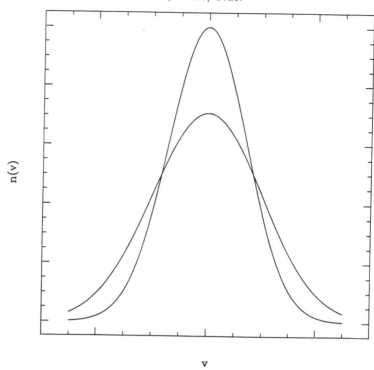

Fig. 1.4 Maxwell's distribution of a component of molecular velocity at two temperatures. The distribution is the same for the component in any direction.

> A remarkable result here presented to us . . . is that if this explanation of gaseous friction be true, the coefficient of friction is independent of the density. Such a consequence of a mathematical theory is very startling, and the only experiment I have met with on the subject does not seem to confirm it.[7]

Later experiments by Maxwell and by M. O. E. Meyer, however, did confirm this startling result. This was probably the first prediction of a previously unknown effect by using the kinetic theory of gases to be experimentally confirmed. It lent considerable support to the kinetic theory.

According to Maxwell's calculation, the viscosity of a gas was proportional to the product of the molecular density n, the mean free path of molecules in the gas, and the mean speed of the molecules. If the molecules were spheres of diameter d, the cross-sectional area presented to a molecule by the other molecules in a unit volume of the gas would be $n\pi d^2$, so the mean free path of a molecule would be inversely proportional to this quantity, making the viscosity independent of the molecular density. A measurement of the viscosity therefore would yield a value for quantity d^2. If the gas were liquefied,

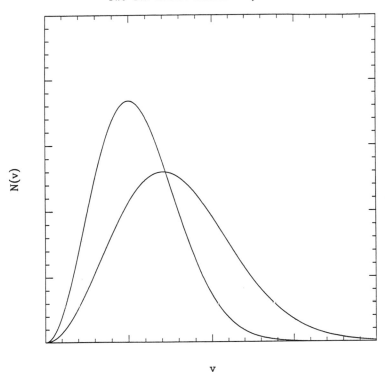

N(v)

v

Fig. 1.5 Maxwell's distribution of molecular speeds for the same two temperatures as in Fig. 1.4.

it would become almost incompressible, and it would be reasonable to assume that here "the volume of space occupied by a body" was a good measure of the amount of matter present. Combining this with the atomist's view that the volume of a sample of matter when the molecules are closely packed was near the number of molecules times the volume of a single molecule, one was led to conclude that the volume of liquid condensed from a unit volume of gas was equal to $n\pi d^3/6$ within a factor near unity arising from the shape of the molecules. Some shapes could be packed more closely than others. In 1866 Johann Loschmidt combined the information from these two sources to arrive at a value for d and for n, and hence for Avogadro's number. His result was $N_A = 0.5 \times 10^{23}$ molecules per mole. An estimate by Maxwell in 1873 was 4.0×10^{23}, and other estimates were given, all of the same order of magnitude. Hard-core skeptics were not convinced of the existence of atoms based on these results because they did not agree with one another very well, and there were not many independent effects involved in the measurements all leading to consistent answers. None of these estimates was based on the observation of individual particles with any atomic or molecular attribute. Collisions of individual molecules with one another provided the mechanism for the existence of viscosity, but there was no proof that this was the only possible explanation.

Maxwell's assumption of the statistical independence of the three components of a molecule's velocity was criticized by Clausius. In another paper, "On the Dynamical Theory of Gases," read in 1866 Maxwell observed:

> In . . . 1860, I have given an investigation of this case [the equilibrium velocity distribution], founded on the assumption that the probability of a molecule having a velocity resolved parallel to x lying between certain limits is not in any way affected by knowledge that the molecule has a given velocity resolved parallel to y. As this assumption may appear precarious, I shall now determine the form of the function in a different manner.[8]

Here he assumed that the probability of a molecule having a collision with another molecule does not depend on what collisions these molecules have had in the past. This is an extremely plausible assumption, and it was not until years later that the possibility of its not being true was pointed out by Paul and Tatiana Ehrenfest, who called it the *Stosszahlansatz* (collision number hypothesis).[9] Boltzmann used this same assumption when he showed that the equipartition of energy covered not only the center of mass kinetic energy and the energy of rotation about the center of mass, but all forms of internal energy that the molecule could possess, such as the kinetic and potential energies of internal vibrations.

Boltzmann believed that the second law of thermodynamics was a consequence of the laws of mechanics applied to large and complicated systems. Over a period of years he gave a series of more and more refined arguments leading to this conclusion. His first line of argument went as follows:

At any one instant of time, say t, the molecules in a gas have velocities distributed according to a function $f(v_x, v_y, v_z, t)$. This means that if there are N molecules altogether, the number having velocities whose components lie between v_x and $v_x + dv_x$, v_y and $v_y + dv_y$, v_z and $v_z + dv_z$ is

$$N f(v_x, v_y, v_z, t)\, dv_x\, dv_y\, dv_z.$$

This velocity distribution will generally change with time as a result of collisions between the molecules. The rate of change can be calculated using the Stosszahlansatz, which Boltzmann did not then recognize as a new assumption going beyond the laws of mechanics. Defining a quantity H (originally denoted by E) as the sum of the values of

$$f \log f$$

over the values of the velocities of all the molecules, he showed that it had the property of never increasing with time as a result of collisions between molecules. Further, the only distribution for which this quantity H did not decrease was the one given by Maxwell, now called the *Maxwell–Boltzmann distribution*.

By identifying the entropy of the gas as proportional to $-H$, Boltzmann seemed to have derived the second law of thermodynamics, for gases at least, from the principles of mechanics, without making any additional assumptions.

That Boltzmann's argument was overly optimistic was pointed out emphatically by J. Loschmidt[10] and much later on other grounds by E. Zermelo, and their criticisms forced him to reexamine the basis for this argument. Loschmidt pointed out that the equations of motion of the system were time reversible. If one motion is a solution of the equations, the motion with all velocities reversed is also a solution. The second motion retraces the path of the first, and if H decreased during one, it increased during the other. Zermelo pointed out that there is a theorem of Poincaré according to which a closed system eventually returns arbitrarily close to its original state. When it does so, the value of H returns to its original value, and any decrease must be balanced by an equal increase. (Zermelo, a student of Planck, considered this an argument for rejecting the entire atomic model of matter.) Boltzmann's mathematics was correct, but his physics was not. His assumption that summing $f \log f$ over all the molecular states of a closed system consistent with what was known about that system at a given time led to a function H analogous to the entropy of that system was untenable. On reversing all the velocities in each of these states, H would retrace its sequence of values, just as Loschmidt said it would. Only if the system were in thermal equilibrium and H were constant at its minimum value would the reversed system not contradict the second law. Loschmidt did not identify the source of the difficulty and correct it; this was done by Boltzmann himself.

In a long and difficult paper published in 1877 entitled "On the Relation Between the Second Law of the Mechanical Theory of Heat and the Calculus of Probabilities with Respect to the Theorems on Thermal Equilibrium" Boltzmann gave a thorough discussion of the way that a statistical theory can give rise to irreversible behavior of physical quantities when these are defined in a statistical way, as the physical properties of large complicated systems inevitably are.[11] Here he introduced the idea of a microstate or *complexion*. A complexion is a specific assignment of position, velocity, and whatever other quantities are needed to determine the state of a molecule to each molecule in the gas, this assignment being consistent with whatever is known about the state of the gas as a whole. For example, the sum of the energies assigned to the individual molecules must add up to the total energy of the gas. Instead of considering a single sample of gas with the appropriate values for the volume, pressure, temperature, and so on, Boltzmann considered an *ensemble* of samples, each one corresponding to one possible complexion consistent with this volume, pressure, temperature, and the like. He then proposed that the probability of the system having a certain value of some physical quantity such as H is proportional to the number of complexions in the ensemble for which that quantity has that value. The average value is then the product of a possible value of the quantity and the probability of its having that value, summed over all possibilities. This required one to be able to *count* complexions in order to evaluate average values.

In order to be counted, complexions must be discrete; one cannot count a continuum. If one of the characteristics of a molecule defining a complexion is the energy, as it nearly always is, Boltzmann had to make this energy take on only values that are an integer times a small unit ε. At the end of the calculation he allowed ε to approach zero, a process that led to no difficulties either mathematically or physically. This way of defining the average value of a dynamical quantity divorces the average from the value of the quantity in any one sample of the system, from any one member of the ensemble taken to represent the gas.

If a gas started out in a state far from thermal equilibrium, such as one in which all the molecules initially were located in one half of a container, there would be relatively few members of the ensemble representing the gas consistent with this state. Call this restricted ensemble C. Because of the motion of the molecules starting from this intial state, the state at later times will not look as special as the initial state does; the molecules will be distributed more widely and without any pattern that can be recognized as having arisen out of the highly special initial state. This later state is represented by complexions belonging to a much larger ensemble W. Note that W contains C, for after all, the molecules might have remained in their original volume or might have returned to it, however unlikely this is. If one then reversed the motions of all the molecules in all the complexions in W, including those in C, only those in C would develop backward so as to represent the special initial state at the initial time; the vast majority would develop in quite different ways not ending up in half the container at the initial time. The value of any physical quantity defined as a sum or an average over the ensemble W would not return in the course of the reversed motion to the value it had coming from the smaller ensemble C at the initial time. The behavior of physical quantities whose values were defined in this way was not reversible, even though the equations of motion specifying the development of any one member of the ensemble were because the ensemble to be used in finding the average changed with the time. At each time this ensemble was made up of all the complexions consistent with the properties of the system known at that time, and the selection of the complexions to include in the ensemble depended only on their possessing these properties at that time and did not depend at all on how they came to have them. In the example of the gas initially concentrated in half its container, all the complexions admitted initially had to fulfill the condition that every molecule was in that half. Since there are as many complexions with any one molecule in one half as with it in the other, all together there are 2^N times as many complexions with the molecules able to be anywhere in the container as there are with all the molecules confined to one half of it. When N is the size of Avogadro's number, 2^N is enormous. The chance for all the molecules in a mole of gas to be simultaneously in the same half of the container is completely negligible. See footnote on facing page.*

Boltzmann's method of assigning values to macroscopic physical quantitites depending on the collective behavior of the molecules comprising the system amounts to a new postulate. The validity of this postulate could, of course, be questioned, and it was. It still provides the fundamental basis for statistical physics.

In the case of gases Boltzmann was able to go further. The function H was defined for any macroscopic state of the gas, so if $-H$ is taken to be the entropy, this provides a definition of entropy for nonequilibrium states and not only for equilibrium states, as in pure thermodynamics. Having a definition for entropy in nonequilibrium states, Boltzmann was able to study the change in entropy as a system evolved in time as a consequence of intermolecular collisions, and he showed that the entropy was stationary—unchanging with time—only when the molecular velocities had the Maxwell distribution. This made the negative of his function H have most of the properties associated with the entropy as defined in thermodynamics, at least for gases. The one property it lacked was an inexorable monotonic increase (a decrease in H) until it reaches its maximum at equilibrium. Boltzmann's H fluctuated, often having a value somewhat above the minimum. It did have the property of being overwhelmingly more likely to be decreasing toward the minimum than increasing further above it. These fluctuations made his theory unacceptable to many of his contemporaries, who believed the entropy could *never* increase. The Ehrenfests showed explicitly how a complicated averaging process led to an H-curve that decreased monotonically until it reached the value it had for a Maxwell–Boltzmann velocity distribution and remained at that value without fluctuations. This averaging involved repeated changes of the ensemble used to define the value of H, as just discussed.

This introduction of ensembles marks the beginning of statistical physics as distinct from kinetic theory. Statistical physics is not a part of mechanics or any other branch of deterministic physics. Rather, it is an extension of it based on statistical hypotheses that appear reasonable and that have led to results in accord with experience. In fact, statistical physics turned out to be more robust than the classical physics that originally underlay it. Observations on a system in thermal equilibrium, blackbody radiation, turned out to be

---

* As an illustration of the size of the numbers entering this discussion, consider the following: If each member of a population of 4 billion has a choice of two equally attractive menu items, the probability of all choosing a particular item—a restaurateur's nightmare—is

$$(1/2)^{4 \times 10^9} = 10^{-1.204 \times 10^9}.$$

Compare this with the probability of 6×10^{23} molecules of a gas being in a specified half of the available volume. This is

$$(1/2)^{6 \times 10^{23}} = 10^{-1.80 \times 10^{23}}.$$

The *logarithm* of the probability of a unanimous menu choice is some fourteen orders of magnitude (fourteen factors of ten) times greater than the *logarithm* of the probability for all the molecules in a mole of gas congregating in one half of the available volume, that is, the probabilities differ by a factor consisting of a one followed by 10^{14} zeros.

irreconcilable with the statistical consequences of any classical theory. When the classical basis for the statistics was replaced by quantum ideas, these observations could and did fit into a new and consistent quantum–statistical scheme.

Notes

1. D. Bernoulli, *Hydrodynamica* (Strasbourg, 1738). This quotation is taken from W. F. Magie, trans. and ed., *A Source Book in Physics* (Cambridge, MA: Harvard University Press, 1963), p. 247.

2. Mary Jo Nye, ed., *The Question of the Atom* (San Francisco: Tomash, 1984), p. 5.

3. Ibid., p. 314.

4. James Joule, *Philosophical Magazine* 4, 211 (1857), excerpted in Magie, *A Source Book in Physics*, pp. 255 ff.

5. Published in *Philosophical Magazine*, January and July 1860; repr. in *The Scientific Papers of James Clerk Maxwell* (New York: Dover, 1965), vol. 1, p. 377.

6. L. Boltzmann, *Wien Berichte* 58, 517 (1868).

7. Maxwell, "Illustrations," in *Scientific Papers*, vol. 1, p. 391.

8. *Scientific Papers*, vol. 2, p. 43.

9. P. & T. Ehrenfest, *Encyklopädie der Mathematischen Wissenschaften*, vol. IV. 2. II D. An English translation is J. Moravcik, trans., *The Conceptual Foundations of the Statistical Approach in Mechanics* (Ithaca, NY: Cornell University Press, 1959).

10. J. Loschmidt, *Wien Berichte* 73, 139 (1876) and 75, 67 (1877); E. Zermelo, *Wiedemann's Annalen* 57, 485 (1896).

11. L. Boltzmann, *Wien Berichte* 76, 373–435 (1877); Fritz Hasenöhrl, ed., *Wissenschaftliche Abhandlungen von Ludwig Boltzmann* (New York: Chelsea, 1968), vol. 2, p. 164.

Amplification

Maxwell's first derivation of the molecular velocity distribution in a gas is perhaps the earliest example of an explicitly statistical calculation in physics that went beyond finding an average. Maxwell made the following argument:[1]

> Let N be the whole number of particles. Let x, y, z be the components of the velocity of each particle in the three rectangular directions, and let the number of particles for which x lies between x and $x + dx$, be $Nf(x)\, dx$, where $f(x)$ is a function of x to be determined.
>
> Now the existence of the velocity x does not in any way affect that of the velocities y and z, since these are all at right angles to each other and independent, so that the number of particles

whose velocity lies between x and $x + dx$, and also between y and $y + dy$, and also between z and $z + dz$, is

$$N^3 f(x) f(y) f(z) \, dx \, dy \, dz.$$

But the directions of the coordinates are perfectly arbitrary, and therefore this number must depend on the distance from the origin alone, that is

$$N^3 f(x) f(y) f(z) = \phi(x^2 + y^2 + z^2).$$

The "distance from the origin" is here the length of the velocity vector, the speed of the particle.

The only function $f(x)$ satisfying this functional equation can be written as

$$f(x) = C e^{-x^2/\alpha^2}.$$

The number C must be such that the number of particles with x-components of velocity anywhere between $-\infty$ and $+\infty$ is N, which requires that

$$f(x) = \frac{1}{\alpha \sqrt{\pi}} e^{-x^2/\alpha^2}.$$

From this Maxwell drew the following conclusions:

1st. The number of particles whose velocity, resolved in a certain direction, lies between x and $x + dx$ is

$$\frac{N}{\alpha \sqrt{\pi}} e^{-x^2/\alpha^2} \, dx.$$

2nd. The number whose actual velocity [speed] lies between v and $v + dv$ is

$$\frac{N}{\alpha^3 \sqrt{\pi}} v^2 e^{-v^2/\alpha^2} \, dv.$$

3rd. To find the mean value of v, add the velocities [speeds] of all the particles together and divide by the number of particles; the result is

$$\text{mean velocity} = \frac{2\alpha}{\sqrt{\pi}}$$

4th. To find the mean value of v^2, add all the values together and divide by N,

$$\text{mean value of } v^2 = \tfrac{3}{2}\alpha^2.$$

This is greater than the square of the mean velocity [speed], as it ought to be.

This derivation contains in an explicit way the assumption that the various Cartesian components of the velocities are uncorrelated with one another. This is the assumption that Maxwell later described as considered "precarious" by some.

The method of averaging required in Boltzmann's statistical approach to gas theory is based on the Hamiltonian form of the equations of mechanics. Each member of the ensemble describing a gas in some state with a definite volume, pressure, temperature, and so forth, is represented in the phase space by a phase point. If the phase points representing the members of the ensemble are thought of as particles of a phase fluid, the velocity of this fluid at any point in the phase space has components $\dot{q}_i$, $\dot{p}_i$ $(i = 1, 2, \cdots, f)$. These time derivatives are given directly by Hamilton's equations. As pointed out earlier in "Classical Mechanic, Amplifications," a consequence of the form of these equations is that the volume of phase space occupied by a set of phase points remains constant as they move. This does not mean that phase points starting out near one another remain close together; over time a compact region can revolve into an extended and highly distorted one, but its volume remains the same. When the total energy E of the system is conserved, the phase point of the system is confined to a $6N - 1$ dimensional "energy-surface" defined by the equation $H(q_i, p_i) = E$ in the $6N$ dimensional phase space. This is analogous to a point in three-dimensional space whose coordinates must obey the equation

$$x^2 + y^2 + z^2 = r^2$$

and which is therefore confined to the two-dimensional surface of a sphere of radius r in three-dimensional space.

In the case of gases it is useful to introduce the phase space for each molecule separately, called by the Ehrenfests the μ-space (mu for molecule), in contrast with the Γ-space (gamma for gas) of the entire gas. For Maxwell's rigid spherical molecules the μ-space describes only the motion of the center of the molecule and has six dimensions, three for the coordinates of the center of mass and three for the components of the center-of-mass momentum. For molecules containing r atoms, the μ-space contains $6r$ dimensions. The probability of a molecule being in the small region $\Delta\tau$ of its μ-space is written as f,

$$f = -\alpha\, e^{-\beta E}$$

where E is the total energy of that molecule in that region. (Because of the possibility of collisions between molecules, the energy of an individual molecule is not conserved, even if that of the entire gas is.) Both α and β are constants; β is inversely proportional to the kelvin temperature T.

The exponential factor, later written as $\exp[-E/kT]$, is known as the *Boltzmann factor* and plays an important role in all of statistical physics. Now k is called *Boltzmann's constant*, so christened by Max Planck.

Note

1. J. C. Maxwell, "Illustrations of the dynamical theory of gases," *Scientific Papers* (New York: Dover, 1965), vol. 1, p. 380.

From Fraunhofer and Kirchhoff: Radiation and Heat

The electromagnetic theory of light as it was understood toward the end of the nineteenth century was enormously successful in describing the propagation of light and other electromagnetic radiation. Thomas Young and Augustin Jean Fresnel had shown that a wave theory was needed to account for the behavior of light. Electromagnetic theory as formulated by Maxwell and Hertz provided a detailed dynamical theory of waves filling all the requirements that optics imposed on them. It gave a picture of how the presence of a material medium could affect the propagation of light, sometimes absorbing it and sometimes reflecting and refracting it. The index of refraction of a transparent medium, defined as the ratio of the speed of light in a vacuum to its speed in the medium, could be explained by the electric field in the light wave polarizing the material constituting the medium. The electric field could move the positive and negative charges inside the material a short distance relative to each other and so induce an electric polarization of the medium. The boundary conditions that govern how the electric and magnetic fields in two media match at an interface between them could be found from Maxwell's equations and the properties of the media. In this way the behavior of optical instruments, the intensity and polarization of light reflected from the boundary between two different media, the dispersion of light into its constituent colors by prisms, and countless other phenomena could be accounted for. What electromagnetic theory, as it was understood in the nineteenth century, could not account for was the way that matter was observed to emit and absorb radiation. The glowing of hot objects, the emission and absorption of line spectra by chemical elements, the weak ionization of gases produced by ultraviolet light, and the photoelectric effect all were unaccounted for by the theory. This was the more puzzling as the emission of radio waves by an antenna such as one used by Hertz in his fundamental experiments did accord with electromagnetic theory. Why was it so hard to explain the emission and absorption of light in the same way?

According to the then current theory, the emission of light of a given frequency required the existence of an electric charge oscillating or an electric current varying with that frequency. The variation with frequency of the index of refraction of many transparent media was ascribed to the existence of electric oscillators of various discrete frequencies in those media. They would be excited by the passage of a light wave and emit secondary radiation of the frequency of the exciting wave that would combine with the original wave to form a new wave of the same frequency traveling through the medium. This new wave would travel at a different, generally lower, velocity than the original wave and would therefore be of shorter wavelength. The degree of excitation of the oscillators in the medium would be greater the closer the frequency of the wave was to that of the oscillators, which would account for the different behavior of light of different frequencies in the material. However, no successful model explaining why these oscillators were present or why they had the frequencies they did was ever suggested.

The well-grounded attitude of nearly all nineteenth-century physicists was that the propagation of light was well understood but that the mechanism of its production and absorption presented many unanswered questions. The problems in understanding observed phenomena were universally ascribed to a lack of knowledge about the interaction of light with matter, and not to the nature of light itself.

One reliable theoretical tool to use in investigating the interaction of light with matter was thermodynamics. Thermodynamics is independent of the particulars of the interactions that produce thermal equilibrium, requiring only that these interactions exist and that they have sufficient time to act. Lack of knowledge of details does not prevent applying thermodynamic methods. The second law of thermodynamics requires the temperature of an isolated system in equilibrium to be everywhere the same and constant in time. Only under these circumstances are we prevented from running a heat engine between a warmer part and a cooler part of the system to extract work from it. In a uniform material, the amount of radiation emitted by any volume per unit time must therefore be equal to the energy absorbed by that volume in that time.

Using this fact in 1860, before the establishment of Maxwell's theory, Gustav Kirchhoff published several results, two of which were especially important.[1] First, the intensity of radiation of any frequency must be the same in all directions; that is, thermal radiation must be isotropic. This is a useful but not surprising conclusion. Of much more influence on the future of physics was the second result concerning the ability of matter to emit and absorb radiation.

Consider a sample of a material that is thick enough to absorb any radiation that enters it. The amount of energy that such a sample at temperature T emits per unit area per unit time in the form of radiation of frequency between v and $v + dv$ can be written as $E(v, T)\, dv$. This $E(v, T)$ is called the *emissivity* of that material. The *absorptivity* of the material, $A(v, T)$, is the fraction of the radiative energy in this frequency range incident on the surface of the sample that is absorbed rather than reflected. Kirchhoff showed that the second law of thermodynamics required the ratio $E(v, T/A(v, T)$ to be independent of the material and therefore to be a universal function of the frequency and the temperature. This theorem was to hold no matter what the form of the interaction between radiation and matter was, as long as there was a way for radiation and matter to come to equilibrium. Kirchhoff stated that the discovery of this function was one of the most important goals of physics. He called a material with $A(v, T) = 1$ a *blackbody* and denoted its emissivity by $e(v, T)$. The radiation emitted by such a material he called *blackbody radiation*. He expected this function to be simple, as basic functions usually are. This is as far as he could go without specifying something about the interaction between light and matter other than its existence.

Maxwell's electrodynamics embodied Faraday's ideas of a stress in the medium filling all space. It required the presence of a *radiation pressure* wherever there was radiation, and in particular, when that radiation was isotropic as thermal radiation in a cavity is, the pressure was one third of the energy density of the radiation. This meant that one could construct a Carnot heat engine with radiation as the working substance and analyze various processes by studying appropriate Carnot cycles. In this way Boltzmann showed that the total energy density of thermal radiation at a kelvin temperature T is proportional to the fourth power of the temperature.[2] Josef Stefan had arrived at this result empirically by analyzing the small amount of available data on the subject.[3] It is now known as the Stefan–Boltzmann law.

Wilhelm Wien extended this approach.[4] By taking into account the Doppler shift in the frequency of radiation on reflection from the moving piston of Boltzmann's Carnot engine, Wien was able to show that the energy density of thermal radiation at any one frequency, denoted by $u(v, T)$, does not depend on the frequency and temperature separately but must, aside from a factor of frequency to the third power, be given by a function of the ratio of the frequency to the temperature. In mathematical language, it must be of the form

$$u(v, T) = v^3 \phi\left(\frac{v}{T}\right),$$

where ϕ is an unknown function. Instead of having to find a function $u(v, T)$ of the two independent variables v and T, one had only the simpler task of finding a function $\phi(v/T)$ of the single variable v/T. Wien did not actually state his result in this manner, but he did offer a way to determine the frequency distribution curve for any temperature from that for another temperature, which amounts to the same thing. Measurements of the radiation from "blackbodies" showed that the function $\phi(v/T)$ has a maximum. If it did not have a maximum, thermal equilibrium could not be achieved because it would take an infinite amount of energy to give equal or increasing amounts of energy to equal frequency intervals as the frequency increased without limit. This would lead to what Paul Ehrenfest christened an *ultraviolet catastrophe*. It followed from Wien's result that the frequency at which this maximum occurred was proportional to the temperature. For this reason it is called the *Wien displacement law*.

Physicists hoped that the form of the function $\phi(v/T)$ could be derived from electrodynamics, and Max Planck embarked on a long series of researches to this end. Familiar radiation processes look irreversible: The moon emits light scattered from the sun in outgoing spherical waves, but no one sees ingoing spherical waves converging back on the moon and being returned to the sun. Planck tried to show that the irreversibility inherent in the second law of thermodynamics arose in the case of electromagnetism from this apparent irreversibility in the emission of electromagnetic radiation. This was essential preparation for Planck's first attempt to derive the form of $\phi(v/T)$. Boltzmann destroyed the prospects of this form of attack on the problem by showing that Maxwell's equations are symmetric under time reversal: To any solution of them describing the scattering of an incident beam of radiation by a system of charges and currents that results in a set of outgoing scattered waves, there corresponds a solution describing a set of incoming waves that are scattered by this system in a way to produce the original incident beam traveling back in the opposite direction. The latter is much harder to prepare than is the former, so it is not seen. Planck had fallen into the same error here that Boltzmann had fallen into earlier in connection with gases. Both were looking for a direct connection between a fundamental time-reversible theory, mechanics or electromagnetism, and the second law of thermodynamics, and eventually both were forced to give up and to recognize that the second law is of a statistical nature; entropy is overwhelmingly likely to increase with time in a large system, but it does not have to increase *always* and under *all* circumstances. Boltzmann had been met with Loschmidt's challenge, Planck with Boltzmann's. Something new had to be added before progress could be made in determining Wien's function $\phi(v/T)$. It was still sought in the interaction of light and matter; no modification of Maxwell's theory was contemplated.

What we are seeing here are attempts to apply known and trusted ideas to an area of physics that was not very well understood. The application was difficult and beset with booby traps into which even the best scientists fell. What distinguished them from their fellow victims was their ability to understand the difficulties they had met with unawares and to overcome them.

The unassailable result of the long effort of applying thermodynamics to the interaction of radiation with matter was Wien's displacement law, not a complete theory of thermal radiation. A framework had been created into which such a theory would have to fit. The freedom to formulate a theory was severely restricted, but it was not reduced to zero. The basis for the future theory was expected to lie in an increased understanding of the mutual interaction of radiation and matter. No one questioned the appropriateness of the known and trusted ideas constituting classical electromagnetism for the derivation of the blackbody spectrum. Statistical elements might have to be introduced, as had been the case in the theory of gases, but there classical mechanics seemed to have survived, and it was expected that here classical electromagnetism would also survive. There was no plausible ground for doubting this.

Notes

1. G. Kirchhoff, *Annalen der Physik* **109** 148 (1860); *Philosophical Magazine* **19**, 275 (1860).
2. L. Boltzmann, *Wiedemann's Annalen* **22**, 291 (1884).
3. J. Stefan, *Wien Berichte* **79**, 391 (1879).
4. W. Wien, *Wiedemann's Annalen* **52**, 132 (1894).

Amplification

Kirchhoff's original thermodynamic argument of 1860 leading to the existence of his universal function giving the ratio of the emissivity of a body to its absorptivity was characterized by Wien as "extraordinarily artificial and difficult," and he described a simpler line of reasoning leading to the same conslusions given some years later by E. Pringsheim.[1] I sketch this simpler argument here. It exploits the ability of radiation to do work directly via the radiation pressure occurring in Maxwell's theory, an approach not available to Kirchhoff at the earlier time because the existence of radiation pressure had not yet been recognized.

When a material body is in thermal equilibrium, all forms of energy that can interact with that body are present. Radiation being such a form of energy, it makes up a definite fraction of the total thermal energy present, and it has a temperature equal to that of the body. The entropy of this radiation has its maximum possible value. Also inside a cavity in this body there is radiation at this same temperature. Pringsheim showed that if this radiation were not uniform in direction, that is, if it were not isotropic, one could easily construct a machine that would produce work at the expense of heat energy without any other change taking place, thereby contradicting the second law of thermodynamics. Consider radiation in the two opposite directions along a line. Place a two-sided mirror in the cavity perpendicular to this line. If more radiation comes from one side than from the other, the mirror will experience a net force due to the greater radiation pressure on one side and will have work done on it if it moves in the direction of this force, with the equilibrium not being disturbed in the process. The second law of thermodynamics says that this is impossible. The radiation in the two directions along a line must be of equal intensity. This argument is easily extended to other directions by considering light incident obliquely on the mirror. Radiation traveling along one line is reflected along another line, so that the radiation in various directions can be compared, and again it must be equal in order to prevent one from obtaining work from the radiation energy without disturbing the equilibrium. By using mirrors that reflect light only in a narrow range of frequencies, this result can be extended to apply separately to each frequency of light. The final result is that thermal radiation of any frequency must be isotropic, the same in all directions.

If the material of which one part of the wall of the cavity is made is replaced by another material at the same temperature, the thermodynamic equilibrium of the walls will not be affected. This change does not affect the radiation of intensity $e(v, T)$ of the radiation of a given frequency coming from a part of the original wall after having either been emitted by that wall or reflected by it and heading in the direction of a part of the new wall. According to the theorem just discussed, the radiation coming in the opposite direction from the new wall must also be $e(v, T)$. The radiation in a cavity at a temperature T therefore does not depend on the nature of the cavity walls; it can depend only on the frequency of the radiation and the temperature of the cavity walls. The function $e(v, T)$ is a *universal* function.

Now consider the radiation of frequency v and of intensity e striking the cavity wall from a certain direction. The wall is made smooth, and so the usual law of reflection is followed. A fraction $A(v, T)$ of this radiation is absorbed by the wall, and the amount $(1 - A)e$ is reflected in another direction. The wall emits an amount of radiation $E(v, T)$ in this same direction, and so the total radiation going in the reflected direction is $(1 - A)e + E$. But this must be equal to e coming in the opposite direction, so that

$$[1 - A(v, T)]e(v, T) + E(v, T) = e(v, T)$$

$$\frac{E(v, T)}{A(v, T)} = e(v, T).$$

The ratio of the emissivity to the absorptivity of any material is a universal function of the frequency and the temperature. Kirchhoff introduced the idea of an ideal body for which

$$A(v, T) = 1$$

so that it absorbs all the radiation incident on it, and he called it a *blackbody*. The radiation emitted by such a body is specified by just the universal function $e(v, T)$ and is called *blackbody radiation*. Kirchhoff thought the determination of this function to be of prime importance for physics because of its universal character.

Kirchhoff's argument, as distinct from Pringsheim's, did not use the existence of radiation pressure explicitly, and neither argument led to any conclusion about the form $e(v, T)$. Boltzmann used the value for the pressure given by Maxwell—namely, one third of the energy density due to all frequencies of radiation—to restrict this function.[2] Using this in the expression for the work done on a piston in a Carnot engine, $dW = p \, dV = u(T) \, dV/3$, he was able to show that the total energy density of radiation, the integral of the energy density per unit frequency interval over all possible values of the frequency, must be proportional to the fourth power of the temperature:

$$u(T) = \int_0^\infty dv \, u(v, T)$$

$$= \sigma T^4.$$

This is the Stefan–Boltzmann law. Because only the total pressure affects the work done, this analysis does not give any information about how the intensity depends on the frequency. It does, of course, require that the radiation pressure integrated overall frequencies be finite. This requirement was so obvious that it was not remarked upon. A little later the consistency of this finiteness with the existence of thermodynamic equilibrium became an important question.

Wien's original derivation of this displacement law involved analyzing what happens when a perfectly reflecting cavity containing thermal radiation is slowly expanded. The details of the calculation need not concern us. On reflection from the moving piston closing the cavity, the light undergoes a Doppler shift to lower frequencies. At the same time the radiation pressure does work on the piston, lowering the total energy in the cavity and hence the temperature. By looking at these processes in detail, Wien established a relation between the change in frequency of the radiation of a given frequency and the change in temperature,

$$\frac{dT}{T} = \frac{dv}{v}.$$

This leads to the form

$$u(v, T) = A v^3 \phi \left(\frac{v}{T} \right),$$

where the factor v^3 is needed to satisfy the Stefan–Boltzmann law. The derivation gives no information about the form of the function ϕ except that it depends on the single variable v/T. The search for the appropriate function was enormously simplified by the limitation to a function of a single variable rather than one of two independent variables.

Notes

1. W. Wien, "Theorie der Strahlung," in *Encyclopädie der mathematischen Wissenschaften*, V 23 (May, 1909); E. Pringsheim, *Verhandlungen des deutsches physikalisches Gesellschaft* (1901), p. 81.

2. L. Boltzmann, *Annalen der Physikalische Chemie* **22**, 291 (1882).

1895

In 1895 physicists saw the world as a place where the classical ideas of the motion of particles and the behavior of fields were well established. There were many unanswered questions about the ultimate structure of matter and the interaction between radiation and matter, but there was little doubt that these questions could be answered within the framework of concepts as it existed at the time. The reality of atoms and molecules was widely though not universally accepted, but knowledge of their number and their structure was very approximate. Only a small fraction of the scientific world was concerned with these questions, and an even smaller fraction saw any profound difficulties in connection with them. There were few theoretical models of atoms and molecules and few pieces of experimental information with which to confront them. Most of the effort in physics research went into the classical subjects of mechanics, sound, optics, heat, and thermodynamics. That whole new areas of physics such as radioactivity, cathode ray and X-ray physics, the quantum of action, and relativity were just ahead was completely unanticipated. Four new constants, e representing the electron charge, h the quantum of action, k, or N_A, specifying the number of molecules per mole, and a newly important constant c, the speed of light in vacuum, would come to dominate physics. In 1895 even the concepts of which the first two are a measure were unknown, and the third had not yet reached universal acceptance. Of course, this had been true of all previous years, too. What makes this particular year special is that it was the year when the rapid development of technology began to make possible new kinds of experiments that probed areas of physics previously inaccessible to anything but speculation. These new capabilities led to the experiments whose interpretation required the introduction of the constants, e, g, and k. Relativity was a separate case. It gave c a meaning going far beyond representing just the speed of another type of waves, electromagnetic waves.

The rest of this book is an account of how these four constants of nature, e, h, k, and c, came to play a dominant role in physics between 1895 and 1925.

CHAPTER 2

Clouds on the Horizon

In the first chapter I gave a survey of the landscape of classical physics. There where clouds on the horizon, but they were few in number and not of ominous darkness. They certainly did not seem to threaten the underpinnings of classical physics. Instead, the clouds appeared mostly in connection with attempts to probe the ultimate nature of matter. In a simple and intuitively attractive way, kinetic theory gave an accurate description of much of the behavior of gases. In some cases, however, gases seemed to behave in too simple a manner. Their molecules seemed not to vibrate when heated, even though the theory required them to do so. The observed value of the molar heat was too small. On the other hand, crystals seemed to act in too complicated a way. Their molar heats should have been independent of the temperature and were so over considerable temperature ranges, but as lower temperatures were attained they were seen to decrease with decreasing temperature for no obvious reason. The whole subject of the emission and absorption of light by matter was a mystery. The characteristic spectra emitted and absorbed by matter, especially gases and vapors, did not fit into any known physical picture. Still, it was not self-evident that these problems were beyond the possibility of resolution within the classical framework, and prolonged efforts were made to do this. In the end, however, no such resolution proved possible. In this chapter I shall describe these clouds as they were perceived at that time.

The Last Stand Against Atoms and Molecules

In a paper read to the Society of German Scientists and Physicians in September 1895, the future Nobel Prize-winning chemist Friedrich Wilhelm Ostwald dismissed the atomic model of matter almost with contempt:

> The [atomic] view required the assumption that when, for instance, all the
> perceptible properties of iron and oxygen had disappeared in iron oxide,
> nevertheless iron and oxygen still existed in the body produced and had only
> assumed other properties. At present we are so accustomed to such an idea
> that it is difficult for us to conceive its strangeness, indeed its absurdity....
> ... Nevertheless it must be remarked that a confirmation of the natural
> deduction from this theory, namely, that all non-mechanical phenomena such
> as heat, radiation, electricity, magnetism and chemical action are actually
> mechanical, has in no single case been obtained.... The assertion that all
> the phenomena of nature can be primarily referred to mechanical ones cannot
> even be designated here as a practical working hypothesis; it is simply
> incorrect.
>
> This error appears more clearly when viewed in the light of the following
> fact. The equations of mechanics all possess the property that they still hold
> good when the sign of the quantity denoting time is changed. That is to
> say, theoretically perfect mechanical processes can take place just as well
> backwards as forwards. In a purely mechanical world there would be,
> therefore, no Before and no After, in the sense of our world; the tree could
> return again to the sapling and the sapling to the seed, the butterfly transform
> itself once more into the caterpillar, and the old man become again a babe.
> For the fact that this does not occur, the mechanical conception of the world
> has no explanation to offer, and can have none on account of the already
> mentioned property of mechanical equations. The evident irreversibility of
> actual natural phenomena proves, therefore, the existence of processes which
> cannot be represented by mechanical equations, and with this statement the
> judgment on scientific materialism is passed. We must accordingly ... give
> up all hope of getting a clear idea of the physical world by referring
> phenomena to an atomistic mechanics.[1]

Similar views were held by Ernst Mach, a physicist and philosopher who was
an authority on the foundations of mechanics, and by a host of others.

Ludwig Boltzmann was reacting to views such as when in 1898 he wrote
the following in the preface to the third edition of his *Lectures on Gas Theory:*

> Just at this time, however, the attacks against gas theory multiply. I am
> convinced that these attacks are based entirely on misunderstanding, and that
> the role of gas theory in science is far from played out. In this book I shall
> try to make intuitively clear a great number of results agreeing with experience
> obtained purely deductively from it by Van der Waals. Very recently this
> same theory has given hints which one could not have obtained in any other
> way. From the theory of the ratio of specific heats Ramsay deduced the atomic
> weight of argon and thereby its place in the system of chemical elements,
> which he showed was the correct one by his later discovery of neon. Similarly
> Smoluchowski established the existence and size of the jump in heat
> conductivity with temperature in very dilute gases from the kinetic theory. I
> would in my opinion be harmful for science if gas theory should be driven
> into oblivion by the temporary success of its opponents' voices, as for
> example, was once the wave theory of light by the authority of Newton.[2]

The great obstacle faced by those trying to convince the skeptics of the reality of atoms and molecules was the lack of phenomena making apparent the graininess of matter. It was only by seeing individual constituents, either directly or indirectly through the observation of fluctuations about the mean behavior predicted by kinetic theory, that the existence of these particles could be shown unambiguously. Nothing of the kind had been seen as yet, as Ostwald so forcefully pointed out in the preceding quotation, although the method that Loschmidt used in estimating Avogadro's number, described earlier, verged on it. Maxwell's velocity distribution for gas molecules had not been verified; the equipartition of energy theorem was consistent with some molar heat data and wildly inconsistent with others; and atomic and molecular weights were measured relative to a standard such as unity for hydrogen, but no reliable connection with laboratory-scale mass units could be made. Detectors for individual particles were unknown, and no atomic model was adequate to suggest the design of one. But all this was about to change at the time these statements were being made.

I shall discuss Ostwald's list of mechanically inexplicable things item by item to see why he put them on it and how they got off it if they did.

1. *Heat.* Ever since Rumford, some scientists had had the feeling that heat must be a form of motion. There had not, however, been any empirical identification of what was moving in thermal motion. The old caloric model had worked well in a limited area of experiment, but by the middle of the century it had been replaced by the recognition, by Robert Mayer and Rudolf Clausius, of heat as a form of energy. This step was regarded as necessary by the chemists and the physicists alike, with a few exceptions such as Ernst Mach. What was not at all clear was the necessity of accepting the idea that heat is a form of *mechanical* energy. Thermodynamics was independent of this assumption, and its success in all its applications made it perfectly acceptable to a large class of scientists as a fundamental theory requiring no interpretation in terms of a model. The analysis of the motion of an anomalously large molecule in equilibrium with an ordinary fluid that would lead to the interpretation of Brownian motion as just such a motion in a way that allowed quantitative comparison of the model with observation was soon to come.

2. *Radiation.* The wave theory of light had long since relegated to history the corpuscular theory of light. The mechanical models of an ether able to support mechanical waves with the properties of radiation were artificial to an extreme and, even then, made the unobservability of the earth's motion through the ether explicable only by conspiracy theories. This had little direct bearing on the atomic controversy, but it did not further the atomic model. A theory of blackbody radiation agreeing with observations and involving discrete energy quanta was at hand.

3. *Electricity.* Electric charge was thought to consist of either one or two conserved imponderable fluids. There was no more reason to think of the electric fluid(s) as consisting of discrete units than there was to think of

ordinary fluids in this way, although some such as Johnstone Stoney did propose the existence of "atoms of electricity." The electron, including the measurement of its finite electric charge, was about to be discovered.

4. *Magnetism.* With the recognized nonexistence of magnetic poles, magnetism became subsidiary to electricity. It had no independent sources of its own, and its association with the spin of the electron lay far in the future.

5. *Chemical action.* It was chemical action in the form of the laws of definite and multiple proportions that gave rise to the modern atomic and molecular pictures in the first place. Here as elsewhere, however, there were only ratios of masses, and there was no chemical reason to insist that the numerator and denominator of these ratios were finite. The detection of individual alpha particles—an ionized form of helium—by scintillation and of individual ionic charges by cloud techniques also was just around the corner.

Although many things were soon to come, in 1895 they had not yet arrived. One can fault the nonbelievers in atomism as having poor scientific taste and being bad guessers, but one cannot show them to have been bad scientists on the basis of what was indisputably known at this time. In short, no body of evidence had accumulated that was so compelling that it could overwhelm the conviction of those who, for whatever reason, rejected the idea that atoms are real. Avogadro's number was not an essential ingredient of physics or chemistry. Most physicists and chemists got along splendidly not knowing its value or even if it had one.

Notes

1. F. W. Ostwald, reprinted in Mary Jo Nye, ed., *The Question of the Atom* (San Francisco, Tomash, 1984), pp. 337 ff.

2. L. Boltzmann, *Lectures on Gas Theory*, 3rd ed. (Leipzig, 1898), preface.

Troubles with the Atomic Description of Heat

By the last decade of the nineteenth century, heat was recognized as a form of energy by all but a very few holdouts. Ernst Mach did not accept it, but his reasons seem to have been that he regarded as meaningless the distinction between conserved quantities such as matter and nonconserved quantities such as the energy of a particular body. Heat as a form of random motion was widely accepted, but even here there were holdouts too among working scientists. The *energeticists* led by G. Helm and Wilhelm Ostwald saw no need for the atomic picture. But even for those who accepted the atomic picture, there were troubling problems.

For Maxwell the largest problem, to which he referred on many occasions, was understanding the specific heats or molar heats of gases.[1] This problem was acute because the atomic model of a dilute gas was so simple that it should have been completely understandable. The calculation of molar heats, which refer to a mole rather than to a unit mass of the gas, seemed too straight-forward to have any hidden or unexpected features. Yet the theory did not account for the experimental results in a satisfactory way.

Maxwell had proved the equipartition of energy among the various modes of motion of a gas in thermal equilibrium at a temperature T, on reasonable statistical assumptions. At first he studied collisions between utterly smooth spherical molecules incapable of any internal vibration; that is, they were rigid. The only force that a colliding molecule can exert on a smooth round one is directed toward its center, so that there is no effect on its rotational motion. The collision can change the energy of the motion of the molecule's center but not its rotational energy. The mean value of the center's kinetic energy coming from motion in any one of the three possible independent directions of motion should be $kT/2$. The molar heat—the amount of heat required to raise by one degree the temperature of a gas composed of N_A such molecules—should, therefore, be $(3/2) \times N_A \times k = 3R/2$, about three calories per degree. Later Maxwell found that for rigid molecules of a more general shape that could be set spinning by collisions, the mean rotational kinetic energy for each of the three independent axes of rotation should also be $kT/2$. Boltzmann then proved this result on still more general assumptions, including any vibrations that could be excited in collisions.[2]

Suppose there are d internal ways that a molecule can store energy, including kinetic energy of rotation around the molecule's center of mass, vibrational kinetic energy, and vibrational potential energy. These are in addition to the three ways it can have kinetic energy associated with the motion of its center of mass, which is the only motion that leads to the gas molecules colliding with the walls and exerting pressure on them. Then the molar heat of a gas that is heated at constant volume should be

$$C_V = N_A \times (3 + d) \times \frac{k}{2} = \frac{3 + d}{2} \times R.$$

Molar heats of gases are difficult to measure accurately. Gases are not very massive, and the heat required to raise their temperature is much less than that required to heat the containing vessel. It is easier to make accurate measurement of the ratio $\gamma = C_P/C_V$ of the molar heat C_P of a gas at constant pressure to the molar heat C_V at constant volume because this ratio enters the expression for the speed of sound in the gas, and the speed of sound is readily determined. Now C_P differs from C_V because when heated at constant pressure a gas expands and does work on the mechanism that keeps the pressure constant, whereas when heated at constant volume it does no such work. Extra heat must be supplied to do this work. For gases that approximate ideal gas behavior, the extra heat increases the molar heat at constant pressure over that at constant volume just by R,

$$C_P = C_V + R.$$

Then the ratio of the two molar heats is

$$\gamma = \frac{C_P}{C_V} = 1 + \frac{2}{3 + d}.$$

A measurement of the ratio γ yields a value of d.

For the so-called permanent gases such as hydrogen, oxygen, and nitrogen that were known when this theory was being developed, the measured value of γ was very nearly 1.40, which corresponds to $d = 2$. It was not clear that at the time that all these gases were diatomic and their molecules therefore not spherical. It was expected that they would be spherical and would have $d = 0$ and $\gamma = 1.67$ or would be unsymmetrical and have three axes around which rotations could be produced by collisions so that they would have $d = 3$ and $\gamma = 1.33$. No gas whose molar heat had been measured yielded a value of γ corresponding to either of these, as should be the case for molecules that cannot vibrate, although by 1895, argon, the first inert gas discovered, and mercury vapor, were found to have γ very nearly 1.67 and were therefore taken to be monatomic. The value 1.40 could be rationalized by saying that the molecules of these gases were symmetrical around one axis so that collisions could not produce rotation around that axis, but it was not satisfying to have this one motion completely unable to partake of thermal motion when all the other motions could. Eventually, in 1895, helium was isolated on the earth and did turn out to behave as a gas composed of rigid spherical molecules should and had $\gamma \approx 1.66$.

None of this was very plausible. If molecules existed, they could not be absolutely rigid. They were supposedly made of atoms and could be broken up by chemical means, so the forces holding the atoms together were finite and should permit vibrations to be excited by collisions. The existence of optical spectra, presumably emitted by atoms and molecules in gases, was well established. These spectra contained light of many frequencies that, according to electromagnetic theory, could be emitted only by electric charges or currents oscillating at those frequencies within the molecule. These oscillations should contribute to the energy of a gas and therefore to its molar heat. They did not, however, which led Maxwell to write in this connection:

> And here we are brought face to face with the greatest difficulty which the molecular theory has yet encountered, namely, the interpretation of the equation $n + e = 4.9$ [$d \approx 2$ in our notation]. . . .
>
> But the spectroscope tells us that some molecules can execute a great many different kinds of vibrations. They must therefore be systems of a very considerable degree of complexity, having far more than six variables [$d = 3$]. Now, every additional variable introduces an additional amount of capacity for internal motion without affecting the external pressure. Every additional variable, therefore, increases the specific heat, whether reckoned at constant pressure or at constant volume.
>
> So does any capacity which the molecule may have for storing up energy in the potential form. But the calculated specific heat is already too great

when we suppose the molecule to consist of two atoms only. Hence every additional degree of complexity which we attribute to the molecule can only increase the difficulty of reconciling the observed with the calculated value of the specific heat.[3]

Crystalline solids provided another case in which the atomic picture almost worked when applied in the simplest way. As early as 1819 Pierre Dulong and Alexis Petit had noted that the molar heat of pure crystals was close to 6 cal/mole for a wide variety of pure materials. This rule was used to correct the atomic weights of several metals, including lead, iron, and mercury, that had been assigned values twice as large as the modern ones because of uncertainty about their valences. Now 6 cal/mole is very nearly three times the gas constant R, so the Dulong–Petit value corresponds to what would be expected from a crystal composed of point masses free to vibrate in three directions, each atom having kinetic energy from possible motions in three directions and potential energy from possible displacements from its equilibrium position in the crystal in those three directions, and nothing else. According to Boltzmann, each of these should have a mean energy of $kT/2$, and so for a mole of material the energy required to increase the temperature T by one degree should be

$$N_A \times 3k = 3R.$$

This is the Dulong–Petit value.

This triumph was not conclusive, for at least two reasons. The rule did not hold with absolute precision. The variations were minor in most cases, but by 1895 there were a few major ones such as diamond, boron, and silicon whose molar heats were too low. Atoms capable of emitting complicated line spectra should have more ways to store energy than point atoms do, and yet here again were some that apparently had even fewer! Worse, by this time it was known that molar heats are not always independent of the temperature, tending to decrease at lower temperatures. This did not fit the equipartition theorem at all. Classical physics never could account for the absence of thermal energy associated with modes of motion whose existence was made clear by other effects such as the emission of line spectra.

In short, by 1895 the atomic theory of heat had met with great successes and also had encountered formidable difficulties whose resolutions were not in sight. It required a separate description of the atoms of each element. There was no prospect of explaining the periodic table of the elements as a consequence of atomic structure because no subatomic constituents out of which to build atoms were known. The existence of a line spectrum characteristic of each element emitted when the element was sufficiently hot was known and used for chemical analysis, but the source of these spectra was totally unexplained. The few regularities recognized in spectra such as Balmer's formula for the wavelengths of light emitted by hydrogen gave no hint of their explanation in terms of a model. In short, the statistical approach based on the model of atoms as particles obeying the laws of classical mechanics had failed to give a consistent description of all observed phenomena, succeeding here and failing there. A period of steady progress was drawing to a close. Much knowledge of the behavior of physical systems had accumulated, and many successes had been achieved. But some of these successes were incomplete, which led to intriguing problems, and there were few clues to how or where to look for new insights that might resolve them. Unanimity on the fundamentals of the atomic model was lacking in that rather small part of the scientific community concerned with questions of basic principles.

Notes

1. "Illustrations of the Dynamical Theory of Gases" and "On the Dynamical Evidence of the Molecular Constitution of Bodies," in *The Collected Scientific Papers of J. C. Maxwell*, ed. W. D. Niven (New York: Dover, 1965), vol. 1, p. 409, and vol. 2, p. 433.
2. L. Boltzmann, *Wien Berichte* (1877), in *Wissenschaftliche Abhandlungen* (New York: Chelsea, 1968), vol. 1, pp. 67, 373.
3. Maxwell, "Dynamical Evidence," vol. 2, p. 433.

The Connection Between Matter and Radiation

We have seen how thermodynamics was applied to the matter–radiation system by Gustav Kirchhoff, Ludwig Boltzmann, and Wilhelm Wien, culminating in the Stefan–Boltzmann law and the Wien displacement law. These well-established principles also were applied to the problem of thermal equilibrium in systems containing both matter and radiation, and important results consistent with what was observed were obtained. Some aspects of this problem had proved resistant to understanding, especially the line spectra associated with atoms and molecules constituting a gas and the shape of the blackbody radiation spectrum.

Atomic spectroscopy began with Joseph Fraunhofer's observation in 1814–15 of a pair of bright yellow lines in the spectrum of a lamp that he was using as a light source while measuring the index of refraction of some glass prisms.[1] These lines were useful to him because they provided a standard color of light with which to measure the index of refraction of different glasses. When dispersing sunlight with a prism and looking closely at the resulting spectrum to see whether these bright lines were also present there, he observed the sequence of dark lines that now bears his name. Labeling them with the letters of the alphabet, two dark lines at the wavelength of the two bright lines in the lamplight received the label *D-lines*, which they have kept ever since.

No interpretation of these observations was offered until the work of

Kirchhoff and Robert Bunsen in 1859. These two investigators studied the line spectra emitted by materials vaporized by a burner with a nonluminous flame developed by Bunsen. They established that these line spectra were characteristic of elements present in the vapor, and they associated the Fraunhofer *D* lines with sodium. They discovered the metals caesium and rubidium through their spectra, which matched no known ones. Even more important, they established that a cool vapor will absorb light of the same color that it emits when heated, so that the light from a bright source viewed through a vapor will show dark lines, whereas the vapor alone would show bright ones. Kirchhoff applied this discovery to the Fraunhofer lines in the solar spectrum, and in 1859 he reported:

> I conclude from these observations that a colored flame in whose spectrum bright sharp lines occur so weakens rays of the color of these lines, if they pass through it, that dark lines appear in the place of the bright ones, whenever a source of light of sufficient intensity, in whose spectrum these lines are otherwise absent, is brought behind the flame. I conclude further that the dark lines of the solar spectrum, which are not produced by the earth's atmosphere, occur because of the presence of those elements in the glowing atmosphere of the sun which would produce in the spectrum of a flame bright lines in the same position. We may assume that the bright lines corresponding with the *D* lines in the spectrum of a flame always arise from the presence of sodium; the *D* lines in the solar spectrum permit us to conclude that sodium is present in the sun's atmosphere.[2]

This work was of tremendous significance for astronomy, as it made possible the identification of chemical elements in the sun and stars. It also led to Kirchhoff's investigation of the relation of emission to absorption in general and to his demonstration that the ratio of a body's emissivity to its absorptivity is the same for all bodies, a theorem that I have already discussed.

The fact that vapors could emit and absorb light of the same definite frequencies led to the suggestion that the atoms of the vapor were resonant systems of some sort, just as a stretched string is resonant. Such a string will emit a definite note when plucked and will become excited if exposed to the sound from a similarly plucked string nearby but not if exposed to the sound of a different note. The study of the spectra of the elements became an important part of physics. Atoms and molecules began to acquire physical properties in addition to their chemical ones, and questions about atomic and molecular structure could be asked, if not answered. (I commented earlier about Maxwell's worries about this structure as it affected the kinetic theory of gases.)

With the discovery of atomic spectra, identifying regularities in the wavelengths emitted was an obvious goal. A vapor or a gas of a single element emitted many spectral lines, some of which were single, some double like the sodium *D*-lines, and some with other patterns. This helped spectroscopists sort out series of lines with similar characteristics. Following a suggestion by G. Johnstone Stoney[3] that the spectra could be better analyzed using wave numbers, that is, the reciprocals of wavelengths, rather than the wavelengths themselves, Walter N. Hartley noticed that the difference in the wave number of triplets in the spectrum of zinc was the same for all members of a series. If light were emitted as a consequence of some periodic electric current in an atom or molecule as sound is emitted by vibrating bodies, light whose frequencies were harmonics of the fundamental frequency of motion should have been present. Any periodic motion can be expressed as the superposition of motions with a fundamental frequency and harmonics with frequencies that are integer multiples of that fundamental, and each of these should radiate at its own frequency. No evidence for the presence of harmonically related spectral lines was found.

The first series to be fitted by a formula giving the wavelengths of its lines was that emitted by atomic hydrogen, which was done by Johann Jakob Balmer in 1885. It was made possible by the accurate measurement of the wavelengths of four lines in the spectrum of atomic hydrogen by Anders J. Ångström. Balmer gave his formula for the wavelength measured in angstrom units as

$$\lambda_m = 3645.6 \times \frac{m^2}{m^2 - 2^2}$$

with m an integer greater than 2. With $m = 3$, 4, 5, 6 this fitted the measurement with remarkable accuracy, differing from them by less than one part in 40,000. Balmer considered this evidence of the high quality of Ångström's measurements. A fifth Balmer line corresponding to $m = 7$ turned up among the lines observed by Hermann Vogel and by William Huggins in hot stars. Within a year fourteen lines of hydrogen were shown to fit Balmer's formula, the higher ones with somewhat greater discrepancies but still accurate. Balmer speculated that the 2^2 in his formula could be generalized to n^2 with n an integer, a speculation that later turned out to have been well founded. The series with $n = 2$ is now known as the Balmer series, and that with $n = 1$ is the Lyman series, whose lines are in the ultraviolet. But no picture of the physics behind this formula emerged for almost twenty years.

Following Stoney's suggestion, Balmer's formula can be written in terms of wave numbers as

$$k_m = 109721 \left(\frac{1}{2^2} - \frac{1}{m^2} \right).$$

This formula was generalized by Johannes R. Rydberg to apply to lithium, sodium, potassium, and the other alkalis.[4] He wrote

$$k_m = n_0 - \frac{R}{(m + p)^2}.$$

The two quantities n_0 and p were the same for corresponding lines of a series but could differ from series to series and from element to element. The constant R was common to all the elements considered. For hydrogen it reduced to

Balmer's formula with $n_0 = R/4$, $p = 0$. Again, no underlying physics emerged from this formula.

Another interaction between radiation and matter was noticed by Hertz during his experiments establishing the existence of electromagnetic waves.[5] His method of detecting these waves was to look for small sparks in a spark gap that was part of a resonant circuit over which the waves passed. Because the sparks were dim, he enclosed the sending spark gap in a cardboard box to shield the observer from the light of the large spark gap sending the radiation though not affecting the electromagnetic radiation he was studying. When doing this he found that the absence of light from the large gap reduced the length of the sparks. Further study revealed that it was the ultraviolet light from the sending spark gap that was most effective in increasing the spark and that it was the illumination of the cathode that produced this effect. The light was producing or freeing negative ions at the cathode. Hertz had discovered the photoelectric effect. The same series of experiments that established the existence of electromagnetic waves also established the effect that eighteen years later would call into serious question the wave nature of light.

The photoelectric effect was difficult to study with precision, but several puzzling features appeared. The most anomalous was the fact that the energy of the negative ions did not seem to depend on the intensity of the light. Any simple mechanical model of the process had the electric field of the light acting on the ionic charge, and the stronger the field was, the stronger this action should have been. The results of Philipp Lenard did not confirm this expectation.[6] The dependence on the color of the light, however, was confirmed: Red light had almost no effect on most surfaces, and the bluer the light was, the more energy was transferred to the ions. These facts were not understood for many years.

A second interaction between radiation and matter was discovered by enclosing a piece of apparatus in a cardboard box. This experiment was carried out in 1895 by Wilhelm Konrad Röntgen and the apparatus enclosed was a gas discharge tube operated by a rather high voltage induction coil.[7] For some reason Röntgen was operating the equipment in a darkened room, and he noticed that a sheet of paper coated on one side with a fluorescent material fluoresced during the discharge. Realizing that this effect must be due to some radiation coming from the discharge, he investigated its properties. It could expose a photographic plate. But its most remarkable property was its ability to penetrate many materials, though not with equal ease. Röntgen had discovered X-rays. He proposed that they were the long-sought longitudinal waves in the ether, an idea that would not, of course, accord with Maxwell's theory, which predicted a purely transverse character of electromagnetic waves. Röntgen's proposal was not widely accepted, but the transverse character of X-rays was not established until much later. The penetrating character of the X-rays made it very hard to perform optical experiments with them, because prisms did not refract them and mirrors did not reflect them appreciably. The discovery caused an enormous stir because of the ability of X-rays to penetrate flesh more easily than bone, so that the human skeleton could be photographed by their means. Röntgen received the first Nobel Prize in physics for this discovery.

There was a considerable body of phenomena involving the interaction of radiation with matter that could be accounted for by classical theory, and another considerable body that had not yet been accommodated by that theory. There was, however, little reason to believe that this latter body never could be accommodated, and so there was little feeling that radical new concepts were required. Indeed, there was confidence that further hard work, both experimental and theoretical, would lead to understanding these as-yet puzzling effects.

Notes

1. J. Fraunhofer, *Gesammelte Schriften*, Munich (1888).

2. J. Kirchhoff, *Berlin Monatsbericht* (October 1859); an English translation appears in W. F. Magie, *A Source Book in Physics* (Cambridge, MA: Harvard University Press, 1963), p. 354.

3. Cited in E. T. Whittaker, *History of the Theories of the Aether and Electricity* (London: Thomas Nelson & Sons, 1951), vol. 1, p. 375; repr. New York: Dover, 1989.

4. J. R. Rydberg, *Philosophical Magazine* **29**, 331 (1890).

5. H. Hertz, *Wiedemann's Annalen* **31**, 983 (1887); H. Hertz, *Electric Waves* (London: Macmillan, 1900).

6. P. Lenard, *Annalen der Physik* **8**, 149 (1902).

7. W. C. Röntgen, *Sitzungsberichte der Würzburger Physikalische–Medizinische Gesellschaft*, December 28, 1895, trans. in *Nature* **43**, 274 (1896).

New Rays Are Seen: Cathode and Canal Rays

The ability of gases to conduct electricity had been noted as early as 1785 by Charles A. de Coulomb, who observed that a charged body lost its charge more rapidly than could be accounted for by losses through its supports.[1] He ascribed the excess rate to the conduction of charge by the surrounding atmosphere, but this effect was not studied systematically until the next century. As Johannes Stark wrote concerning this field of research when it was well developed, "In order that while doing experimental research we may avoid purposeless groping and getting lost in inconsequential details and secondary phenomena, we require theoretical viewpoints which allow us to see a definite goal and the direction to approach it."[2] Here such theoretical viewpoints would be lacking for many years.

The sequence of studies of the conduction of electricity through gases that led to the discovery of the electron began with Julius Plücker. These studies became possible only when vacuum pumps were developed that could reduce the pressure in the gas to much less than a millimeter of mercury. A partially

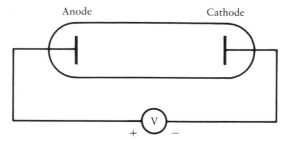

Fig. 2.1 The simplest form of gas discharge tube. The appearance of the gas during a discharge depends strongly on the pressure of the gas and on its chemical identity.

evacuated tube was fitted with two internal electrodes, as illustrated in Fig. 2.1; and a large potential difference was impressed across them. In early experiments this was provided by an induction coil; later, large batteries that could supply a steady potential difference were used. A vivid description of the woes involved in building and maintaining such a battery is given by Hertz.[3] Plücker noted the presence of a greenish yellow fluorescence of the glass envelope near the cathode, the negative electrode.[4] Johann Wilhelm Hittorf showed that an object placed between the cathode and the glass caused a shadow in the fluorescence on the glass and so established the presence of some sort of radiation emitted by the cathode.[5] And Eugen Goldstein found that the shadow of a small object in front of a large cathode was well defined.[6] A small object in front of an extended luminous surface does not cast a shadow on a wall, just as a person does not cast a shadow on a cloudy day. Each point of the wall receives light from all parts of the luminous surface except those hidden from view at the point by the small object, because every part of such a surface emits light in all directions. Goldstein's observation demonstrated that each point of the tube's fluorescing wall received radiation from only a small area of the cathode, an area that could be hidden from that point by the small object. Therefore this radiation was emitted from each part of the cathode in a definite direction, and not in all directions, as is light. Goldstein coined the name *cathode rays* for this radiation.

When the tube was modified so that the positive electrode was off to one side, as in Fig. 2.2, under proper conditions a bright fluorescence in the glass could be seen at the end of the tube. The location of the fluorescence changed when a magnet was brought nearby, the change being consistent with the idea that the cathode rays were composed of a stream of negatively charged particles acted on by a magnetic field, but difficulties with this view soon became apparent.

A convenient way to study the cathode rays was found: The cathode was operated at a large negative potential and the anode at ground. The anode was provided with a hole through which some of the rays could pass into a space free of the electric field used to produce them, and because the gas

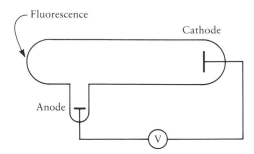

Fig. 2.2 The simplest gas discharge tube showing cathode rays. The pressure must be low for fluorescence to appear at the end away from the cathode.

pressure in the evacuated tube was very low, these rays could travel a distance of many centimeters without becoming diffused by scattering by the residual gas. The electron gun had been invented. With cathode ray tubes of this kind, William Crookes studied the behavior of the cathode rays as the pressure within the tube was changed.[7] The rays appeared only when the pressure was low, 0.077 mm Hg or lower, with their deflection by a magnetic field growing smaller and the definition of the fluorescence growing better as the pressure was reduced further. This could be accounted for by picturing the rays as composed of negatively charged particles coming from the cathode, their speed increasing as the pressure was lowered.

Hertz gave an account of the confusion of trying to understand what happened in a gas discharge:

> As is well known, the cathode rays spread outward in straight lines, approximately perpendicular to the cathode and without reference to the position of the anode. According to the density of the gas, they proceed in the medium for a few millimetres, centimetres, or even up to lengths of the order of a metre. In air they are blue, but at low densities their luminosity is exceedingly feeble; they are then most noticeable on account of the phosphorescence which they excite when they strike the glass. If a magnet is brought near the tube they appear bent much as an elastic wire attached to the cathode and traversed by a current would become bent under the influence of the magnet. This is universally regarded as an electromagnetic action, and, excepting that passing doubts were expressed, the view that used to be held by physicists was as follows: The cathode rays indicate the path of the current, and their blue light arises from the glowing or phosphorescence of the gas-particles under the action of the current. As a fuller knowledge of the facts was attained this view appeared less probable, and more recent experimenters express themselves very reservedly as to the relation between the cathode rays and the actual process of discharge.[8]

Hertz and others attempted to deflect the discharge sideways by placing electrodes beside the tube, but their attempts failed. Hertz then placed plates inside the tube and applied a potential difference between them but still could not produce a deflection of the rays. All this put the negatively charged particle picture in serious doubt. Hertz then mapped the magnetic field in a discharge at low-enough pressure to show the cathode rays.[9] He found that the current needed to produce that field was not the cathode rays, that in places the current was almost at right angles to the rays. Hertz thereby concluded that it was likely that the magnet's action on the rays was not an electromagnetic one but something else. He also noted the rapidity with which the rays responded to changes in the magnetic field and found it too rapid to be associated with the current. But at that time, the ideas of currents carried by particles much lighter than chemical atoms and of the possible ionization of gas molecules had not yet occurred to anyone, and without them these phenomena were incomprehensible. Indeed, many workers in the field considered the discharge to be a phenomenon in the ether.

When electric fields perpendicular to the direction of motion of the rays were produced by electrodes inside the tube and there was a sufficiently good vacuum there, electric deflection of the fluorescence was observed. With the plates outside the tube, ions formed in the gas accumulated on the glass, and their charge neutralized the charge on the deflection plates and canceled the field in the interior of the tube, thereby preventing any electrical field from acting on the rays. Similarly, with the plates inside the tube but too high a pressure of gas in it, enough ions collected near the plates to neutralize the charge on them. But because the behavior of gas ions in a discharge was not known at that time, our explanation of these phenomena could not have been given at the time of Hertz's work.

With both the magnetic and the electric deflection of the rays established, the radiation could again be pictured as consisting of a stream of particles carrying a negative charge. This was confirmed by Jean-Baptiste Perrin, who collected the rays in an insulated metal cylinder and measured the charge that accumulated.[10] But both the mass m and the charge $-e$ of the individual particles composing the rays remained unknown.

Rays of positively charged particles were also discovered. The initial discovery seems to have been made by Goldstein, who studied the glow in the gas immediately alongside the cathode. Its color depended on the kind of gas in the discharge tube, unlike that produced by the cathode rays that depended only on the kind of glass forming the tube wall. On drilling holes or canals in the cathode, Goldstein saw this glow appear at the back of it, and at low-enough pressure it extended some distance into the space behind it, indicating the presence of a new kind of radiation in the vicinity of the cathode. Perrin, using the same method as for the cathode rays, showed that these *canal rays* carried a positive charge. Both kinds of rays, cathode and canal, appeared to consist of beams of charged particles, but their behaviors were quite different, and they arose in different parts of the gas discharge. The former were the same for all discharges, but the latter produced glows of different colors in different gases, and the former had a much longer range than the latter. There was no symmetry between the negatively and the positively charged rays other than their corpuscular nature. The nature of these radiations was the topic of future investigations, with revolutionary consequences.

Notes

1. C. A. de Coulomb, *Mémoires de l'Académie des Sciences* (1785), p. 612.
2. Johannes Stark, *Die Elektrizität Gasen* (Leipzig: Johann Ambrosius Barth, 1902), p. 23.
3. H. Hettz, *Wiedemann's Annalen* **19**, 782 (1883), repr. in *Miscellaneous Papers by Heinrich Hertz*, trans. D. E. Jones and G. A. Schott (London: Macmillan, 1896), p. 224.
4. J. Plücker, *Poggendorf's Annalen* **107**, 77 (1859), **116**, 45 (1862).
5. W. Hittorf, *Pottendorf's Annalen* **136**, 8 (1869).
6. E. Goldstein, *Sitzungsberichte der Berliner Akademie* (1876), p. 284.
7. W. Crookes, *Philosophical Transactions of the Royal Society* **170**, 135 (1879).
8. H. Hertz, *Miscellaneous Papers*, p. 238.
9. Ibid., p. 787.
10. J. Perrin, *Comptes rendus* **121**, 1130 (1895).

An Elusive Ether

When the wave theory of light was accepted early in the nineteenth century, the mechanically inclined scientists of the time felt the need of a material medium in which these waves would constitute a disturbance. They first considered this medium to be a fluid. The great speed of light required this fluid to be almost weightless and almost incompressible, and the large distances that light could travel required it to be without friction. The discovery of the polarization of light required the medium to be able to support transverse waves. A fluid can support only longitudinal, compressional waves, and so an almost rigid medium, the ether, became necessary. Ordinary matter had to be able to move through this ether, at least slowly, because the earth's population was not immobilized in it like flies in amber.

By our starting date of 1895, optics was a precise science, and optical measurements were among the most accurate that could be made. The wavelength of the light emitted by the elements in electrical discharges in gases could be measured with a precision requiring many decimal places for its expression. As early as 1892 Albert A. Michelson had devised a method of measuring the length of the standard meter in terms of the wavelength of light. To do this, light of a well-defined and reproducible wavelength must be used. Michelson chose three lines in the spectrum of cadmium vapor produced under precisely specified conditions, one red, one green, and one blue. (The red line was used by others as a standard wavelength for many years afterward.) Using an interferometer Michelson effectively counted the number of wavelengths between the marks at the ends of a replica of the standard meter. He gave his results in these terms, and not as the wavelength of the light in meters, his purpose being to establish a more permanent standard of length than a bar that might be damaged or destroyed. These results were 1553163.5 wavelengths of the red line, 1966249.7 wavelengths of the green line, and 2083372.1 wavelengths of the blue line of cadmium in one meter, all the wavelengths determined in air at 15°C and 760 mm Hg.[1]

The universal view of those who studied optics was that these waves had to be mechanical waves in a material medium, the luminiferous ether. In the words of Heinrich Hertz,

> What, then, is light? Since the time of Young and Fresnel we know that it is a wave-motion. We know the velocity of the waves, we know their length, we know that they are transversal waves; in short we know completely the geometrical relations of the motion. . . . It is therefore certain that all space known to us is not empty, but is filled with a substance, the ether, which can be thrown into vibration. But whereas our knowledge of the geometrical relations of the processes in this substance is clear and definite, our conception of the physical nature of these processes is vague, and the assumptions made as to the properties of the substance itself are not altogether consistent. At first, following the analogy of sound, waves of light were freely regarded as elastic waves, and treated as such. But elastic waves in fluids are only known in the form of longitudinal waves. Transversal elastic waves in fluids are unknown. They are not even possible; they contradict the nature of the fluid state. Hence men were forced to assert that the ether which fills space behaves like a solid body. But when they considered and tried to explain the unhindered course of the stars in the heavens, they found themselves forced to admit that the ether behaves like a perfect fluid. These two statements together land us in a painful and unintelligible contradiction, which disfigures the otherwise beautiful development of optics.[2]

An elastic solid can support transverse or shear waves, but it can also transmit longitudinal or compressional ones. These generally travel at different speeds but should not be completely independent of one another. For example, when a transverse wave in one medium obliquely strikes the interface with a different medium, it should give rise to both transverse and longitudinal waves in the second one. But this did not happen with light waves. No longitudinal waves were ever seen, and no known physical effects could be associated with them. All the intensity of a light wave was contained in the two transverse polarization states just described; nothing was left over for the third, longitudinal polarization. It was conceivable that such waves existed and had not been detected. Shortly after the discovery of X-rays it was suggested that they were the missing longitudinal waves, but this idea did not survive the experimental tests and had to be abandoned. A similar suggestion concerning the cathode rays met the same fate.

The difficulties arising in astronomy appeared insurmountable. Ether must fill the space between us and the stars, for starlight reaches us. As early as

1728 James Bradley had noticed changes in the apparent position of stars over the course of a year that could be explained simply by the corpuscular theory of light then prevalent.[3] In its annual motion around the sun, the earth very nearly reverses its velocity every six months. The direction from which a light corpuscle appears to approach the earth depends on both the velocity with which the corpuscle left the star from which it came and the velocity of the earth. The direction from which rain strikes one depends on the direction of the raindrop's motion through the air (downward) and on one's own motion through the air. One gets wet in front while moving forward, not while standing still. Expressed mathematically, one must compound the velocity of the corpuscles and the velocity of the earth in order to determine the direction from which starlight strikes the earth. The reversal of the earth's velocity relative to the sun every six months then changes the result of this compounding and makes the apparent angular position of a suitably located star change over the course of a year before returning to its original place. Broadly observed this *aberration*, and the angular displacement was consistent with the speed of light and the speed of the earth in its orbit around the sun. It is what we now call a *first-order* effect; it is directly proportional to the ratio of the speed of the earth to the speed of light, v/c, which is about one part in ten thousand. The maximum angle of aberration is about twenty seconds of arc.

As Hertz pointed out in the earlier quotation, early in the nineteenth century the corpuscular explanation became irrelevant along with the corpuscular theory of light. The wave theory provides a perfectly straightforward explanation of aberration, but only if the earth moves through the ether without disturbing it at all. Under these circumstances, one compounds the velocity of the wave with that of the earth, just as one compounded the velocity of the light particles with that of the earth before, and one gets the same aberration. However, according to this view, there would be an *ether wind* at the surface of the earth that should affect the behavior of light in terrestrial optical instruments.

A number of tests of this ether wind hypothesis were made over the century. The earliest was done by François Arago.[4] Using a prism, he measured the refraction of light from a star known to be subject to aberration. His result was that the refraction was precisely what should be produced by a light source at the apparent position of the star by a stationary prism; the presumed motion of the prism through the ether had no effect on its optical properties.

The difficulty was avoided by a suggestion of Fresnel's. He proposed that the density of ether in a transparent material of index of refraction n is increased over the density in vacuum by the factor n^2, a density increase that would account for the diminution of the speed of light in the material by the factor n.[5] He proposed further that this excess ether is carried along with the material when it moves and that the velocity of light in the moving material is to be compounded from its velocity c/n when the material is at rest with the velocity of the center of mass of the original ether and the excess ether.

The resultant velocity is then

$$\frac{c}{n} + \frac{n^2 - 1}{n^2}\, v.$$

This, he showed, would just cancel any first-order effect of motion through the ether on optical experiments. This effect is known as *Fresnel drag*. *Second-order* effects, proportional to $(v/c)^2 \approx 10^{-8}$, were too small to be detected in Fresnel's day.

An even more sensitive test of this proposal also was suggested by Fresnel and carried out much later by George Biddell Airy. This was to observe the angle of stellar aberration using a telescope containing water. The angle of aberration is essentially the ratio of the speed of the earth to the speed of light down the tube of the telescope. If the telescope is filled with water, the speed of light down the tube is reduced by the index of refraction of water, about 1.6. This should cause the angle of aberration to increase by 60 percent. But Airy found no change in the angle. Late in the century, Fresnel drag was demonstrated experimentally by Hippolyte I Fizeau, who compared the speeds of light in two streams of water moving in opposite directions and found a positive effect in agreement with Fresnel's prediction.[6]

A second explanation of aberration was offered much later by George Gabriel Stokes.[7] He showed that if the ether moved around the earth in a vortex-free way and was at rest at the surface of the earth, stellar aberration would not be affected by the motion, and of course, neither would the behavior of light at the surface. Lorentz pointed out that Stokes's two conditions were mathematically contradictory for an incompressible ether, the kind contemplated by Stokes, and so this attempt failed.[8] Max Planck showed that these conditions could be satisfied to an arbitrary degree of precision by an ether compressed like a gas by the gravitational attractive force of the earth, but that in order to account for experimental results, the compression would have to increase the density of the ether at the surface of the earth by a factor of over 50,000, and this compression must have no effect on electromagnetic phenomena. Though not impossible, this was hardly an attractive "explanation" of aberration. Such behavior was more mysterious than the aberration it was designed to explain. An ether wind at the surface of the earth and Fresnel drag therefore became the accepted picture.

Testing this picture presented enormous difficulties. It was designed to account for astronomical observations depending on the motion of the earth through the ether, at least to first order in v/c, and in this it succeeded. Second-order effects on aberration were too small to be detected. Lorentz showed that terrestrial experiments to detect the ether wind must always involve light either following a closed path and returning to its starting point or following two different paths between two points. No "one-way" experiment involving light going from a point A to a point B by a single path would suffice. In a two-path experiment it does not matter which way the light travels;

it can be reversed in direction without affecting the outcome. This means that the outcome cannot depend on the sign of v, and the largest effect is again of order $(v/c)^2 \approx 10^{-8}$. For a long time such experiments appeared beyond reach.

It was Michelson who determined how second-order effects could be detected:

> The length of a light-wave is, however, so small that one hundred million of them make up a distance of 50 m, and if an interferometer be so arranged by repeated reflections from appropriately placed mirrors, the actual dimensions of the apparatus need not be very great in order to obtain a displacement easily measurable if second-order effects are appreciable.[9]

In other words, in an instrument involving two light paths of 50 m, a path difference between them of one part in 10^8 would be about one wavelength of visible light, big enough to be detected by its effect on the interference of two light waves. This instrument was, of course, the famous Michelson interferometer.

In Michelson's experiment, an incident beam of light was split into two by a half-silvered mirror, the waves of the two half-beams traveling back and forth in directions at right angles to each other, which were then recombined by another such mirror. Their manner of interference when joined revealed any difference in travel time for the two paths. If one path were parallel to the ether wind and the other across it, the former should take longer to cover than would the latter by an amount of order $(v/c)^2$. On rotating the apparatus so as to interchange the two paths, the interference fringes should be seen to move, but no such motion of the fringes was detected. The interferometer used in his experiment in 1881 was too short to make the negative result completely convincing. But in 1887 its repetition with Edward W. Morley on a larger scale was definitive: There was no shift.[10]

This amazing negative result was explained on the basis of a new, second-order conspiracy theory proposed independently by George F. FitzGerald and by Hendrik Antoon Lorentz.[11] They proposed that the dimension along the direction of motion of any material body moving through the ether was contracted by the factor $\sqrt{1 - (v/c)^2}$, the FitzGerald–Lorentz contraction. This shortened the time required for the light following the parallel path just enough to yield no effect of the ether wind. The failure to detect even a second-order effect of the earth's motion through the ether did not cause any questioning of the need for the ether. The experts on electromagnetism such as Lorentz, Wien, Planck, and Poincaré considered the ether to be there of necessity but just to be hidden from detection. Indeed, Poincaré went so far as to declare that this must happen to all orders in v/c, that the conspiracy must be perfect, without suggesting that the ether concept was therefore superfluous. Michelson himself as late as 1927, when describing some additional experiments by Lodge and others, remarked:

It must be admitted, however, that these experiments are not sufficiently conclusive to justify the hypothesis of an ether which is entrained with the earth in its motion. But then how can the negative results be explained?[12]

The problem with the ether was not that the theory predicted wrong results, because as interpreted, it could account for the known phenomena as long as one was willing to encumber it with additional hypotheses such as Fresnel drag and the FitzGerald–Lorentz contradiction. Rather, the difficulty with the theory was largely aesthetic. It incorporated the idea of an absolute space and then went on to prevent its identification by any electromagnetic experiment. The fact that physical processes take place in the same way in all inertial reference frames seemed to be the result of a series of coincidences or conspiracies and not a consequence of some essential feature of the theory. These difficulties were entirely different from those in the areas of atomic physics, in which there were blatant contradictions between what was to be expected on the basis of theory and what was actually observed to happen.

Notes

1. A. A. Michelson, *Studies in Optics* (Chicago: University of Chicago Press, 1927).

2. Heinrich Hertz, "On the Relations Between Light and Electricity," in *Miscellaneous Papers* (London: Macmillan, 1896), p. 313.

3. J. Bradley, *Philosophical Transactions of the Royal Society* 35, 637 (1728).

4. F. Arago, *Comptes rendus* 8, 326 (1839).

5. A. Fresnel, *Annales de chimie* 9, 57 (1818).

6. H. Fizeau, *Annales de chimie* 62, 385 (1859).

7. G. Stokes, *Philosophical Magazine* 87, 9 (1845).

8. H. A. Lorentz, *Encyklopädie der Mathematischen Wissenschaften* V 13, 104 (1903); H. A. Lorentz, *The Theory of Electrons* (New York: Dover, 1952), chap. 5.

9. A. A. Michelson, *American Journal of Science* 22, 20 (1881).

10. A. A. Michelson and E. Morley, *American Journal of Science* 34, 333 (1887); A. A. Michelson and E. Morley, *Philosophical Magazine* 24, 449 (1887).

11. G. F. FitzGerald, *Science* 13, 390 (1889); H. A. Lorentz, *Versl. K. Ak. Amsterdam* 1, 74 (1892); H. A. Lorentz, *Collected Papers*, eds. P. Zeeman and A. D. Fokker (The Hague: Nijhoff, 1936), vol. 7, p. 219.

12. Michelson, *Studies in Optics*, p. 155.

CHAPTER 3

The Entrance of the Electron

T he discovery of the electron provided physics with the first completely (almost) characterized atomic particle. Even those who believed in the existence of atoms had only vague ideas of what atoms consisted. How big were they? How many of them were there in a mole of any material? What were they made of, if that question had a meaning? The electron's properties made it possible to begin answering these questions in a way free of large-scale speculation but based on obvious interpretations of unobvious experiments. Shortly after the identification of the electron Max Planck provided an altogether distinct way of determining Avogadro's number and the electron charge. His argument, based on his theory of blackbody radiation, was so complicated and obscure that few tried to understand it and even fewer took his values of these basic quantities into account at all. Planck's approach is described in detail in Chapter 4. In this chapter I shall describe the course of events leading up to J. J. Thomson's announcement of the existence of a universal constituent of matter, the electron. At the end I include still another way of finding Avogadro's number using Einstein's study of Brownian motion. It came some five years later, but it had such a direct connection with the existence of molecules that it persuaded most of even the most stubborn skeptics who were unwilling to understand the significance of the electron for the problem. This would be sufficient reason for including it. But a better reason is its beauty and simplicity. It does not quite fit the title of this chapter, but it is relevant to the larger aspect, the very existence of atoms and molecules as real physical particles.

e/m for the Cathode and Canal Rays

By 1895 the cathode rays had been identified as a stream of matter carrying negative charge, and the canal rays as a stream of matter carrying positive charge. What had not yet been learned was the nature of this matter: Was it a continuous fluid, or did it have a discrete structure atomic in character? How was it related to other more familiar forms of matter? The first step taken in answering these questions was to measure the ratio of the charge e associated with the matter constituting these rays to the mass m of that matter. This ratio might depend on the particular source used to generate the rays or on some other features of the experiment, as the value of the corresponding ratio in electrolysis did. If the rays did consist of a stream of particles of one or more kinds, the difficulty remained that no experiment measuring the motion of a charged particle of mass m under the influence of forces proportional to its charge e could determine more than the ratio e/m. This is because the acceleration of the particle is, according to Newton, the ratio of the force on the particle to its mass, and when the force is proportional to the charge, this ratio does not depend on the charge and the mass separately but only on the ratio of e to m. This ratio was known for materials freed at the electrodes in electrolysis. There it depended on the kind of material released, being smaller for heavier substances; its value for hydrogen was the largest found. The amount of charge needed to release one gram of hydrogen in the electrolysis of water had been determined to be about 96,000 coulombs, an amount of charge called the faraday because it was Michael Faraday who had established the quantitative laws governing this process. The charge-to-mass ratio of the cathode and canal rays was expected to be in the same range as that found in electrolysis. The conduction of electricity in gases was widely thought to take place by mechanisms similar to conduction in liquids, whatever these might be. The year 1897 was the year of e/m for the cathode rays. During its course three investigators published independently measured values for this quantity.

All three investigators used the deflection of a beam of cathode rays in a magnetic field.[1] The force exerted on a moving charged particle by a magnetic field is proportional to the size of its charge, to its velocity, and to the strength of the magnetic field through which it is moving. The direction of the force is at right angles to both the particle velocity and the field. It was this last feature, which had so surprised Oersted early in the century, that led to a deflection of the rays rather than to their speeding up or slowing down. When the charge e is measured in electrostatic units and the magnetic field B in gauss, the force F in dynes on a particle moving with speed v across the field is given by

$$F = \frac{evB}{c},$$

where c is the speed of light. If this force acted for only the short time

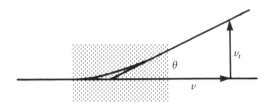

Fig. 3.1   The change in direction of an electron's motion on crossing a magnetic field.

$\Delta t = L/v$ required for the particle to cover the distance L over which the field was present, according to Newton it would produce a small transverse velocity component v_t, that is, one at right angles to the original direction of motion, given by

$$v_t = a_t \Delta t = \frac{FL}{mv}$$

$$= \frac{eBL}{mc}.$$

This changed the direction of motion without changing the speed of the particle (see Fig. 3.1). After passing through the region containing the magnetic field, the particle was deflected by an amount proportional to this transverse velocity component, with the angle of deflection θ for small deflections being given by

$$\theta = \frac{v_t}{v} = \frac{e}{mv} \frac{LB}{c}.$$

A measurement of this deflection θ gave the value of e/mv in terms of experimentally known quantities. To get a value of e/m it was necessary to determine v. This could be done either by measuring the time required for the particle to cover a known distance or by some less direct method.

The first of our three investigators, E. Wiechert, pioneered in applying the time-of-flight method, now a mainstay of high-energy physics, to particle physics; the other two, W. Kaufmann and J. J. Thomson, measured the speed of the cathode ray particles in other ways.

Wiechert's method was difficult to carry out with the technology available at the time but was very ingenious.[2] He arranged his discharge tube so that the rays emerging from an electron gun could be deflected by a steady magnetic field in a region near the electron gun and by the fields produced by each of two loops of wire. The rays had to go through two apertures, one near the cathode and the other farther down the beam, to produce fluorescence at the far end of the tube. The first loop was between the electron gun and the first aperture, the second after the second aperture. Whether or not the rays got through was detected by the presence or absence of fluorescence (see Fig. 3.2).

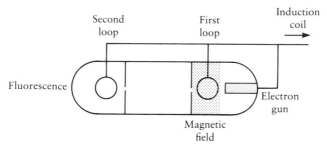

Fig. 3.2 Wiechert's time-of-flight tube for measuring the charge-to-mass ratio of the electron.

The steady field would deflect the rays out of the first aperture, so there would be no fluorescence unless its effect was annulled by that of some other field. If the right amount of current went through the first loop, its field would cancel the effect of the steady field in the region in front of the first aperture, and then the rays would go through the two apertures and produce a sharp spot of fluorescence on the glass. If a current were going through the second loop also, then the rays emerging from the second aperture would be deflected, and the fluorescent spot would move away from the straight line through the two apertures.

The observation was made by using high-frequency (by 1897 standards) current to furnish a series of pulses, each of which would do three things:

1. Excite the discharge momentarily.
2. Cause a current pulse to flow through the first loop.
3. Cause a similar current pulse to flow through the second loop.

Using a low frequency with correspondingly long pulses, Wiechert adjusted the fields and currents so that a spot of fluorescence appeared at the end of the tube. He used long pulses so that the current in the loops would not change appreciably during the time it took the cathode rays to travel from the gun to the end of the tube. The fluorescence consisted of a series of flashes, one from each pulse, but these occurred too rapidly to be resolved by the eye. Each flash came from the arrival of cathode rays that went through the magnetic field of the first loop when the current in it had just the right value to counter the effect of the steady magnetic field. This was adjusted to be the peak value of the current in the pulse. The spot was displaced from the straight-line path through the apertures because when the rays went past the second loop, current from the long pulse was still flowing through it.

Wiechert then increased the frequency of the current, shortening the pulses, until the rays no longer were deflected in their passage between the second aperture and the glass. This lack of displacement of the pulses of cathode rays meant that when they got to the second loop there was no current flowing through it. The transit time of the rays down the tube was then the time it look for the current in the second loop to decrease from its peak value to zero, one quarter of the period of a complete oscillation, aside from the small correction coming from the time it took the current pulse itself to travel to the second loop. The current pulse traveled down the wires at nearly the speed of light, so its transit time was small compared with that of the rays. In this way Wiechert measured a speed of 3×10^9 centimeters per second. Then measuring the deflection produced by a known magnetic field gave him a value of e/m of 2×10^7 emu/gram. After improving the apparatus, two years later, Wiechert determined that the value of e/m lay between 1.55 and 1.01×10^7 emu/gram or 4.7 and 3.04×10^{17} esu/gram.[3]

Wiechert also got an estimate of v by another, independent method. The energy of the cathode rays traveling through the rarefied gas in the tube is nearly conserved. The particles are accelerated in the first place by the potential difference ΔV between the cathode and the anode of the electron gun. This gives each of them a kinetic energy proportional to its charge e,

$$\tfrac{1}{2}mv^2 = e\Delta V.$$

If little of this kinetic energy is lost by the rays in going through the residual gas in the tube, in conjunction with the measured deflection of the rays this gives an accurate way of estimating the speed v of the particles as they go down the tube. Wiechert used this to obtain an upper limit on the value of e/m of 4×10^7 emu/gram. The unknown energy losses of the particles due to their motion through the gas made this value less certain than the one obtained by the time-of-flight method.

W. Kaufmann used magnetic deflection and the same energy conservation method as did Wiechert in his measurement of e/m for cathode rays. He varied the material of the cathode and the gas used in the discharge generating the rays, finding little difference in the results. His first value was approximately 10^7 emu/gram.[4] More precise measurements of the magnetic field along the path of the particles led to the value 1.77×10^7 emu/gram or 5.3×10^{17} esu/gram.[5]

J. J. Thomson at the Cavendish Laboratory of Cambridge University was the first to use the fact that the velocity of electrons could be measured by observing their deflection in crossed electric and magnetic fields.[6] The force exerted on a particle in the beam by an electric field is independent of the particle's speed, being simply the product eE of the particle's charge (measured in electrostatic units) and the strength of the electric field, and depending on whether e is positive or negative, it is in the direction or opposite to the direction of the electric field. The force exerted by a magnetic field was, as we have seen, proportional to the particle's speed. The fact that the electric force did not depend on the speed and the magnetic one did offered a way of measuring the speed of the particles in the beam by comparing these forces. If both electric and magnetic fields were present, the net force on the particle was the sum of the forces due to these fields separately. After he measured the deflection of the beam produced by a magnetic field and so determined the value of e/mv, Thomson passed the beam through both a magnetic and an electric field arranged to produce deflections in opposite directions. By adjusting the strength of the magnetic field until there was no net deflection of the beam, he found the condition under which the electric and magnetic forces were equal,

$$eE = \frac{evB}{c},$$

so that

$$\frac{v}{c} = \frac{E}{B}.$$

The speed of light was known and the fields E and B were controlled by the experimenter, so this determined the speed v and enabled him to obtain a value for e/m. His 1897 mean value for e/m was 2.3×10^{17} esu per gram or 0.77×10^7 emu per gram (see Fig. 3.3).

The numerical value of the charge-to-mass ratio of the cathode rays was astounding. It was a thousand times larger than the value occurring in electrolysis. On the particle picture of the rays this enormous difference between the two values could be produced by either a large charge or a small mass for the cathode ray particles in comparison with the hydrogen ion in electrolysis. In addition, the value of the charge-to-mass ratio was the same no matter what material made up the source of the cathode rays. The cathode rays had a universality not present in electrolysis, in which this ratio depended strongly on the material being electrolyzed, and Thomson emphasized the conclusion:

> Thus on this view we have in the cathode rays matter in a new state, a state in which the subdivision of matter is carried very much further than in the ordinary gaseous state: a state in which all matter—that is, matter derived from different sources such as hydrogen, oxygen &c.—is of one and the same kind; this matter being the substance from which all the chemical elements are built up.

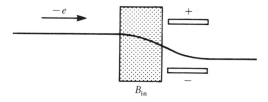

Fig. 3.3 The principle of Thomson's double-deflection apparatus. The upward magnetic deflection of the electrons was countered by the downward electric deflection. Actually the magnetic and electric fields were not separated as in the figure but occupied the same space, so the sideways displacement shown did not occur.

This "new state of matter" was a constituent of the matter of all chemical identities, a universal constituent. For a short time Thomson believed that the large value of e/m was partly due to a value of e larger than that of hydrogen in electrolysis and not entirely due to a small value of m. This opinion was based on some rather speculative considerations of the dielectric constants of gases. This question of the size of e, and therefore of m, for the cathode rays was an urgent one, and it was soon addressed. Only when a finite value for the smallest observable charge was found would the atomicity of charge, and therefore of the cathode rays, be firmly established.

The existence of an atomicity of electricity had been suggested much earlier by Hermann von Helmholtz, who even gave an estimate of the atomic unit of charge and by G. Johnstone Stoney.[7] Helmholtz did this by dividing the value of the faraday obtained from electrolysis by the value of Avogadro's number obtained from rather crude estimates based on kinetic theory. Stoney had no estimate of the mass associated with this "atom of electricity," which he named the *electron* in 1894.

Using similar experimental methods, that same year W. Wien measured the charge-to-mass ratio for the canal rays.[8] These positively charged rays were more difficult to work with because their source was less copious and their range in the residual gas of the discharge tube was shorter than that for cathode rays. Wien obtained a value much smaller than that for the cathode rays, being comparable to that found in electrolysis. The ratio also depended on the materials present in the tube. The canal rays were, unlike the cathode rays, not universal in nature. A measurement of e for canal rays would now lead directly to a measure of the mass of a chemical atom or ion.

In 1896, the year before measurements of e/m for free electrons in the cathode rays were made, Pieter Zeeman at the University of Leiden in the Netherlands made observations that H. A. Lorentz interpreted as giving e/m for the particles responsible for the atoms' emission of spectral lines.[9] Zeeman observed that when sodium atoms are placed in a strong magnetic field, the D-lines of the spectrum they emit when excited were broadened by an amount proportional to the strength of the field. This *Zeeman effect* could have been taken to be a splitting of each line into several lines, but Zeeman was not able to observe anything beyond a general broadening of the line.

Zeeman's observation of the broadening of spectral lines led Lorentz to a value of e/m for the electric charges in the atom presumed to be responsible for emitting the lines. According to electromagnetic theory, a light wave of a given frequency would be emitted by an electric charge oscillating with that frequency. To account for the spectra emitted by atoms excited by an electric discharge, Lorentz assumed that there were charged particles within the atom that were bound to equilibrium positions by elastic forces. These oscillators would radiate light of their frequency whenever they were provided with energy in some way. This model of an atom had to contain a great many such particles because there were a great many spectral lines emitted by all kinds of atoms.

Lorentz's argument concerning the broadening of the sodium spectral lines was based on this model. He considered a particle oscillating at a frequency v_0 about its equilibrium position in the atom. In the absence of an external magnetic field, this particle could oscillate along any of the three directions in space with this frequency. These three modes of oscillation are independent of one another and can be excited at the same time. An equivalent way of describing these modes of motion is as a combination of an oscillation along one line and of two rotations in opposite directions around that line with angular velocity $\omega_0 = 2\pi v_0$. Lorentz used this latter picture and chose the line to be in the direction of the external magnetic field.

Because a charged particle could move freely along a line of magnetic field, the frequency of the motion along this direction was not affected by the field. The two rotational modes of motion involved motion across the direction of the magnetic field, and so magnetic forces arose in connection with these modes. The angular velocities of the two circular motions were affected, and the frequencies of radiation emitted in their course would be changed, one increased and the other decreased. The broadening of the spectral line that Zeeman observed gave a measure of how much the two altered frequencies differed from the unaltered one and so provided a way of evaluating the charge-to-mass ratio. The value that Thomson obtained was not very different from the estimate of this ratio obtained from the broadening that Zeeman saw. This strengthened Thomson's conclusion that there were light charged particles, electrons, of the same kind in atoms of all the elements and that they could be extracted from the atom by sufficiently large forces.

A further check on the validity of Lorentz's interpretation of the broadening was the observation that the light from the high- and low-frequency edges of the broadened line observed along the direction of the field was circularly polarized in opposite directions. This should be the case according to Lorentz's model. Looking along the field one would see only the two circular motions of slightly different frequency, because motion along the line of sight does not produce any radiation in this direction. In the spectroscope the light resulting from one of these motions would be slightly displaced from that resulting from the other. The light at one edge of the broadened line would come primarily from just one of these circular motions, whereas that at the other edge would come primarily from the other. This would cause the light at the two edges to be circularly polarized in opposite directions.

The success of Lorentz's beautiful model was short lived, however, for when observations with higher resolution could be made it was seen that many lines split into more than three, and the harmonic binding model could account for only three. In fact, even the two sodium D-lines that Zeeman observed acted anomalously according to this model. One line was seen to split into four and the other into six components. Fortunately, Zeeman could not resolve them in his first experiments. The anomalous Zeeman effect had to wait twenty years for quantum mechanics and the discovery of electron spin for its complete explanation.

Neither Zeeman nor Lorentz suggested that the initial success of this theory of line broadening or splitting meant that these elastically bound particles were able to exist as free particles outside the atom. Indeed, the harmonic-force law used would not allow the particle to escape no matter how high its energy; it was confined to the atom. The existence of a kind of particle extractable from atoms of any element was first proposed by Thomson.

Notes

1. An account of all these methods is given by J. Townsend, *Electricity in Gases* (Oxford: Oxford University Press, 1915). A more detailed account is given in P. Lenard's article in Wien-Harm's *Handbuch der Experimentalphysik* (Leipzig, 1927), vol. 14.

2. E. Wiechert, *Göttingen Nachrichten* 38, 1 (1897).

3. E. Wiechart, *Wiedemann's Annalen* 69, 739 (1899).

4. W. Kaufmann, *Wiedemann's Annalen* 61, 544 (1897).

5. W. Kaufmann, *Wiedemann's Annalen* 62, 598 (1897).

6. J. J. Thomson, *Philosophical Magazine* 44, 299 (1897).

7. E. T. Whittaker, *A History of the Theories of Aether and Electricity* (New York: Dover, 1989), p. 354.

8. W. Wien, *Annalen der Physik* 65, 440 (1898).

9. See H. A. Lorentz, *The Theory of Electrons* (New York: Dover, 1952), chap. 3.

Amplification

Circular motion with angular velocity ω_0 required a centripetal force $m\omega_0^2 r$ to be exerted on the moving particle. In the case of an elastically bound electron rotating around its position of equilibrium in an atom in the absence of an external field, this force was provided by the atom, but in the presence of an external magnetic field perpendicular to the plane of rotation, the magnetic Lorentz force $evB/c = eB\omega_0 r/c$ either reinforced or countered the atomic force, thereby slightly altering the frequency. If ω was the perturbed angular velocity with the magnetic field present, then

$$m\omega^2 r = m\omega_0^2 r \pm \frac{eB}{4\pi mc}.$$

The plus sign was for rotation in one direction, the minus sign for rotation in the other. Then for weak magnetic fields,

$$\omega = \omega_0 \pm \frac{eB}{2mc}, \qquad v = v_0 \pm \frac{eB}{4\pi mc},$$

and a measurement of the line broadening or splitting yielded a value of e/m. Zeeman's observations led to a value of 3×10^{17} esu/gm = 10^7 emu/gm.

The Ionic Charge

The existence of a definite value of e/m for the cathode rays demonstrated by J. J. Thomson, Emil Wiechert, and Walter Kaufmann and the existence of distinct values of this ratio for canal rays of various sorts as demonstrated by Wien's measurements could easily be understood for the corpuscular picture, but they did not establish it beyond the possibility of attack. No observation of individual particles was involved in any of these experiments. To those working in the field of the conduction of electricity in gases, the atomic picture was the only reasonable one, but in the larger community of scientists, this was not the universal view. One thing that could settle the question and establish the atomic theory beyond doubt would be finding an effect necessarily produced by a small number of these purported atoms or ions, preferably just one. Demonstrating the discreteness of the charges in a beam of cathode rays, or of canal rays, or of the ions released at an electrode during electrolysis would accomplish this purpose. The next goal of Thomson's group in Cambridge was determining the charge on a single ion. If they found a smallest ion charge, the precise electrical neutrality of matter made it almost certain that the magnitude of the charges on any positive or negative ions would have to be an integer multiple of it. If two incommensurable ionic charges existed, a molecule containing them both could not be neutral.

The first measurement of an ionic charge was made by John S. Townsend a student of Thomson at Cambridge, in 1897.[1] The idea of the method was somehow to associate an ion with a small but directly observable neutral particle, a water droplet. By producing a large number of these charged droplets in a reproducible way, Townsend could measure their total charge by standard laboratory means. If he could also count them, he would have his desired result. Droplets are visible, and although the number needed to carry a measurable charge could not be counted directly, a little ingenuity permitted him to get a good estimate of their number.

The first problem was to get droplets most of which carried a single ionic charge. Townsend had observed that there is a fog in the bubbles of oxygen formed on the electrode of an electrolytic cell where water was being decomposed. It was known that charged particles act as nuclei for the condensation of saturated vapor into droplets. The presence of this fog signaled the existence of oxygen ions inside the bubbles. Some of the oxygen released in electrolysis was released as molecular oxygen ions rather than as neutral molecules. These ions remained charged even after the bubble burst at the surface of the electrolyte, and so electrolysis provided a source of oxygen ions. Townsend proceeded in the following way:

Oxygen gas freed in electrolysis and containing an unknown number of molecular ions was passed through a vessel containing water vapor in equilibrium with water at a fixed temperature so that the vapor was saturated. A droplet of water condensed from the saturated vapor onto each oxygen ion, and a cloud of these droplets was formed. Townsend then passed this cloudy gas through an insulated series of bulbs containing sulfuric acid, an effective drying agent, which removed the water vapor and the cloud. The increase in the mass of the bulbs was measured to give $W_1 + W_2$, the mass W_1 of the vapor plus the mass W_2 of the cloud. The density of water in saturated vapor at a given temperature was well known, and so the mass of the vapor could be determined from the temperature and the volume of gas. This then led to the value of the mass W_2 of the cloud.

Next the size of the individual water droplets had to be measured. The radius of the droplets was found by observing the rate of fall of the top of the cloud in air. (This was done on a separate cloud prepared in a similar way.) Stokes' law of fall gave the force F required to move a sphere of radius a through a fluid of viscosity η at constant speed v as

$$F = 6\pi\eta a v.$$

Here the force causing the droplets to move was the force wg provided by gravity:

$$F = wg = \frac{4\pi}{3} a^3 \rho \times g,$$

so that equating these two expressions gives

$$v = \frac{2a^2 \rho g}{9\eta}.$$

Determining v by watching the top of the cloud descend and knowing the values of the viscosity η of air, the density of water ρ, and the acceleration of gravity g, the radius a could be found. The density and the radius determine the mass, so finally Townsend knew the mass of an individual droplet. The number N of droplets was now given by the total weight of the cloud, measured after the cloud was condensed, divided by the weight of an individual droplet:

$$N = \frac{W_2 - W_1}{w}.$$

A typical number arrived at in this way was 10^7, a number much too large to have been arrived at by directly counting the droplets.

The number of droplets in a cloud having been determined, the next thing was to determine the total charge on that cloud. To do this, the dried gas was passed into an insulated receiver, and the charge q_2 on this receiver as well as the charge q_1 on the drying bulbs was measured. The total charge of the ions originally obtained from the electrolytic cell was then the sum of the charge collected in the drying tubes and that collected in the final receiver. It was assumed that all the ions present had charges of the same sign and that each droplet had formed about just one ion, so that the number of ions and the number of droplets would be the same. On this basis the charge per ion was simply the total charge collected divided by the number of droplets originally formed in the cloud:

$$e = \frac{q_1 + q_2}{N}$$

which yielded the result that the ionic charge was

$$e = 3 \times 10^{-10} \text{ esu} = 1.0 \times 10^{-19} \text{ coulomb.}$$

If there were ions of the opposite sign present, then the total charge on the droplets would be the difference between the charge of the positive and negative ones. This would lead to an underestimate of the charge of each ion. The presence of some negative ions was suspected on the basis of electrical conductivity measurements on the gas. The conductivity was greater than that expected from the charge per ion just obtained. On making a correction for this, the next year Townsend published the value

$$e = 1.7 \times 10^{-19} \text{ coulomb.}$$

This is remarkably close to the currently accepted value of about 1.6×10^{-19} coulomb.

In this first measurement of e there was still no observation of individual particles with that charge. The graininess of the charge was associated with the graininess of the cloud, and the degree of the cloud's graininess was determined by the effect of viscosity on the fall of the cloud, not by the fall of any individual droplet. This is reminiscent of the determination of Avogadro's number by Loschmidt based on Maxwell's expression for the viscosity of a gas, as described earlier. There the viscosity gave a bulk measure of the probability of intermolecular collisions, and for this reason the method was not as compelling as some skeptics wanted. In Townsend's work it was the motion of the cloud as a whole rather than the motion of individual droplets that was observed, a fact that may have diminished the impact on the larger scientific community. Still, it was more direct evidence for the graininess of matter than that provided by the viscosity of a gas.

Townsend assigned no estimated uncertainty to his value, but if he had it would have been rather large. Important sources of error were the assumption of one ion per drop, the correction for the presence of ions of the opposite sign, and the neglect of the effect of evaporation and air currents on the rate of the cloud's fall. The great accuracy of Townsend's result in 1898 must be regarded as fortuitous.

This method was much modified and supposedly improved over the next few months. Thomson had found that the newly discovered X-rays[2] and ultraviolet light[3] caused gases to become conducting, and he proposed that this was due to the production of ions in the gas by these radiations. He determined the charge on ions produced in this way rather than by the obscure mechanism involved in the electrolytic process. The drops containing the ions were formed in the newly developed cloud chamber where a volume of saturated air is rapidly expanded, supersaturating the vapor and causing rapid condensation on any ions present.[4] Here the uncertainties of Townsend's method were also present, perhaps in accentuated form due to the rapid expansion of the cloud chamber. The result of the work done at Cambridge up through 1900 was a best value of the ionic charge in the range 6.5–6.8 × 10^{-10} esu or 2.2–2.3 × 10^{-19} C, much farther from the present-day value than Townsend's second value. The value of Avogadro's number corresponding to this value of the ionic charge and the faraday was

$$N_A = \frac{F}{e} = 4.3 \times 10^{23}.$$

This value was of the same general size as the estimates coming from the kinetic theory of gases, and the similarity of results from such different kinds of measurements lent considerable weight to the atomic picture of matter.

Several advances in experimental method were made over the next few years. Townsend had measured the charge on a cloud by collecting it on an electrometer. He had measured the size of the drops in the cloud by observing their rate of fall through air under gravity and had weighted the cloud by condensing it. The second and third measurements gave him the number of drops. If each drop carried one charge, the ratio of the total charge to the number of drops would be the ionic charge. Thomson's measurements involved this same assumption. C. T. R. Wilson introduced a new method of determining the charge of the cloud drops formed around each ion. Instead of measuring the charge of a whole cloud with an electrometer. Wilson provided a known vertical electric field that could be applied to the cloud and observed the rate of fall of the top of the cloud when the field was on or off. The difference between these two rates gave him a direct measure of the charge on the drops he was looking at, and he could select those least affected by the field and hence carrying the least charge, that is, a single ion's charge. His values ranged from 4.4 to 2.0 × 10^{-10} esu, with a mean value of 3.1 × 10^{-10} esu = 1.0 × 10^{-19} C.

The ultimate drop method was devised later by Robert A. Millikan. He replaced the cloud of water drops of his predecessors with oil drops, which have the great advantage of not evaporating nearly as quickly as water drops do. He observed the motion of individual drops under the influence of gravity and a vertical electric field through a telescope. The electric field could be made strong enough to balance the effect of gravity. Exposing the drops to X-rays, Millikan could observe sudden changes in the motion of a drop as it gained or lost a unit of charge. One of the principle ways in which a drop could change its charge was through the photoelectric effect in which X-rays expel electrons from atoms or molecules, these electrons perhaps being picked up by other drops. Thus Millikan was assured that what he measured was the electron charge. His final (1917) value of the size of the electron charge was

$$e = 4.774 \pm 0.005 \times 10^{-10} \text{ esu}$$

$$= 1.591 \pm 0.002 \times 10^{-10} \text{ coulomb.}[5]$$

Actually, the value of the elementary charge with the smallest experimental uncertainty that had been published by 1900 was that of Max Planck.[6] As discussed in the next chapter, Planck's analysis of blackbody radiation led directly to a value for Avogadro's number of 6.2×10^{23} molecules per mole. Reversing the procedure of getting Avogadro's number from the faraday and the ionic charge, Planck got the ionic charge from the values of the faraday and Avogadro's number! No one understood Planck's argument leading to his value for Avogadro's number, and so no attention was paid to his value, $e = 4.69 \times 10^{-10}$ esu $= 1.56 \times 10^{-19}$ C, close to Townsend's corrected value but not so close to the value that Thomson's group thought best, until 1908 when Ernest Rutherford cited it when it supported the value he obtained from scattering experiments.[7] The best value in 1988 was

$$e = 4.803\,206\,8 \times 10^{-10} \text{ esu} = 1.602\,177\,33 \times 10^{19} \text{ C},$$

with an uncertainty of 0.3 parts per million.[8]

Notes

1. J. S. E. Townsend, *Proceedings of the Cambridge Philosophical Society* 9, 244 (1897).

2. J. J. Thomson, *Philosophical Magazine* 46, 528 (1898).

3. J. J. Thomson, *Philosophical Magazine* 48, 547 (1899).

4. C. T. R. Wilson, *Philosophical Transactions of the Royal Society* A 189, A 192, 403 (1899).

5. R. A. Millikan, *Philosophical Magazine* 19, 209 (1910); R. A. Millikan, *The Electron* (Chicago: University of Chicago Press, 1917).

6. M. Planck, *Annalen der Physik* 4, 564 (1901).

7. E. Rutherford, *Proceedings of the Royal Society* A 81, 162 (1908), included in vol. 2 of his *Collected Papers* (London: Allen & Unwin, 1963).

8. *Particle Physics Data Booklet* (Amsterdam: North-Holland, 1990).

A Universal Constituent of Matter

J. J. Thomson's 1897 conjecture that the cathode ray particles were universal constituents of matter rapidly received support from a variety of sources. Thomson based his conjecture on the identity of the value of e/m found in the cathode rays coming from electrical discharges in tubes containing a variety of gases and having cathodes made of a variety of metals. Strong support came from the way that it made the presence of the electrons needed to account for the Zeeman effect a general feature of all atoms. The scale of the splitting of spectral lines emitted by atoms in a magnetic field was consistent with the value of e/m as determined in the cathode ray experiments even if the number of components into which a given line split was later seen not to be that predicted by Lorentz's harmonically bound electron model. No plausible model for the Zeeman effect based on the charge-to-mass ratio found in electrolysis could be given. A magnetic field could not affect the motion of such massive particles enough to alter their frequency of oscillation that much. This support came from the past, Zeeman's observation and Lorentz's explanation of it having preceded Thomson's experiments; Zeeman and Lorentz had not, however, suggested the existence of a universal constituent on the basis of their work.

Hertz's discovery of the photoelectric effect while he was doing the experiments that established the existence of electromagnetic waves has already been mentioned. Hertz showed that the ability of a spark gap to support a potential difference large enough to cause a long spark was much reduced if the negative pole of the gap was illuminated with ultraviolet light. This work was done in 1887, just ten years before Thomson's measurements of e/m for the cathode rays. The proof that the photoelectric effect was due to the ejection of electrons from the cathode by the ultraviolet light was found when Thomson measured the value of e/m for the ejected particles and found the cathode ray value. Electrons had to be present in the material of the photocathode.

The ability of metals to conduct electricity had long been a subject of speculation. Whether the conduction was due to the motion of positive or of negative charge, and what the nature of the charges of either sign was, could not be determined. In electrolytic conduction, the charge was carried through the electrolyte by ions with definite chemical identity, as was shown by the chemical activity at the electrodes where material went into or out of solution in proportion to the amount of charge that passed through the electrolytic cell. Metallic conduction, on the other hand, did not involve the transport of matter in this way. A wire could transport any amount of charge without being altered in any perceptible way. This led Thomson to propose that the positive and negative charges in metals were different in nature, with the charge of one sign being associated with the fixed chemical atoms in the solid material and the charge of the other sign being free to move through the material. The comparison of metallic conduction with that in gases led to the idea that it was the negative charge that was free to move in metals, just as the cathode rays were relatively free to move in ionized gases. When the electron was identified in the cathode rays, it was a small step to suggest that there are electrons in metals free to move and to carry current. Lorentz credited E. Riecke, P. Drude, and J. J. Thomson with ideas that he put together in developing his theory of electrons as the active particles in the electromagnetic behavior of both metals and insulators.[1]

One other occurrence of electrons deserves to be mentioned in this connection. The discovery of radioactivity will be described in a later chapter, but at about the same time that these developments were taking place, the emission of particles of negative charge by radioactive materials was discovered. Their charge-to-mass ratio was measured by Ernest Rutherford and found to be the same as for Thomson's electron. The atom had not yet been divided into a nucleus and an electron cloud, and so the source of these electrons was not distinguished from that of other electrons that had been seen. They were more energetic, but this high energy could not be interpreted.

In the very early years of the twentieth century, the electron had become a well-established component of all matter in the gaseous and solid forms and, by extension, also in liquids. How it was incorporated in matter was unknown but was a subject of intense interest and speculation. The means to investigate this question in the laboratory had not yet been developed. It was, of course, taken for granted that electrons have a continuing existence. The idea that they might be created and annihilated was not to appear for another thirty years, and so the highly energetic electrons found in radioactivity were also considered to be preexisting constituents of matter. This preexistence might lead one to say that the electron was even more universal then than it is now, for now we consider the electrons emitted in radioactive decay to be created in the decay process and not already present in the atom before the decay.

Note

1. H. A. Lorentz, *The Theory of Electrons* (New York: Dover, 1952), p. 10.

Brownian Motion: A Determination of k

Because Albert Einstein's theory of Brownian motion was so influential in persuading holdouts against accepting the reality of atoms, it deserves to be looked at in some detail. The central idea of Einstein's approach to the theory of Brownian motion was to recognize that there was no line of demarcation between a "particle" suspended in a fluid and a large solute molecule in that fluid regarded as a solvent.[1] The particle might be a colloid particle, a pollen grain, or some other object visible through a microscope. If it was suspended in the solvent, all that was known about the behavior of solutes in solvents should apply. There was, however, a well-founded theory of the motion of a body through a fluid that had been developed by G. G. Stokes and that had already been used by J. S. E. Townsend, J. J. Thomson, and C. T. R. Wilson in their determination of the ionic charge described in the second section of this chapter, namely, Stokes' law of fall. What Einstein did in his 1905 paper on this subject was to put these things together in a masterful way to arrive at a visible and inescapable consequence of the thermal motion of molecules.

Einstein's discussion was divided into three parts. The first stated the known properties of dilute solutions of large molecules, in particular the law of osmotic pressure on a semipermeable partition in the presence of external forces on the molecules. The second examined the behavior of spherical objects acted on by an external force in a viscous medium and related this to diffusion under the influence of the thermal motion of those spherical objects regarded as solute molecules. It was here that the gap between phenomena on an atomic scale and those on a directly observable scale was bridged. The third examined the diffusion process and deduced the observable, long-time behavior of either a large solute molecule or a spherical object diffusing in a fluid. Following Einstein, I shall consider these three processes separately.

Osmotic Pressure

Einstein considered a cylinder filled with a solvent and divided into two parts by a movable piston that was permeable to that solvent. On one side of the piston a small quantity of a substance whose molecules were too large to get through the piston, for which the piston was impermeable, was thought to be dissolved. The solute molecules were able to move through the solvent on that side even though their motion was retarded by the viscosity of the solvent, but they were unable to enter the other side because of the presence of the piston. Consequently, they exerted an osmotic pressure on the piston, just as molecules of a gas exert a pressure on any containing wall. It had been established by J. H. van t'Hoff that this osmotic pressure p in a solution occupying a volume V at temperature T was governed by the ideal gas law,

$$pV = \frac{n}{N_A} RT,$$

where n was the number of solute molecules, N_A was Avogadro's number, and R was the universal gas constant. Although n and N_A were not known separately, their ratio was the number of moles of solute, which could be accurately known. This permitted the expression of the pressure in terms of the number v of solute molecules per unit volume,

$$p = v \frac{RT}{N_A}.$$

Next Einstein imagined a force K toward the piston to be applied to each solute molecule. (A familiar example of such a force is the force of gravity acting on particles suspended in the air.) This caused the pressure of the solute molecules to be bigger closer to the piston than farther away. The increase in pressure over a short distance just balanced the total force per unit area exerted by the external force on the solute molecules in a layer of this thickness. But at constant temperature the change in pressure is proportional to the change in density of the particles producing the pressure. The effect of applying the external force K to each molecule was, then, to produce a concentration of solute molecules that increased on approaching the piston, just as the density of the atmosphere increases on approaching the earth's surface. This completed the osmotic pressure part of the argument. It was completely macroscopic because it involved only directly observable quantities, the pressure, the temperature, the gas constant, the number of moles of solute present, and the total external force applied to the solute in a layer of the solution. The force per molecule and the number of molecules present were used in describing the physical situation but did not enter the final result.

Molecular Motion

It was in this section that Einstein confronted the macroscopic picture with the molecular one. If the solute molecules were spherical particles, the force K applied to each one would give it a velocity v in the direction of the force determined by Stokes' law,

$$K = 6\pi\eta av.$$

The right side of this equation gave the viscous force on a particle of radius a moving with velocity v through the fluid whose viscosity is η. The force K produced a current of particles in the direction of the force. If the system was in equilibrium, there could be no net current, and so this current had to be canceled by another one arising from some other cause. The only available cause was the thermal motion of the solute molecules. This thermal motion would cause no current if the density of the solute were uniform, as the direction of thermal motion is random. If, however, the density were not uniform, a current would flow from regions where the density was high to those where the density was low, in the way illustrated in Fig. 3.4. Throughout the illustrated container the thermal motion of the solute molecules was random in direction. When the density of molecules was greater on the right than on the left, however, on the right there would be more solute molecules with thermal velocity to the left than there were molecules on the left with thermal velocity to the right. There was therefore a net diffusion current from right to left. The size of this current depended on the density gradient and the properties of both the solvent and the solute and was characterized by a parameter D called the *diffusion constant*. It was this molecular diffusion current that countered the drift current to produce a steady equilibrium condition. The necessity of these two currents' canceling each other in a steady-state situation produced a relationship between the Stokes equation containing the macroscopic coefficient of viscosity and the diffusion process resulting from the molecular constitution of the solution and characterized by

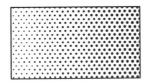

Fig. 3.4 Molecules diffuse from regions where they are dense to regions where they are rare.

the molecular diffusion constant D. By establishing this relationship Einstein derived one of the first *fluctuation–dissipation* theorems ever found, a connection between a molecular quantity D describing an effect of the random thermal motion of molecules and a macroscopic quantity η describing the irreversible dissipation of energy through friction when a spherical object moves through a viscous fluid.

Observable Molecular Motion

If a small drop of solute were put into the solvent, it would not remain a well-defined drop for long. Its boundary would become diffuse, the solute would slowly distribute itself through the solvent in a way governed by a differential equation called the *diffusion equation*, and it would eventually become uniformly distributed. At intermediate times the concentration of the solute around the location of the original drop would be given by the curves of Fig. 3.5. The peak of this curve moves out with time, not in proportion to the time, but in proportion to the square root of the time.

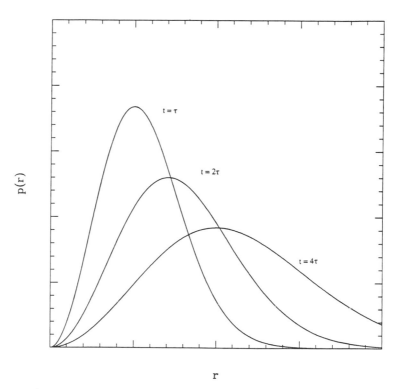

Fig. 3.5 The probability of a solute molecules being at a distance r from its starting point at times τ, 2τ, and 4τ.

If one were to follow a particular molecule during this diffusion process, the probability of it being found at any given distance from the position of the original drop would be proportional to the concentration of solute at that distance and is therefore given by the same curve as is the concentration. This is the point where Einstein applied the theory just developed to the Brownian motion. One could observe the distance r between the locations of a small particle suspended in a fluid as seen at two times separated by an interval t. But r was not the distance that the particle traveled during this interval, because the particle followed a very zigzag path as a consequence of collisions with fluid molecules; rather, it was the straight-line distance between two points on this crooked path. Regarding this particle as a molecule diffusing through the fluid, Einstein predicted that the mean of the *square* of this distance r measured in many pairs of observations separated by the interval t should be proportional to the length of that interval. Roughly speaking, the distance between the starting point and the point at which the particle is located at a later time should increase as the square root of the time and not in direct proportion to the time. His prediction included the value of the constant relating distance and time in terms of the particle size, the viscosity of the fluid, the temperature, and Avogadro's number. It was

$$\sqrt{\overline{r^2}} = \sqrt{t}\sqrt{\frac{RT}{N_A}\frac{1}{\pi\eta a}}.$$

Everything in this equation except Avogadro's number N_A could be measured, so Einstein urged that experiments be done to confirm the theory and to determine Avogadro's number in this way.

The most complete experimental confirmation of Einstein's theory was made by Jean Perrin, who in 1909 wrote a comprehensive review of the whole subject and the determination of Avogadro's number.[2] Perrin's best value for Avogadro's number was 7.05×10^{23}. It is interesting to note that Perrin included the Planck value obtained from the blackbody spectrum but assigned it the enormous uncertainty of lying between 6.0 and 8.0×10^{23}, much larger than that assigned by Planck. This may reflect his lack of understanding of the Planck and the Rayleigh theories of blackbody radiation as late as 1909, or perhaps Perrin felt that his value was better than Planck's, so that the uncertainty in Planck's result must have been greater than Planck believed. At the end of this long review Perrin wrote:

> **Molecular Reality.** I think it impossible that a mind, free from all precon-
> ception, can reflect upon the extreme diversity of the phenomena which thus
> converge to the same result, without experiencing a very strong impression,
> and I think that it will henceforth be difficult to defend by rational arguments
> a hostile attitude to molecular hypotheses, which, one after another, carry
> conviction, and to which at least as much confidence will be accorded as to
> the principles of energetics. As is well understood, there is no need to oppose
> these two great principles, the one against the other, and the union of Atomists
> and Energetics will perpetuate their dual triumph.

Perrin found it necessary even at the late date of 1909 to stress that his work, together with Einstein's theory, added to the evidence that atoms and molecules exist. By this time the dispute was almost settled, but there were still a few reluctant resisters to be won over.

Notes

1. A. Einstein, *Annalen der Physik* 17, 549 (1905); an English translation is given in A. Einstein, *Investigations of the Theory of the Brownian Movement*, trans. A. D. Cowper (New York: Dover, 1956).

2. J. Perrin, *Brownian Motion and Molecular Reality*, repr. in Mary Jo Nye, ed., *The Question of the Atom* (San Francisco: Tomash, 1984).

Amplification

Einstein's first paper on Brownian motion, like his other two major papers of 1905, needed very little calculation to achieve its remarkable ends. Here I write out the steps described in the previous section in mathematical language, mainly to show just how simply a great idea was formulated.

Osmotic Pressure

From the ideal gas law that applies to dilute solutions,

$$pV = \frac{n}{N_A} RT,$$

where p is the osmotic pressure exerted by the solute, not the total pressure in the solution, one immediately obtains

$$p = \frac{nRT}{VN_A} = v \frac{RT}{N_A},$$

where v is the number of solute molecules per unit volume. At a constant temperature, then, any variation in pressure is accompanied by a variation in the density of the molecules,

$$dp = \frac{RT}{N_A} dv.$$

The external force K on each molecule produces an increase in pressure in a distance dx equal to the total force per unit area exerted on all the molecules in a layer of thickness dx. Thus the force K produces a pressure change over the distance dx of

$$dp = vK\,dx.$$

This pressure variation due to the force K therefore produces a density gradient

$$\frac{dv}{dx} = \frac{vN_A K}{RT}.$$

Molecular Motion

According to Stokes' law, the force K applied to a spherical molecule of radius a makes it move through a viscous fluid of viscosity η with a speed v such that

$$K = 6\pi\eta a v.$$

This motion of molecules constitutes a current s given by

$$s = vv$$

$$= v\frac{K}{6\pi\eta a}.$$

By the definition of the diffusion constant, the diffusion current is given by

$$s = -D\frac{dv}{dx}$$

$$= -D\frac{vN_A K}{RT}.$$

In a steady state these two currents must add up to zero, which requires that

$$D = \frac{RT}{N_A}\frac{1}{6\pi\eta a}.$$

The force K and the velocity v do not appear in this equation; the application of the force was simply a device to investigate the relation of diffusion to viscosity in a special case in which everything could be easily evaluated.

Observable Molecular Motion

As a consequence of diffusion, the number density of solute molecules obeys the diffusion equation

$$\frac{\partial v}{\partial t} = D \frac{\partial^2 v}{\partial x^2}$$

if the density varies only in the x-direction and is uniform in the other two directions. This equation gives the rate at which the density at a point with coordinate x changes in time in terms of the nonuniformity of the density distribution. It was even then a well-known equation, being the one that describes the diffusion of heat through a thermally conducting material. A solution of this equation that describes a distribution of solute molecules that for small times is very concentrated in the plane defined by $x = 0$ is

$$v(x, t) = \frac{1}{\sqrt{4\pi Dt}} \exp\left(-\frac{x^2}{4Dt}\right).$$

For fixed t this is the famous bell-shaped Gaussian curve which gets lower and wider as t gets larger.

Because the solute molecules in a dilute solution move independently of one another, the density of molecules in a given plane x is proportional to the probability of any one molecule being in that plane. If, therefore, the molecule is directly visible because it is extremely large, a suspended particle visible in a microscope for example, this function gives the probability that it will have moved a distance $|x|$ in time t in either direction away from the plane $x = 0$. The mean value of x over many trials will be zero because movements in the two opposite directions are equally likely. The mean value of x^2, however, does not vanish because x^2 is never negative. It is given by

$$\overline{x^2} = 2DT.$$

The result quoted in the main text and tested experimentally by Perrin is for the root mean square of the distance of the position at time t from the initial position when motion in all three directions is considered. It is given by

$$\overline{x^2 + y^2 + z^2} = \overline{r^2} = 6DT$$

$$\sqrt{\overline{r^2}} = \sqrt{t}\sqrt{\frac{RT}{N_A}\frac{1}{\pi \eta a}}.$$

CHAPTER 4

A Chaotic Period Begins

The way that the quantum entered physics was a model of obscurity. Blackbody radiation was not a topic of central concern at the end of the nineteenth century. It had attracted some attention from some of the leading physicists of the time, including Gustav Kirchhoff, Ludwig Boltzmann, Wilhelm Wien, Ernst Pringsheim, and a few others. The problems of making accurate measurements had long been a challenge to experimentalists, Samuel Langley in the United States and a group in Germany among whom Heinrich Rubens, Ferdinand Kurlbaum, Otto Lummer, and Pringsheim were prominent. We have seen how their efforts led to the Stefan–Boltzmann law and Wien's displacement law, both of them sound thermodynamic results restricting but not determining the actual distribution of energy over frequency or wavelength. In this chapter we will follow the efforts of three people to derive this distribution from what two of them optimistically regarded as first principles. Wien gave an admittedly *ad hoc* derivation that led to a result that agreed closely with experiments until the middle of the year 1900. Max Planck gave a more pretentious derivation of Wien's result that he believed for a while to be soundly based on thermodynamics. Lord Rayleigh wrote a brief note applying statistical mechanics to the problem, invoking in a low-key way the principle of the equipartition of energy in connection with the modes of the radiation field, in particular the low-frequency end of the spectrum. His result made no sense at the high-frequency end of the spectrum, but it would agree with the soon-to-be-acquired data at the low-frequency end.

Into this gap came the German experimentalists, who in 1900 established that the Wien law was not valid at low frequencies and high temperatures. They also established that Rayleigh's law was good there but paid no attention to this fact.

Planck was then forced to recognize that his previous derivation of the Wien law contained arbitrary assumptions, and he found a way to make different ones that produced a formula that fitted the new data within experimental uncertainty over the entire known frequency range. He then felt the need to justify his assumptions, to render them less arbitrary. To accomplish this he found that he was forced, for the first time, to adopt Boltzmann's statistical approach to thermodynamics. To count the possibilities that this entailed, he introduced an energy quantum. This quantum was associated with the oscillators that he used to mock up the material with which the radiation was in equilibrium and that determined the temperature of that equilibrium rather than with the radiation field itself. His inability to let it approach the value zero but his necessity to give it a nonzero value at the end of his theoretical development quietly admitted a totally new concept into physics, the quantum of action.

This is clearly a long and complicated story. It is the subject of this chapter.

Particles and Fields

At the end of the nineteenth century there were two large divisions of physics that had very different underlying structures. The older of these was classical mechanics, the science of the motion of material bodies, and the younger one was electrodynamics, the study of the effects produced by electrified and magnetized bodies, including the production and detection of electromagnetic waves.

Classical mechanics dealt with discrete physical objects, bodies that could be separately described and whose positions and motions could be specified by using a finite set of numbers. For example, the position of an extended body of fixed shape, a rigid body, could be completely described by giving the three coordinates of some one point of the body and by giving the orientation of the body about that point, which required three more numbers for its specification. Its motion could be completely specified by giving the rate at which these six numbers changed with time. Classical mechanics accounted for the influence of one body on another by introducing action at a distance, in practice if not necessarily in principle.* Sometimes this was done by giving the force that the bodies exert on each other, as in Newton's law of universal gravitation or Coulomb's law giving the attractive or repulsive force between

---

* Newton wrote the following to Richard Bentley concerning action at a distance: "That gravity should be innate, inherent and essential to matter so that one body may act upon another at a distance through a vacuum without the mediation of any thing else by and through which their action or force may be conveyed from one to another is to me so great an absurdity that I believe no man who has in philosophical matters any competent faculty of thinking can ever fall into it. Gravity must be caused by an agent acting constantly according to certain laws, but whether this agent be material or immaterial is a question I have left to the consideration of my readers." See R. S. Westfall's biography of Newton, *Never At Rest* (Cambridge: Cambridge University Press, 1980), p. 505.

two electrically charged bodies. It could be done by specifying a potential energy depending on the relative positions of the bodies. The mechanism by which one body's influence was transmitted to another did not come into consideration. This influence was completely specified by the description of the interacting bodies and their positions and perhaps their velocities. No finite time interval was allowed for this influence to propagate from one body to another; the influence was considered to act instantaneously. The propagation of anything between bodies implied the existence of that thing at intermediate locations, and this contradicted the concept of action at a distance.

The Faraday–Maxwell picture of electromagnetic phenomena was different from action at a distance. In their picture the electric and magnetic fields had an existence apart from the bodies that generated them or on which they acted. They had their own dynamics, which affected the way in which one charged or magnetized body influenced another. In particular, the transmission of such influences was not instantaneous but, rather, was at a finite speed not exceeding the speed of light, so a finite time was required for the effect of a change at one place to appear someplace else. The fields were truly continuous and had values at every point of space, so they were not localized in the manner of particles. No finite set of numbers could describe a field completely. The very large value of the speed of light—the speed characterizing the rate at which electromagnetic influences were transmitted between bodies—obscured the presence of this independent dynamics. Only when bodies were at astronomical separations or when very short time intervals could be distinguished did the lapse of time between the occurrence of a change at one point and the appearance of its effect at another point become apparent.

Einstein began his famous paper introducing the light quantum with the following paragraph:

> There exists an essential formal difference between the theoretical pictures physicists have drawn of gases and other ponderable bodies and Maxwell's theory of electromagnetic processes in so-called empty space. Whereas we assume the state of a body to be completely determined by the positions and velocities of an, albeit very large, still finite number of atoms and electrons, we use for the determination of the electromagnetic state in space continuous spatial functions, so that a finite number of variables cannot be considered to be sufficient to fix completely the electromagnetic state in space. According to Maxwell's theory, the energy must be considered to be a continuous function in space for all purely electromagnetic phenomena, thus also for light, while according to the present-day ideas of physicists the energy of a ponderable body can be written as a sum over the atoms and electrons. The energy of a ponderable body cannot be split into arbitrarily many, arbitrarily small parts, while the energy of a light ray, emitted by a point source of light is according to Maxwell's theory (or in general according to any wave theory) of light distributed continuously over an ever increasing volume.[1]

One could simply say that there were these two different aspects of nature and not make an issue of it. This was, in fact, what nearly all physicists did. Why should everything be of the same character? A particle was a particle and a field was a field, and why should they not coexist as separate entities? One reason to think that they could not was the overwhelming success of thermodynamics and its statistical foundation. Thermodynamics was based on the law of conservation of energy and the law of increase of entropy, laws that did not depend on the structure of the underlying system. Energy and entropy could be assigned to systems of particles and also to fields, and so the energy and entropy of a field were subject to the two laws of thermodynamics. This fact was used by Kirchhoff, Boltzmann, and Wien, among others, in ways we described earlier, and the results were in accord with experiments. The culmination of this long development as applied to light was the Wien displacement law we discussed earlier.

The kinetic theory of gases as elaborated by Boltzmann had made it seem that the thermodynamic behavior of systems of particles could be understood with the help of a statistical model in which the second law was given a probabilistic interpretation. In this statistical model the principle of equipartition of energy was valid. It was inevitable that one should want to extend this statistical treatment of thermodynamics to the electromagnetic field. This is where the qualitative difference between systems of discrete particles and continuous fields had a devastating effect.

The equipartition theorem was originally conceived by Maxwell and Boltzmann in connection with mechanical systems, in particular molecular gases. Its mathematical formulation, starting from Hamilton's form of the equations of motion and some statistical assumptions, could be extended to Maxwell's electromagnetic theory, although Rayleigh was one of very few who realized this. As Rayleigh pointed out, the fields inside a cavity of any finite size, however large, can be considered as made up of a discrete set of *modes*, each one having its own characteristic frequency of oscillation varying from a lowest value corresponding to a wavelength about twice the linear dimension of the cavity up to arbitrarily large values corresponding to arbitrarily short wavelengths. In this way the fields could be pictured as composed of an infinite set of harmonic oscillators of frequencies ranging from a lowest one, depending on the size and shape of the cavity, to higher and higher frequencies with no upper limit. Rayleigh considered this situation in a short note published in 1900 entitled "Remarks upon the Law of Complete Radiation." We shall return to this note in more detail later on, but here it is interesting to quote him on the subject of the equipartition of energy:

> Speculation on this subject [complete or blackbody radiation] is hampered by the difficulties which attend the Boltzmann–Maxwell doctrine of the partition of energy. According to this doctrine every mode of vibration should be alike favoured.[4]

This immediately led to the ridiculous situation that it would be impossible to have radiation in a cavity in equilibrium with the walls. There was an unlimited number of modes of the field of shorter and shorter wavelength, and so it would take an unlimited amount of energy to give each one any finite amount of energy, however small. We might ask if the case of an acoustical cavity is not as bad, there being an unlimited number of modes of the sound field to supply with energy. The answer is no. Sound waves consist of alternate compressions and rarefaction of the gas in the cavity. The gas consists of molecules, and in the cavity there are only a finite number of them, so they have a finite separation. It is impossible to have alternate compressions and rarefaction between adjacent molecules! There is, therefore, no unlimited number of short-wavelength acoustic modes in the cavity. If the gas were a continuous, structureless substance, then the same difficulty would arise with sound waves as arises with electromagnetic waves. Referring to the equipartition principle or "doctrine," Rayleigh remarked,

> Although for some reason not yet explained the doctrine fails in general, it seems possible that it may apply to the graver modes

These remarks concerning the application of the principle of equipartition of energy to the modes of the electromagnetic field attracted no immediate attention from the theorists worrying about the radiation problem.

It was only some years after the correct formula for the spectrum of thermal radiation had been discovered by Planck later in 1900 that the equipartition paradox attracted much attention. The whole theory of the mechanical interpretation of heat was opened up for new study. The final result was the emergence of quantum theory, a theory in which the distinction between particles and fields becomes blurred and almost disappears and the equipartition theorem becomes an approximation valid only under appropriate circumstances.

Notes

1. A. Einstein, *Annalen der Physik* **17**, 132 (1905); an English translation appears in Mary Jo Nye, ed., *The Question of the Atom* (San Francisco: Tomash, 1984), p. 461.

2. Lord Rayleigh, *Philosophical Magazine* **49**, 539 (1900).

Blackbody Radiation According to Wien

By the late 1890s the experimental work of many people, especially F. Paschen, O. Lummer, E. Pringsheim, and H. Rubens, had revealed a lot about the spectrum of the light emitted by a hot blackbody. The variation of the intensity of the radiation with color, the frequency distribution of the radiation, was seen to have the same general form as the Maxwell distribution of the speeds of molecules in a gas. At low and at high frequencies it was small and it had a

maximum somewhere between. The location of the maximum was seen to
change toward higher frequency with increasing temperature. The color at the
lowest temperature giving visible radiation was a dim red, and as the radiating
body got hotter, the radiated light got brighter and whiter, the amount of
yellow, green, and blue radiation increasing faster than the amount of red.
We have seen that in 1893 Wien had shown, using thermodynamics and the
electromagnetic theory of light, that if the blackbody spectrum has a maximum,
the frequency of light at which this maximum occurs must be proportional to
the absolute temperature of the radiating body. This displacement theorem
meant that the ratio of the frequency at which the maximum occurred at a
given temperature to this temperature measured on the absolute scale—or the
product of the corresponding wavelength and temperature—had to be constant.
By 1895 the value of this constant had been determined by Paschen to be 2700
when the wavelength was measured in microns (millionths of a meter) and
the temperature was measured in kelvins (degrees absolute).[1] There was no
reason to doubt this law on either experimental or theoretical grounds. Experiments were consistent with its most striking prediction, and its derivation was
a classic of thermodynamic reasoning. It was expected, therefore, that the
spectrum would be of the form required by that theorem. If at temperature
T the energy of radiation in a unit volume with frequency in a unit frequency
interval centered on the frequency v is denoted by $u(v, T)$, it should be possible
to write this as the product of the cube of the frequency and a function of the
ratio of the frequency to the temperature. Symbolically,

$$u(v, T) = v^3 \phi\left(\frac{v}{T}\right),$$

where $\phi(v/T)$ was an unknown but presumably well-behaved function of the
single variable v/T trailing off at high values of v/T. The decrease of the
intensity toward low frequencies required by observation would be provided
by the factor v^3 in front as long as $\phi(v/T)$ did not increase too fast here. The
least-known part of the spectrum was that at low frequency or long
wavelength, a region known as the *far infrared*. In 1895 there were no very
sensitive detectors available that worked in this region, and at even the highest
temperatures that could be used, the radiation here was very weak. At the
high-frequency end of the spectrum the energy decreased rapidly. This was
the situation when Wien proposed the spectral distribution bearing his name.[2]

   Wien based his argument for the distribution law on Kirchhoff's thermodynamic theorem requiring the spectrum of thermal radiation at a given
temperature to be independent of the kind of matter with which the radiation
is in equilibrium. He chose as a material system a gas whose molecules had
speeds given by the Maxwell–Boltzmann velocity distribution. As we have
seen, this distribution was well founded in statistical physics, even though it
was not required by thermodynamics and as yet no direct experimental
confirmation of it existed. Wien then assumed that the frequency and intensity
of the light that a given molecule could emit depended only on the speed of

that molecule and that the intensity of radiation at any frequency v was some function of the frequency times the number of molecules that were moving with the appropriate speed as given by Maxwell. Neither of these assumptions could be derived from basic principles, but they were not totally implausible and their consequences seemed worth exploring. Insistence on their consequences being consistent with the displacement law led to the Wien law: The energy density per unit frequency of thermal radiation was given by

$$u(v, T) = c_1 v^3 e^{-c_2 v/T}.$$

Because wavelengths, rather than frequencies, were the quantities measured, the energy density per unit wavelength interval was more often used. It was given by

$$u(\lambda, T) = \frac{\alpha}{\lambda^5} e^{-\beta/\lambda T}.$$

The constant c_2 (or β) allowed fitting the maximum given by the formula to the frequency at which the maximum brightness was observed to appear, and the constant c_1 (or α) allowed fitting the absolute brightness of the radiation to that observed. Values of these constants accepted at the beginning of 1900 were $\alpha = 1.24 \times 10^{-5}$ cgs units and $\beta = 1.4435$ cgs units.

Plots of this function for several temperatures are shown in Fig. 4.1. Their similarity to the Maxwell–Boltzmann speed distribution was not coincidental. The function $\phi(v/T)$ left unspecified by the displacement law was simply the exponential function appearing in this equation. For a fixed temperature, when the frequency was large the exponent was large and negative. This made the exponential function decrease, and it decreased so much faster than the factor v^3 in front increased that it made the radiation get very weak at high frequencies. Again for fixed temperature, as the frequency approached zero the exponent approached zero, so that the exponential function approached unity. The factor in front then made the radiation weak at low frequencies. In between these limits the predicted energy density per unit frequency interval had a maximum.

The only place that the temperature appeared in Wien's formula was in the exponent, where it was in the denominator. If the intensity of radiation of any fixed frequency were calculated from this formula for increasing temperatures, the exponent would approach zero, and the exponential function itself would approach the value unity. This meant that at any fixed frequency the energy density of radiation per unit frequency interval would approach the constant value $c_1 v^3$ as the temperature rose. The formula predicted that the radiation of any particular color or frequency would not exceed this limit, no matter how hot the radiation body got.

Wien's law was found to fit the observations made before 1900 very well when appropriate values were assigned to the constants c_1 and c_2, but these observations did not include accurate ones at very low frequency. For the temperatures attainable at this time, the predicted cap on the intensity at any

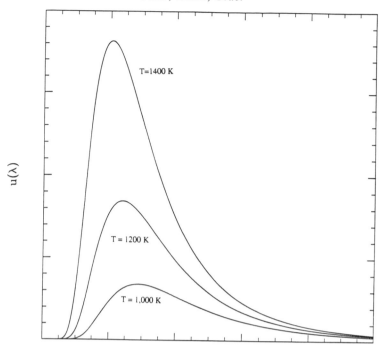

Fig. 4.1 The blackbody spectrum according to Wien for three temperatures, 200°C, 700°C, and 1200°C.

given frequency could be approached only at very low frequencies for which there were no sensitive detectors, and so this feature of the Wien law remained untested until the summer of 1900.

Lord Rayleigh did not like the prediction of a limit to the intensity of radiation with increasing temperature required by Wien's law and remarked:

> Upon the experimental side, Wien's law ... has met with important confor-
> mation. Paschen finds that his observations are well represented if he takes
>
> $$c_2 = 14,455$$
>
> θ [T] being measured in centigrade degrees and λ in thousandths of a
> millimetre (μ). Nevertheless, the law seems rather difficult of acceptance,
> especially the implication that as the temperature is raised, the radiation of
> given wavelength approaches a limit. It is true that for visible rays the limit
> is out of range. But if we take $\lambda = 60 \mu$, as (according to the remarkable
> researches of Rubens) for the rays selected by reflexion at surfaces of Sylvin,
> we see that for temperatures over 1000° (absolute) there would be but little
> further increase of radiation.[3]

Being an expert on the theory of sound and its production by resonant systems, Rayleigh naturally thought of any radiation in a region bounded by walls in terms of the resonant modes defined by those walls. The longest wavelength modes for either sound or electromagnetic waves, those with the lowest frequencies, have wavelengths comparable to the separation of the walls, and different ones of these modes are readily distinguished from one another. The high-frequency modes are much more numerous and much harder to distinguish from one another. Rayleigh gave as an example a stretched string whose ends were fixed. Here the modes were such that a whole number of half-wavelengths had just to fit into the length L of the string and their frequencies were integral multiples of the frequency v_0 of the mode with $\lambda/2 = L$:

> Let us consider in illustration the case of a stretched string vibrating transversely. According to the Boltzmann–Maxwell law the energy should be equally divided among all the modes, whose frequencies are as 1, 2, 3, ... Hence if k be the reciprocal of λ, representing the frequency, the energy between the limits k and $k + dk$ is (when k is large enough) represented by dk simply.

In other words, the number of modes with frequencies specified by integers lying in a range dk, say from 96 to 105, is simply proportional to the extent of that range, here 10. For vibrations in three dimensions Rayleigh showed that the number of modes with frequencies in the range from k to $k + dk$ is proportional to $k^2 dk$. This made the density of high-frequency modes much greater than that of the low-frequency ones:

> If we apply this result to radiation, we shall have, since the energy in each mode is proportional to θ,
>
> $$\theta k^2 dk \qquad (3)$$
>
> or, if we prefer it,
>
> $$\theta \lambda^{-4} d\lambda \qquad (4)$$
>
> . . . The suggestion is that (4) rather than, as according to (2) [Wien's law],
>
> $$\lambda^{-5} d\lambda$$
>
> may be the proper form when $\lambda\theta$ is great.

According to Rayleigh's suggestion, the radiation of a given wavelength (color) should increase indefinitely in energy in proportion to the temperature, which he denoted by θ, and not reach a limiting value. For high frequencies, Rayleigh's law clearly could not be right, for there was no limit to the number of high-frequency modes, each requiring its share of any energy supplied to the system, so that it would take an infinite amount of energy to increase the system's temperature by a nonzero amount. To patch up this difficulty he kept the exponential appearing in Wien's law as an *ad hoc* factor to make the radiation decrease at high frequencies and so to give the spectrum a maximum. His final suggestion differed in form from Wien's law simply in setting the unspecified function ϕ equal to T/ν times the exponential function instead of just the exponential function alone. This conformed to the displacement law, and it did give the equipartition result at low frequency and high temperature.

Rayleigh's formula did not fit the data where in 1900 they had been accurate, at the higher frequencies (shorter wavelengths) for which good sources and detectors were available and Wien's law did fit. Consequently neither the formula nor the reasoning behind it receives much attention. About the only mention of it in the literature on blackbody radiation was in the 1900 experimental paper by Rubens and Kurlbaum that showed that the Wien law, as well as some other proposed laws, did not work at long wavelengths and high temperatures.[4] Here Rayleigh's law with the exponential was plotted along with others and proved to work where Wien's did not, but because it did not work at the other end of the spectrum where Wien's did, it was not considered a candidate to replace Wien's. The physical argument Rayleigh was making using the equipartition of energy principle was totally ignored.

By the turn of the century Max Planck had long been concerned with thermodynamics and radiation. He was well versed in the classical approach to thermodynamics. He had prepared Kirchhoff's lectures on the *Theory of Heat* for publication, and so he was familiar with the kinetic theory of gases, though not in sympathy with it. He regarded the second law of thermodynamics as an exact law never to be violated, not as a statistical law having only very improbable exceptions. For Planck the second law meant that in naturally occurring situations, entropy *always* increased. A reversible process, one in which the entropy remains constant, he regarded as an idealization that could never quite be achieved in reality, but in *no* process occurring in an isolated system on any scale was it possible for the entropy of that system to decrease. His student E. Zermelo was one of those who attacked Boltzmann's statistical mechanics, considering its requirement that the second law be given a statistical interpretation as a refutation of the atomic hypothesis. Planck himself apparently conceded the existence of atoms and molecules but did not accept the statistical methods of Maxwell, Kelvin, and Boltzmann as an appropriate way to investigate the effect of their existence on the bulk properties of matter, gases in particular.

Planck set himself the task of deriving the Wien law for blackbody radiation from Maxwell–Hertz electrodynamics and thermodynamics with no admixture of statistical ideas. During the 1890s he had been preparing for this task by studying the behavior of radiation with a view to finding irreversible processes that were a direct consequence of the Maxwell–Hertz theory.[5] In the scattering of an electromagnetic wave by a particle of mass m and charge e, he thought he had found such a process. A charged particle subject to an incident electromagnetic wave is forced to oscillate by the electric field of that wave. It then acts like an antenna sending out scattered radiation of the frequency of the incident wave in all directions . The radiated energy comes from the incident wave, which consequently loses energy. To Planck, the mechanism transferring energy from the incident wave to the scattered radiation looked similar to the mechanism of viscous friction. A sphere moving through a viscous fluid loses mechanical energy, just as the incident electromagnetic wave loses some of its electromagnetic energy, and the fluid gains this energy in the form of heat, just as the scattered wave gains the energy lost by the incident wave. There were two features, however, that distinguished the electromagnetic from the mechanical case. In the mechanical case the form of the energy was changed from kinetic energy to heat (which Planck did not want to regard as molecular kinetic energy), whereas in the electromagnetic case the energy remained electromagnetic. The parameter characterizing the rate at which a moving sphere lost its mechanical energy and transferred it to the viscous medium was the coefficient of viscosity. This depended strongly on the medium, and in general it could not be calculated from the fundamental equations of mechanics but had to be measured experimentally in each case. In the electromagnetic case, Maxwell's theory made the amount of energy lost by the incident wave in the form of scattered radiation directly calculable from the properties of the incident wave and of the scattering particle, its charge and its mass. Planck could write down the expression for this so-called radiation damping coefficient, the analog of the coefficient of viscosity in mechanics. If the incident wave is of frequency v, the radiation damping coefficient is $2\pi e^2 v^2/(3mc^3)$. Because of these differences, especially the second one, he thought that there was a fundamental distinction between the two cases and that the scattering of radiation provided an example of irreversible behavior originating in the fundamental equations of the system, Maxwell's equations. He could then hope to *derive* the second law of thermodynamics as it applies to electrodynamics from the fundamental equations of the theory.

It was Boltzmann who showed that this was not the case. He was the first to show that the form of Maxwell's equations remains unchanged if the sign of the time and the directions of all currents and magnetic fields are reversed.[6] Just as the time reversibility of Hamilton's equations prevents any distinction between past and future based on the fundamental equations of mechanics, the time reversibility of Maxwell's equations precludes the description of irreversible phenomena using the fundamental equations of electrodynamics. To every solution of Maxwell's equations describing the scattering of an incident electromagnetic wave by a charged particle, there corresponds a solution with the scattered waves converging back on the scatterer and combining with the reversed unscattered wave to reconstitute the incident traveling in the opposite direction. Planck's system of a wave incident on a charge particle initially at rest is easy to conceive of and relatively easy to prepare in nature. The time-reversed state with what survived of the incident wave traveling in the opposite direction and combining with the reversed scattered wave in just such a way as to constitute incident wave and nothing else is harder to conceive of and is almost, if not entirely, impossible to prepare in nature. But practical difficulty was irrelevant; the question was one of principle, not convenience. Planck was faced with a difficulty much like the one that had faced Boltzmann after Loschmidt pointed out the time reversibility of Hamilton's equations, thereby invalidating his proof of the second law of thermodynamics on the basis of mechanics alone. To avoid it, Planck was forced to reformulate his ideas.

In order to study blackbody or cavity radiation, Planck, like Wien before him, took advantage of Kirchhoff's result that this radiation is independent of the material with which it is in interaction and whose energy determines the temperature of the radiation. Where Wien had chosen moving molecules, Planck chose stationary oscillators of all frequencies. These oscillators were in contact with waves traveling in all directions in the cavity and undergoing many reflections from the cavity walls. At the position of one of these oscillators the total electric field acting would be fluctuating as a result of the interference of the many waves present. Some of these fluctuations would be very rapid, with frequencies comparable to the frequencies of the waves themselves, and some would be much slower. Planck assumed that the rapid fluctuations averaged out and did not produce any long-term effects. He retained only the effects produced by the slow fluctuations in order to calculate the way that the radiation changed with time. The hypothesis that this was justified he called the hypothesis of *natural radiation*. All the long-term effects such as the approach to equilibrium, he assumed, were produced by these slow changes or fluctuations.

It is clear that Planck did not recognize this introduction of natural radiation as analogous to Boltzmann's introduction of ensembles in his discussion of the irreversible behavior of gases. The ensemble that Boltzmann had introduced was an ensemble of gases sharing the same composition, mass, volume, temperature, and so forth, but not sharing the same state of molecular motion. He had introduced it in order to calculate averages of physically interesting quantities over all members of the ensemble. Planck was studying his oscillators individually, not en masse, but in introducing natural radiation he had done some averaging over possible states of the radiation field. This appeared to be quite different from bringing in ensembles of oscillators. Planck used his concept of natural radiation in connection with the behavior of a single one of his oscillators, not a vast array of them, and drew attention to this difference between his theory and Boltzmann's. In order to find the expression for the entropy of this oscillator, he did consider the way that the energies and entropies of *many* of them with the same frequency should be combined to yield the total energy and entropy of them all, but his way of combining them was simply to multiply the energy or entropy of one oscillator by the number of oscillators, a straight addition involving no element of randomness. Finally, with the help of his hypothesis of natural radiation, Planck arrived at a connection between the entropy S and the average energy U of his oscillators. He wrote this as

$$S = -\frac{U}{a} \log \frac{U}{ebv}.$$

Here a and b are constants that could be adjusted to fit experiments, and *e* is the base of the natural logarithms that he introduced to make later formulas simpler. From this he could derive a relation between their average energy and the temperature when the entropy had reached a maximum and the oscillators were in thermal equilibrium. At this point the electromagnetic field had disappeared from his result, with only the oscillators being involved. A ghost of the field remained unnoticed, however, in the effect on the oscillators of the hypothesis of natural radiation used in the calculation.

Planck's next task to relate the average energy of his oscillators to the average energy of the radiation at their frequency. In an earlier study Planck had found this connection. The two are proportional to each other, with a proportionality constant depending on the frequency. Then knowing the equilibrium energy of his oscillators, he also knew the energy of the radiation of that frequency. In this way he arrived at the equilibrium energy distribution of radiation in thermal equilibrium with his oscillators and therefore, according to Kirchhoff, in equilibrium with anything else. He believed he had *derived* the blackbody spectrum without resorting to Boltzmann-like probability considerations. His result was the Wien law.

Planck pointed out the limitation of the intensity of light of a given color with increasing temperature that is contained in the Wien law. Hertz had produced electromagnetic waves of high frequency if thought of in terms of oscillating electric circuits but of extremely low frequency if thought of in terms of light or heat radiation, and there had been no evidence at all of any limitation on the amount of the energy that these waves could have. Was this in contradiction of the Wien law with its low limiting value of radiation at low frequencies? Planck explained the absence of this limitation in the case of Hertzian waves by saying that they were highly ordered waves and did not constitute *natural radiation*. Therefore the Wien limitation did not apply to them.

For a few months Planck and most of the rest of the scientific community interested in such matters believed that the issue of the spectrum of blackbody radiation was settled. He had based the Wien spectrum on what appeared to be firm thermodynamic ground, and he expressed the opinion that the range of validity of the Wien distribution and that of the second law of thermo- dynamics were the same. He could not have made a stronger statement of confidence in his result. At that time the Wien law represented the experimental data well, although there were some discrepancies. In his renowned *Theory of Optics*, whose introduction is dated January 1900, Paul Drude described the Wien law, on the grounds of Planck's derivation as "universal." In a foot- note, however, to his giving the value of β as 1.4435 in cgs units he reported,

According to Beckmann (diss. Tübingen, 1898) and Rubens (*Wied. Ann.*, **69**, p. 576, 1899) the constant c_2 [our β], when calculated from the emission of waves of great length, is considerably larger. According to this Wien's law is not rigorously correct.[7]

This was a precursor of what was to happen in the summer of 1900, when the "emission of waves of great length" was examined with a precision previously unavailable. Planck's argument contained a fault. The fault slipped, and its slipping shook the classical edifice.

Notes

1. F. Paschen, *Göttingen Nachrichtungen* (1895).

2. W. Wien, *Annalen der Physik* **58**, 662 (1896).

3. Rayleigh, *Philosophical Magazine* **49**, 539 (1900).

4. H. Rubens and F. Kurlbaum, *Berlin Berichte* (1900), p. 929.

5. M. Planck, A Series of Five Communications to the *Prussian Academy of Sciences* entitled "On Irreversible Radiation Processes," extending from February 1897 to May 1899; *Annalen der Physik* **1**, 69 (1900).

6. L. Boltzmann, "On Irreversible Radiation Processes," *Berlin Berichte* 660 (1897).

7. P. Drude, *Theory of Optics* (Leipzig, 1900). The quotation is from the English translation by C. Mann and R. A. Millikan (New York: Dover, 1959), p. 526.

Amplification

Wien's argument leading to his blackbody spectrum was a simplification of and improvement on a proposal by V. Michelson that had led to an expression with more than two parameters. Wien suggested that the wavelength of radiation emitted by a molecule in thermal equilibrium with other molecules of a gas at a temperature T depended solely on the speed v of the molecule and not on its direction of motion, so that he could write

$$\lambda = g(v^2).$$

This equation could be solved for v^2 to give

$$v^2 = f(\lambda),$$

where $f(\lambda)$ was a function to be found. He further suggested that the intensity of radiation at a given wavelength was proportional to the number of molecules having the corresponding speed, which was given by the Maxwell speed distribution and to a function of the wavelength alone, $F(\lambda)$. He then used the Maxwell distribution, according to which the probability of a molecule having speed v is proportional to $v^2 \exp[-\alpha v^2/T]$, to obtain

$$u(\lambda, T) = F(\lambda)\, e^{-f(\lambda)/T}.$$

Now according to Wien's displacement law, the exponent could depend only on the product λT (or the ratio v/T) so that $f(\lambda)$ must be simply β/λ. Further, the function $F(\lambda)$ had to be α/λ^{-5} in order to have the total radiative energy proportional to the fourth power of the temperature as required by the Stefan–Boltzmann law. This was the Wien spectrum written in terms of wavelength rather than frequency:

$$u(\lambda, T) = \frac{\alpha}{\lambda^5} e^{-\beta/\lambda kT}.$$

The Planck derivation was more complicated than this, and here only one key step will be described, a step that showed his conviction that everything concerning thermal equilibrium was definite and sharply defined, allowing no room for statistical uncertainty.

Planck believed that if a system had an energy that differed by the amount ΔU from the value it would have at equilibrium at the ambient temperature, then the energy of the system would change with time according to the equation

$$\frac{dU}{dt} = -2\sigma v \Delta U.$$

This equation described an exponential approach to the equilibrium energy at a rate given by $2\sigma v$. It was an equation of first order in the time and therefore not invariant under a reversal of the sign of the time. It arose from Planck's use of the hypothesis of natural radiation. After some mathematical manipulation and requiring that the entropy have maximum as a function of the energy U, he arrived at the result that

$$\frac{\partial^2 S}{\partial U^2} = -f(U),$$

where $f(U)$ was a positive function of the energy. This equation was taken to apply to a single oscillator and to a set of n similar oscillators in the same cavity. Since the n oscillators were identical and in the same cavity field, Planck assumed that they all would act in precisely the same way and that the energy and entropy of the set would be the sums of the separate energies and entropies, all of which would be alike. He could then write the equation connecting the total entropy to the total energy as

$$\frac{\partial^2 (nS)}{\partial (nU)^2} = -f(nU).$$

But n is a constant and produces simply a factor $1/n$ on the left side, so he arrived at the functional equation

$$\frac{1}{n} f(U) = f(nU).$$

This has the unique solution

$$f(U) = \frac{\text{const}}{U}.$$

The entropy could now be found as a function of a function of U, and this led straightforwardly to the Wien law.

Planck's procedure could be defended from a completely deterministic view of thermodynamics, though even there it was not watertight. From a statistical point of view it was clearly wrong, and wrong even from an experimental point of view. It was known in 1900 by those working in the field of electrical discharges in gases that not all atoms exposed to the same beam of X-rays responded to this radiation in the same way; a few atoms were ionized, but most were not. This was a puzzle to J. J. Thomson and Einstein, which the latter addressed in his famous paper introducing the light quantum. The field of natural radiation was not uniform in a microscopic sense over the entire cavity and was not exactly the same at all the locations of Planck's oscillators. The fluctuations that Planck said averaged out need not do so exactly, and if the energies of the separate oscillators differed at all, the sum of the energies at any one time would not be simply n times the energy of an individual one, as was assumed at a crucial point in the argument. We cannot fault Planck for having proceeded in this way, as it was consistent with his view of physics, a view well supported by enormous successes in the past. More important to our opinion of Planck is the fact that when he found this view untenable, as he soon did, it was he and not his critics who broke the ground that led to the new, quantum view.

The Experimental Verdict

During the last decades of the nineteenth century, there were great advances in experimental techniques that could be used to improve both the production and the detection of blackbody radiation. Kirchhoff had defined a blackbody as one that absorbed all the radiation incident upon it and reflected none. A body could be black for radiation of certain frequencies and far from black for radiation of others. The blackness of a body could be estimated quite well for visible radiation by measurements using the eye as detector. If a surface looked really black when illuminated by bright white light, it was absorbing all the visible light incident on it. As one went down in frequency below the visible range, the eye had to be replaced by some other detector. The standard way of detecting infrared radiation was to measure the heat produced when the radiation was absorbed and this required being able to know how much of the incident radiation the detector absorbed. Preferably the detector should have a black surface to do the absorbing, and so one was led back to the original problem, determining how black a body really was.

Kirchhoff had indicated how to build an ideal blackbody. One should form a cavity in a good conductor of heat and drill a small hole in the wall. The walls of the cavity are maintained at the desired temperature. The interior of the cavity should be shaped so that any radiation entering the hole is necessarily reflected away from the entrance so that it will have to undergo many reflections, each with the possibility of some being absorbed, before it can possibly reach the hole and escape from the cavity. Such a source, if well designed and constructed and with a well-controlled temperature, could be taken to be a source of blackbody radiation. It was necessarily large and massive. It had to be used in an environment much cooler than itself so that any radiation present assuredly came from the blackbody and not from something else in the neighborhood. Sources built according to Kirchhoff specifications were actually constructed in Berlin in the last decade of the century.

Detectors were a more delicate problem. To detect small amounts of radiation, they had to be small and light and to have a black absorbing surface. The best instrument was the bolometer developed by S. P. Langley.[1] This involved a thin and narrow strip of blackened platinum placed in the radiation to be measured, which had usually been transmitted through a prism or a grating to separate the various wavelengths. Its narrowness made it receive radiation of a narrow range of wavelengths or frequencies, and its thinness made it light so that it took very little energy to raise its temperature. This strip was connected to a resistance-measuring device, a Wheatstone bridge, in which the detecting galvanometer measured a current arising from the difference in resistance between the detecting strip and a standard strip kept at a constant temperature. The blackness of the absorbing strip could not be ensured by using Kirchhoff's idea, and its determination was a difficult part of carrying out the measurements.

Another problem was separating the various frequencies of radiation, a separation needed to determine a frequency spectrum. At the long wavelengths involved, a grating spread the radiation out so much that the intensity at the detector was reduced below the detector's capability to measure it accurately, and also different orders of spectra overlapped so that the radiation detected was not necessarily of a single frequency. The dispersion of materials such as rock salt that transmit radiation of these frequencies quite well was not accurately known, and so the frequency of the radiation bent through a given angle by a rock salt prism and then detected was not known accurately.

Through much hard work the difficulties of making measurements in the far infrared were overcome. By 1886 Langley could state that the wavelength at which thermal radiation had its maximum intensity decreased with increasing temperature, a result not immediately accepted by all. By 1893, when Wien published his displacement law, the constancy of the product of $\lambda_m T$ of the wavelength at the maximum and the temperature, was fairly well established. By 1896, when Wien presented his proposed spectrum of blackbody radiation, it was possible to use experimental data from below the maximum up toward the high-frequency end of the spectrum to evaluate the parameters appearing in the law. Up to the beginning of 1900 the accumulating data were consistent with the law, and the uncertainty in the values of the parameters appearing in the law was decreasing. When Planck published his derivation of the Wien law in 1900, that law seemed to be well established.

Isolating infrared radiation of a well-defined range of frequencies presented difficulties, especially in the very far infrared, where an acid test of the theoretically predicted blackbody spectrum could be made. To verify the prediction that the intensity of radiation at a fixed frequency would approach a limit with increasing temperature, a narrow band of frequencies had to be isolated out of the very broad spectrum emitted by a blackbody. Rubens and Nichols developed a technique of producing a beam of highly monochromatic long-wavelength radiation by multiple reflections from the faces of rock salt or fluorite crystals.[2] Each of these materials possessed one or more atomic resonances at an infrared frequency, and for radiation in the immediate neighborhood of this frequency, the reflectivity of the usually transparent material approached that of a metallic mirror, whereas for other frequencies it reflected very little of the incident radiation, like a piece of glass reflecting visible light. For rock salt the strongest reflection occurred for wavelengths near 54.0 μ (microns or millionths of a meter), and for fluorite it occurred at 22.9 and 32.8 μ. By reflecting the radiation coming from a blackbody cavity several times from surfaces of rock salt crystals onto the sensitive element of a bolometer, Rubens and Kurlbaum could measure the change of the intensity of this far infrared "residual radiation" as the temperature of the emitting body was changed.[3] That is, they could look at the dependence of the intensity on T for fixed frequency v. For the first time they could reach well-defined values of the ratio v/T so small that the limiting intensity predicted by the Wien law should be approached and the intensity should no longer increase in proportion to the temperature.

When these new methods were applied to the problem of blackbody radiation, the saturation predicted by the Wien law was found to be absent. For radiation of wavelength 51.2 μ, the increase of the intensity with increasing temperature should have begun to taper off markedly when the temperature reached 1500°C. But this did not happen; the intensity continued to increase proportionately to the temperature beyond the saturation value. This had been indicated a little earlier by somewhat less definitive observations, especially those by Lummer and Pringsheim.[4] By October 1900 the validity of the Wien law was shown to be restricted to not-too-long wavelengths (too-low frequencies) and not-too-high temperatures, at which the exponential factor was appreciably smaller than unity. Rubens and Kurlbaum published a figure in which the behavior of the intensity at 51.2 μ as observed and predicted by various theories was shown, the theories including that of Rayleigh with the *ad hoc* exponential (see Fig. 4.2). Rayleigh's came very close to fitting the

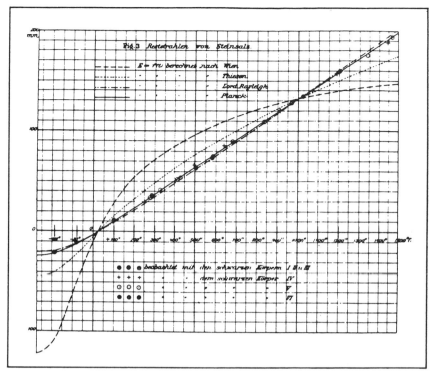

Fig. 4.2 Rubens and Kurlbaum's plot of their results for the dependence of the intensity of blackbody radiation at a wavelength of 51.2 microns on the temperature, and plots of various theoretical predictions. [From *Annalen der Physik* **4**, 649 (1901).]

data. This agreement was not especially noted, however, as the theory was disqualified by its predictions at larger values of v/T. The basis of the theory in equipartition was not mentioned.

In the fall of 1900 the situation regarding blackbody radiation can be summarized as follows:

1. The law governing the total amount of energy radiated per unit area per unit time by a blackbody had been well confirmed and the corresponding Stefan–Boltzmann constant had been determined. The Wien displacement law had also been confirmed over a considerable range of frequencies, and the value of the constant $\lambda_m T$ was known. There was no surprise here because the theoretical basis of these laws was well established and not subject to doubt.

2. The Wien law had been suggested by Wien, had been given theoretical backing by Planck, had been verified under the more easily attained conditions of short wavelength and moderate temperature, but had then proved inadequate under the experimentally more difficult conditions of long wavelength and high temperature.

3. Rayleigh had been disturbed by the saturation predicted by the Wien law at long wavelengths and high temperatures and had suggested a theory that incorporated the equipartition of energy for long wavelengths (low frequencies) but that retained the exponential decay at short wavelengths of the Wien law so as to keep finite the total radiated energy. The result was a modification of the Wien law. His proposal was shown to be inadequate at short wavelengths by the data that confirmed the Wien law at short wavelengths and moderate temperatures, even though it did fit the new data at long wavelengths and high temperatures.

What was needed was an interpolation between Wien and Rayleigh, although it was not then thought of in this way, as Rayleigh's equipartition argument was being ignored. The new formula was provided by Planck on a fairly *ad hoc* basis. Planck then succeeded in giving a physical basis for his formula. This work of Planck is the subject of the next section.

Notes

1. S. P. Langley, *Chemical News and Journal of Science* **43**, 6 (1881).
2. H. Rubens and E. F. Nichols, *Wiedemann's Annalen* **60**, 418 (1897).
3. H. Rubens and F. Kurlbaum, *Berlin Berichte* (1900), p. 929; *Annalen der Physik* **4**, 649 (1901).
4. O. Lummer and E. Pringsheim, *Verhandlungen der deutsche Physikalische Gesellschaft* **1**, 23, 215 (1899); **2**, 163 (1900).

Blackbody Radiation According to Planck; the Appearance of *h*

On the first of June 1899 Planck had submitted the fifth in his series of communications on "irreversible radiative processes" to the Prussian Academy of Sciences in Berlin. It was in this communication that he derived the Wien law describing blackbody radiation and concluded that "the limits of validity of this law, if such exist at all, coincide with those of the second law of thermodynamics." He had said that this made the precise experimental test of the Wien law of the greatest interest of principle. His basis for the Wien law was not a rigorous, step-by-step proof on the model of Euclid, but it was nonetheless a serious and persuasive argument. To use Sir E. T. Whittaker's adjective, it was not *coercive*, but it made the road to any other result look very forbidding.

In the latter half of 1900 Planck obtained the results of the test he had urged, and they showed, at first tentatively and then unambiguously, that the Wien law was not valid at small values of v/T, at low frequencies and high temperature. First it was Paschen, then Lummer and Pringsheim and finally Rubens and Kurlbaum, who showed that the intensity of radiation of a given color does not approach the limiting value predicted by Wien as the temperature increases but grows beyond that value in proportion to the temperature.[1] To accomplish this it had been necessary to develop blackbodies of the kind suggested long before by Kirchhoff, uniformly heated cavities with small holes designed so that any radiation entering encountered many opportunities to be absorbed before it could again leave the cavity, to refine the methods of measuring the high temperatures needed, and to have accurate and sensitive detectors in the far infrared.

This demonstration that nature did not behave according to his expectations forced Planck to go back over his arguments and find out where he had erred. According to his own account given to the Prussian Academy of Sciences in October 1900, he found a possible complication that he had not anticipated, one that opened up a new range of possibilities.[2]

In his derivation of the Wien law Planck had assumed that if the energy of one of his oscillators differed from the equilibrium energy corresponding to the temperature of the system, it would approach that equilibrium value smoothly and steadily at a rate calculable from electromagnetic theory. This energy loss was a manifestation of Planck's irreversible process coming from Maxwell's equations, no longer directly but indirectly via the hypothesis of natural radiation, and it entailed an increase in the total entropy of the oscillator and the radiation. In order to evaluate this entropy increase, he had assumed that if n of his oscillators—all formed in exactly the same way and all exposed to the same radiation field in the same cavity—started off with energies deviating from the equilibrium value by the same amount, they all would approach equilibrium at the same rate. The energy of the n oscillators would be just n times the energy of one; the deviation of the total energy of these oscillators from the equilibrium value would be just n times the deviation of any one of them; the change in the total energy in a time interval would be just n times the change in the energy of any one during this time interval; and the total entropy of the oscillators would be just n times the entropy of any one of them. He had assumed further that the rate of change in the entropy of these n oscillators depended on the total energy and the total deviation of the energy from its equilibrium value in just the same way that the rate of change in entropy of a single oscillator depended on its energy and its energy deviation from its equilibrium value. From these assumptions, which he considered almost self-evident although not consequences of the second law of thermodynamics, he had shown that the entropy S of each oscillator must depend on the energy U of the oscillator in a particular way, namely,

$$S = -\frac{U}{a} \log \frac{U}{ebv}.$$

This equation for the entropy had led him directly to the Wien law. The argument leading to this equation, therefore, had to be analyzed, and the place where it had led him astray had to be found.

By October 1900 Planck had decided that the change in total entropy of his set of n oscillators could also depend explicitly on the energy of the individual oscillators and not just on the total energy of the set, "even if it would not be easily understandable." The explicit dependence of the total entropy on the energy of an individual oscillator meant that somehow not all these oscillators were behaving completely coherently, remaining in lockstep as they approached equilibrium. To one believing in the complete regularity of nature, this would indeed "not be easily understandable." In any case, this made his earlier line of argument inapplicable and opened up new possibilities.

Among these new possibilities Planck found one that intrigued him particularly. It was a relation between the change in entropy and the change in energy of an oscillator in equilibrium with the radiation field that accomplished the following: "It gave a very simple logarithmic expression for the dependence of the entropy of an irradiated monochromatic resonator on its vibrational energy." This property was a requirement if Planck's new distribution was to decrease exponentially at high frequencies, as the Wien law and the observed spectrum did. Also, it was almost as simple as his original one; it approached the same form as his original one in the region where the original one worked; and it yielded something new where the original one did not work. This new relation led him to the distribution of energy over a wavelength given by what is now called the Planck law,

$$u(\lambda, T) = \frac{C\lambda^{-5}}{e^{c/\lambda T} - 1}.$$

The corresponding distribution over frequency was

$$u(v, T) = \frac{c_1 v^3}{e^{c_2 v/T} - 1}.$$

These differed from the corresponding Wien laws only by replacing the decreasing exponential function in the numerator with an increasing exponential minus one in the denominator. It was not evident that this apparently minor change heralded the beginning of a new era in science.

Immediately after this distribution was proposed, Rubens and Kurlbaum carefully compared their results with this formula. They found that by appropriately choosing the values of the parameters c and C, or equivalently c_1 and c_2, they could fit the formula to their measurements within experimental error. This agreement has persisted ever since over an ever-increasing range of temperatures and frequencies. In 1990, Cosmic Background Explorer (COBE) satellite measurements of the background radiation in space presumably left over from the Big Bang showed that this radiation follows the Planck law very exactly, the temperature being 2.735 ± 0.06 kelvins or −270.41°C.[3] The measurements extended from 0.05 to 1 cm or from 500 to 10,000 μ, covering both sides of the peak of the curve. Ninety years earlier Rubens and Kurlbaum had been forced to work hard to reach the wavelength of the residual rays of rocksalt, 51.2 μ, in their measurements establishing the Planck law.

Planck had found the right formula, but he had no good grounds to believe it to be correct other than the very strong one that it agreed with observation. He had arrived at it by picking one out of a large number of possibilities. The physical basis for its selection was missing. He needed to find out what distinguished this one from the others and to demonstrate that the others arising along with this one could be excluded.

To do this, Planck found himself forced to go beyond the bare statement of the laws of thermodynamics that throughout his scientific career he had championed as sufficient. There may be some hints that Planck had begun this process before 1900, but he did not go beyond them explicitly until this time.[4] He retained his model of the matter with which the radiation was in equilibrium as consisting of harmonic oscillators of all frequencies. His concern was with this set of oscillators. He did not follow Rayleigh's example and use the modes of the radiation field in a cavity as the oscillators to deal with, and therefore he did not recognize any need to reconsider the application of thermodynamics to Maxwell's electromagnetic field, but only its application to his oscillators. The relation of thermodynamics to the radiation field itself was left for Einstein to examine five years later.

Planck's object was to find the entropy of a set of oscillators of a specified frequency bathed in a field of thermal radiation of a given temperature. Because in his first way of getting the correct distribution, the entropy of a set of n of these oscillators could depend on the energy of an individual oscillator, he could not assume that all the oscillators behaved in exactly the same way. They could have different energies even after having started out with the same energy. Since all the oscillators were identical and in the same radiation field, this had to mean that some random element entered the situation. For the first time in his career Planck felt obliged to introduce a statistical interpretation of entropy, stating further on in the paper cited earlier:

> Entropy means disorder, ... The constant energy of the stationary vibrating resonator can thus only be considered to be the time average, or, put differently, to be an instantaneous average of the energies of a large number of identical resonators which are in the same stationary radiation field, but far enough from one another not to influence each other directly. Since the entropy of a resonator is thus determined by the way in which the energy is distributed at one time over many resonators, I suspected that one should evaluate this quantity by introducing probability considerations into the electromagnetic theory of radiation, the importance of which for the second law of thermodynamics was originally discovered by Mr. L. Boltzmann. This suspicion has been confirmed; I have been able to derive deductively an expression for the entropy of a monochromatically vibrating resonator and thus for the energy distribution in a stationary radiation state, that is, in the normal [blackbody] spectrum.

The derivation was contained in a paper presented to the Prussian Academy of Science in December 1900 and submitted for publication to the *Annalen der Physik* in January 1901.[5] It described a new approach. Before, Planck had been concerned with just one oscillator at a time, and now he was considering an ensemble of them. Following Boltzmann he took the giant steps of equating the time average of the energy of one oscillator to the instantaneous average of the energies of an ensemble of similar oscillators with a given total energy, and of relating entropy to probability.

In order to find the probability of a particular distribution of an amount E of energy among N oscillators, Planck had to be able to count the number of possible distributions of that energy over those oscillators. One cannot count the number of ways of dividing a continuous quantity into even two parts; two divisions are distinct no matter how little the boundaries between the parts differ. What one can do is to break up the quantity into a finite number of identical bits that can be counted, and then the number of ways of distributing these bits among discrete possible recipients is finite. Boltzmann had used this device in gas theory when he wanted to find the probable distribution of kinetic energy among the molecules. Kinetic energy depends on the momentum of a molecule in a continuous way, so Boltzmann had divided this energy up into many equal elements ε and had then counted the ways of distributing them among the molecules. The more ways there were to achieve a state in which a molecule has a given energy, the more probable it was that the molecule did have that energy. As long as it was sufficiently small, the size ε of his energy element turned out not to influence any physical result. After having done the counting Boltzmann could go to the limit in which ε approached zero. An important result arrived at in this way was the Maxwell–Boltzmann velocity distribution and its generalization to the internal motions of molecules.

Planck proceeded to follow this path. The problem he set himself was to assign probabilities to the various ways of distributing a given amount $E = NU$ of energy among a set of N oscillators of frequency v, U being the average energy of any one of the oscillators. Because the oscillators were identical and in the same cavity, their average energies would be equal even though their instantaneous energies would not be. He divided this energy into P equal elements or quanta of size ε so that $E = P\varepsilon$. He then counted the number of ways of distributing this energy $P\varepsilon$ over the N oscillators. There is a simple combinatorial formula for the total number W of ways:

$$W = \frac{(P + N - 1)!}{P!(N - 1)!}.$$

Having this number, Planck then, in the spirit of Boltzmann, assigned an entropy S_N to the state of the system in which the set of oscillators had this energy and wrote

$$S_N = k \log W,$$

where k is Boltzmann's constant. Planck was the first to write the expression for the entropy in this fashion, including the proportionality constant k. The value of k is fixed in principle by applying this formula to an ideal gas and counting, just as Boltzmann had done. The result of this is that k is the ratio of the ideal gas constant R to Avogadro's number N_A or is the gas constant per molecule rather than per mole. Avogadro's number was not well known at this time, so k could be considered a parameter to be fitted to the results of measurements of blackbody radiation. Its evaluation would then lead to a value of N_A.

After considerable mathematical manipulation, Planck arrived at an expression for the average entropy of one of his oscillators, more complicated looking than the expression for the entropy he had arrived at earlier in obtaining the Wien law, but in fact it was just the difference of two terms, each of which had the same form as the earlier one. A most important feature of this expression was that the energy U entered it only in the form of the ratio of U to the quantum size ε. This relation between the energy and the entropy of the oscillators could be transformed into one between the average energy of an oscillator U and the temperature T. The result was

$$U = \frac{\varepsilon}{e^{\varepsilon/kT} - 1},$$

and this bore a strong resemblance to the form of the spectrum he was trying to derive.

Having the average energy of an oscillator of frequency v as it depends on the temperature, Planck could use a relation between this average energy and the average energy and the average energy of the field mode with this frequency he had already derived and used in connection with the Wien law. It was that these two average energies are proportional to each other with a proportionality constant depending on the frequency. The result was that

$$u(v, T) = \frac{8\pi v^2 \varepsilon}{c^3} \frac{1}{e^{\varepsilon/kT} - 1}.$$

This is an expression for the frequency distribution in blackbody radiation and had therefore to conform to Wien's displacement law, which required the exponent to be a function of the single variable v/T. This made ε proportional to v. For this reason Planck introduced the time bomb of an equation

$$\varepsilon = hv$$

specifying the size of his energy quantum. This quantum was altogether different from Boltzmann's ε because Planck could not allow it to approach zero as Boltzmann had allowed his ε to do. The constant h was one of the two constants in the expression for the spectrum, and its value had to be determined by matching the expression to the observations of the spectrum. The other one was k. The proportionality constant h was Planck's constant, or the quantum of action.

As Planck pointed out in his first lecture on his derivation, the fact that the constant k appearing in the spectrum and in the expression $k \log W$ for the entropy of Planck's oscillators is the ratio of the universal gas constant to Avogadro's number offers a way to determine Avogadro's number from measurements of the blackbody spectrum, since the macroscopic gas constant R is known. In this way Planck arrived at the value $N_A = 6.2 \times 10^{23}$ molecules per mole. As remarked in our discussion of the ionic charge in Chapter 3, this was by far the most accurate value available in 1900, but one that aroused no interest and was never used because of the obscurity of its derivation. Planck combined this value of N_A with that of the faraday to get a value of the ionic charge $e = 4.69 \times 10^{-10}$ esu, also by far the most accurate estimate at that date. He regarded these determinations as very important, even devoting a separate note to them.[6] But they were ignored. For example, three years later J. H. Jeans, who was an authority on kinetic theory and was soon to be interested in blackbody radiation, wrote a paper entitled "The Determination of the Size of Molecules from the Kinetic Theory of Gases" in which he discussed methods of estimating Avogadro's number (or rather Loschmidt's number L, the number of molecules per cubic centimeter under standard conditions of temperature and pressure).[7] Among them was the method that Loschmidt had used earlier. Jeans considered these methods quite inaccurate, preferring the electrical method based on the measurement of the electron charge by Thomson, Townsend, and others. The former gave him L in the range 4.6 to 4.92×10^{19}, corresponding to N_A being in the range 10.3 to 11.0×10^{23}. The size of the charge he used for the electrical evaluation was in the range 3.0 to 3.4×10^{-10} esu. He made no mention of Planck's much superior values of 1900.

The other constant in the radiation law, h, had no such clear meaning. It related energy and frequency in a completely novel and incomprehensible way. The immediate reaction was to consider it a formality necessary at the current stage of understanding but sometimes that would go away in the end. It was associated not with the radiation field but with the Planck oscillators that were only stand-ins for the real material of which blackbodies were made. Only later did the far-reaching implications of what turned out to be a universal association between energy and frequency begin to be appreciated.

It is strange that even after having adopted Boltzmann's statistical concept of entropy, a concept based on the kinetic theory of gases, Planck did not take the next step and consider the implications of the principle of equipartition of energy for blackbody radiation. He seems simply to have regarded gas theory as irrelevant to the problem, even though he had used it in assigning the value of R/N_A to k, the coefficient of log W in his expression for the entropy. It has been remarked that we are fortunate in his not having applied the principle too early, as it would have cut short the argument that led to the correct law, but one might expect him to have noticed that the low-frequency end of his spectrum did conform to the principle, especially as Rayleigh's work was cited in the Rubens and Kurlbaum paper showing the data in this range.

Notes

1. For a detailed account of the sequence of measurements at first confirming the Wien law and then, as the range of parameters was extended to lower frequencies and higher temperatures, finding systematic deviations from it, see H. Kangro, *Early History of Planck's Radiation Law* (London: Taylor and Francis, 1976).

2. *Planck's Original Papers in Quantum Physics* (New York: Wiley, 1972).

3. J. Mather et al., *Astrophysical Journal Letters* 354, L37 (1990).

4. See, for example, T. S. Kuhn, *Black-Body Theory and the Quantum Discontinuity, 1894–1912* (New York: Oxford University Press, 1978), pp. 96 ff.

5. M. Planck, *Annalen der Physik* 4, 553 (1901).

6. Ibid., p. 564.

7. J. H. Jeans, *Philosophical Magazine* 8, 692 (1904).

Amplification

In his earlier work Planck had been led to the Wien law by deriving an expression for the entropy of an oscillator of the form

$$S = -\frac{U}{a} \log \frac{U}{ebv}.$$

It was this form that he needed to modify in order to account for the new experimental data at low frequencies. The new possibility that intrigued him was of the form

$$S = \alpha\left[\left(1 + \frac{U}{\beta}\right)\log\left(1 + \frac{U}{\beta}\right) - \frac{U}{\beta}\log\frac{U}{\beta}\right].$$

Applying the same steps to this expression as he had applied to the earlier one led him to the new, Planck distribution. In his presentation of this expression for the entropy, Planck made a cryptic parenthetical remark that its logarithmic form suggested an application of probability theory. This may have been stimulated by its similarity to Stirling's approximation for factorials, which certainly loom large in probability calculations, namely,

$$\log(n!) \approx n(\log n - 1).$$

Whatever the source of this remark, two months later Planck gave a combinatorial derivation of this expression for the entropy.

The manipulations Planck used to get from the Boltzmann-like combinatorial expression for the entropy of an ensemble of oscillators to his result are not without interst. The first was to make advantage of the largeness of the integer N representing the number of oscillators of frequency v and the integer P, giving the number of elements into which the energy E of these oscillators was divided for the purpose of counting. They were related by

$$E = NU = P\varepsilon$$

with U the average energy of any one of the oscillators. Their largeness permitted the use of Stirling's approximation for the factorials:

$$P! \approx \left(\frac{P}{e}\right)^P, \qquad \log P! \approx P \log P - P$$

and similarly for $(N - 1)!$ and $(P + N - 1)!$ Using these in the expression for W led to

$$\log W \approx (P + N) \log(P + N) - P \log P - N \log N,$$

where the difference between $N - 1$ and N was neglected. Now setting

$$S_N = k \log W$$

and replacing P by NU/ε Planck got

$$S_N = kN\left[\left(1 + \frac{U}{\varepsilon}\right)\log\left(1 + \frac{U}{\varepsilon}\right) - \frac{U}{\varepsilon}\log\frac{U}{\varepsilon}\right].$$

This gave the desired expression for entropy of the set of N oscillators. Its validity depended, of course, on the validity of the prescription used for counting the ways of distributing the energy E among the N oscillators. The counting could have been done differently, by distinguishing between oscillators or between quanta, for example. It is, however, of the form that Planck knew led to the experimentally correct spectrum, so there was no motive for him to question it.

What Planck wanted was the entropy S of a single oscillator. This is the quantity he had worked with in his derivation of the Wien law. The obvious way of getting it was simply to divide the preceding expression by the number N of oscillators, giving the same entropy of each. In this way he obtained the expression

$$S = \frac{k}{\varepsilon}\left\{\left(\frac{U}{\varepsilon} + 1\right)\log\left(\frac{U}{\varepsilon} + 1\right) - \frac{U}{\varepsilon}\log\frac{U}{\varepsilon}\right\}$$

for the entropy of a single oscillator as it depends on the average energy of that oscillator.

Planck now needed to bring in the temperature. To do this he considered the change in entropy caused by a change in average energy when no external work is done. Under these conditions the entropy, energy, and temperature are related by

$$dS = \frac{dU}{T}, \qquad \frac{dS}{dU} = \frac{\dot{=}Q1}{T},$$

because here $dU = dQ$ by the first law of thermodynamics. This immediately led to the equation for the average energy of one of Planck's oscillators:

$$U = \frac{\varepsilon}{e^{\varepsilon/kT} - 1}.$$

Next he considered other sets of oscillators with different frequencies v', v'', and so on and with different energies E', E'', and so on to be distributed over them in elements of size ε', ε'', and so on. Taking all these sets together, he could find the entropy corresponding to any distribution of energy among all these oscillators and also could find the distribution that made this entropy the maximum possible. He did not give the details of this calculation. In this way he found the equilibrium state of the oscillators, and so he arrived at the distribution of blackbody radiation:

$$u(v, T) = \frac{8\pi h v^3}{c^3} \frac{1}{e^{hv/kT} - 1}.$$

His earlier constants c_1 and c_2 were now expressed in terms of two new constants, h and k, the latter of which led to the value of Avogadro's number.

This argument eventually, when it became important to understand it in the light of further applications of the quantum idea, led to endless and involved discussions. We will examine a few of them in the next chapter. Detailed accounts are given by Kangro, by Klein, and by Kuhn.[1]

Note

1. H. Kangro, *Early History of Planck's Radiation Law* (London: Taylor and Francis, 1976); M. J. Klein, *Archives for the History of the Exact Sciences* 1, 459 (1961); T. S. Kuhn, *Black-body Theory and the Quantum Discontinuity, 1894–1912* (New York: Oxford University Press, 1978).

CHAPTER 5

Allergic Reactions to h

T he quantum of action, as Planck called his new constant, was not greeted with enthusiasm. Planck himself seemed to regard it as relating only to the oscillators with which he represented the matter defining the temperature of blackbody radiation. He certainly did not associate it with the radiation field. In this chapter I shall describe some of the early responses to Planck's theory. The response to his formula was one of unqualified approval because it worked so well then, as it still does. The response to his theory, however, took a while to develop because it was found to be difficult and obscure. The most violent reaction was that of J. H. Jeans, who insisted that the only possible value of h was zero. Paul Ehrenfest concluded that Planck's quantization was a purely formal device, useful for calculation but otherwise insignificant. Einstein introduced his light quantum without the benefit of h or Planck's formula, restricting his attention solely to the region where the Wien law held. His energy quanta had the size $(R/N)\beta v$ rather than hv as Planck's had, β being one of the constants in the original form of Wien's law. Eventually Planck tried to limit its effect even in connection with his oscillators, clearly hoping that it could be made to go away. This rejection or reluctant acceptance persisted up to the time that Niels Bohr introduced his model of the hydrogen atom. This interim period is our subject here.

An Initial Lack of Interest

The initial reaction to Max Planck's papers of 1900 was confined almost entirely to the October one in which he had proposed his formula for the blackbody spectrum without giving a compelling argument for it. It had been shown immediately to fit the experimental data better than did any other proposed formulas such as those of Wien, Rayleigh, and other more *ad hoc* ones, being in fact exact as far as could be determined. It was in the December lecture[1] and the paper elaborating it[2] that the quantization of the energy of the material oscillators was introduced to produce thermal equilibrium of the cavity itself. The equilibrium of the radiation with the cavity then produced the Planck law. These steps in arriving at the law received little attention. Although the experimentalists greeted the formula with enthusiasm, incorporating it in the experimental handbooks starting as early as 1902, its derivation apparently did not concern them, and they did not comment on the radical nature of the assumption needed for its derivation. Planck had found himself forced to use Boltzmann's method of counting possible states of his system and, therefore, to introduce finite energy quanta to have something to count. He appeared to be more struck with his need to use Boltzmann's methods than he was with the fact that having done so he could not, as Boltzmann had done, go to the limit as this energy quantum became indefinitely small and so have the energy become a continuous variable. That was strange, but Planck associated the phenomenon with the interaction between matter and radiation, a subject full of obscurities.

An early contributor to the puzzlement over the quanta was H. A. Lorentz. He had developed the theory of electrons to a high degree, attributing the electrical and magnetic properties of materials to the presence of electrons in them. In conductors there were electrons free to move through the material, and in insulators the electrons were bound to fixed centers and able to move only in their immediate vicinity. On the basis of this model Lorentz calculated the radiation emitted by electrons in thermal equilibrium in a thin sheet of metallic conductor:

We find

$$F(\lambda, T) = \frac{8\pi k T}{\lambda^4}.$$

It is very remarkable that this result is of the form (129) [satisfies the Wien displacement law] and that it agrees exactly with that of Planck [in the long-wavelength limit].[3]

Among those to take up the matter was Lord Rayleigh.[4] In 1905 he completed the calculation he had sketched in 1900, evaluating the number of modes in a cavity with frequencies between v and $v + dv$ and ascribing to each mode the average energy kT as required by the equipartition theorem. Rayleigh pointed out that

a very remarkable feature of Planck's work is the connection which he finds between radiation and molecular constants. . . . Though I failed to notice it in the earlier paper, it is evident that [it] leads to a similar connection.

Uncharacteristically, Rayleigh made a mistake and found eight times as many modes as there are. Because of the factor eight error in his mode count, he obtained for Boltzmann's constant k—or equivalently Avogadro's number and consequently for the electron charge e—a value that was off by this factor. Puzzled, he failed to understand how his argument, based on Boltzmann's ideas, could differ from Planck's, which also was based on them:

> A critical comparison of the two processes would be of interest, but not having succeeded in following Planck's reasoning, I am unable to under-take it.

The rest of the theoretical community shared this response to Planck's work. Jeans pointed out to Rayleigh his slip, and so the puzzle about Boltzmann's method leading to two different values of Boltzmann's constant or its equivalent Avogadro's number was resolved, but the lack of understanding of Planck's reasoning remained.

Jeans had been worrying about the old difficulty in the kinetic theory of gases that had bothered Maxwell so much, the incomprehensible smallness of the observed molar heats. He proposed a way out of the difficulty by denying the existence of thermal equilibrium between the theoretically required internal motions of molecules and the rigid motions, linear and rotational, even after long times. Strictly speaking, any system requires an infinite amount of time to attain equilibrium. When a hot body and a cold body are put in thermal contact, the temperature difference between them decreases more or less exponentially in time,

$$T_{\text{hot}} - T_{\text{cold}} \propto e^{-\lambda t}.$$

The rate constant λ depends on the strength of the interaction between the two bodies that produces equilibrium; the weaker this interaction is, the smaller λ and the longer the time will be needed for equilibrium to be achieved. If, for example, heat were added to a gas in equilibrium at a low temperature, according to Jeans the added heat would increase the energy of those forms of motion that interacted strongly with the walls of the vessel defining the temperature and would not appreciably increase the forms of motions with no direct coupling to the walls. Over a long period of time the energy should become distributed over these forms of motion also, and the temperature of the gas should decrease as the energy once associated with the center of mass motion of the molecules was shared with internal motions of some kind. No such gradual change of temperature of a gas after a change of its energy content had ever been observed, but as Rayleigh remarked, it would have been a very difficult observation. This resolution of the molar heat difficulty was not widely accepted. Jeans pursued this line of thought, however, and proposed that blackbody radiation was produced by molecules of the gas only during collisions, and because these were rare events in a dilute gas, it would take millions or billions of years to achieve equilibrium between the gas and radiation.[5] If this were true, then clearly the equipartition theorem would not be relevant, and neither would Planck's theory, based on equilibrium as it was. Jeans insisted that Planck should have let his constant h, and hence the size of his energy quanta, approach zero so as to retain equipartition, since he was calculating the equilibrium state of the system that Jeans denied would be reached in practice. Jeans dismissed as fortuitous the agreement with observation that Planck had achieved by keeping h finite. This proposal also was never widely accepted, even though Jeans persisted in advancing it as late a 1909.[6]

The coefficient of kT in the low-frequency limit of the law of blackbody radiation was the same in Lorentz's, in Planck's, and in Rayleigh's formulas as corrected by Jeans, despite the apparently different starting points of their derivations. Rayleigh invoked the principle of equipartition of energy for modes of vibration of the ether; Lorentz involved it for the motion of free electrons in metals; and Planck did not invoke it at all.

One of the few physicists to react substantively to Planck's theory paper before Einstein's work of 1905 and 1906 was Ehrenfest. In a 1905 paper entitled "On the Physical Assumptions of Planck's Theory of Irreversible Radiative Processes,"[7] he carefully analyzed Planck's procedure of arriving at his spectral distribution and concluded that Planck's model of linear oscillators interacting with the electromagnetic field could not lead to a unique spectral distribution. Planck provided no way for radiation or oscillators of different frequencies to exchange energy, and so an energy distribution among the frequencies differing from the one he derived would be permanent and there would be no tendency to establish a unique spectrum. This follows from the fact that the equations of motion of both the field and the oscillators are strictly linear. The sum or difference of any two solutions of such equations is itself a solution, and so there is no influence of the motion corresponding to one solution on the motion corresponding to another solution. Taken literally, Planck's model would reach some state that did not change with the time, but this state would depend on initial conditions and would not be a unique equilibrium state. Ehrenfest categorized Planck's introduction of energy quanta, an essential step in his arriving at his spectral distribution, in the following way:

> The hypothesis [of energy quanta], which in the present form is clearly meant only formally, requires a still further reduction. As far as I can see, Boltzmann's theory [of gases] lacks any analog for it.

Ehrenfest concluded a paper published the next year by asking:

> By what means does the Planck theory render the infinitely numerous ultraviolet overtones of the cavity harmless, so that they do not suck away all the energy as with Rayleigh and Jeans, but rather make the spectral curve fall off so strongly toward the ultraviolet? One recognizes easily how the assumption $[h\nu]$ for the size of the energy atom can work in this sense: According to Eq (9) $[\varepsilon_0 = h\nu]$ the Planck energy atoms grow above all limits with growing ν. This circumstance does not itself give rise to any serious physical objection; for as an estimate on the basis of the numbers given by Planck shows, ε_0 for visible light has a value of the order of magnitude of the mean kinetic energy of a molecule for $T = 1000$. However, it seems to be

meaningful methodologically for comparing the Planck theory with that of Rayleigh–Jeans and that given by H. A. Lorentz.[8]

To summarize the initial reaction to Planck's proposal of his law of blackbody radiation:

1. It was regarded as a very accurate, perhaps exact, specification of the spectrum of blackbody radiation.
2. The derivation was regarded as obscure and not very important by most of those few interested in the whole subject.
3. Planck himself and others regarded the introduction of the energy quanta for the Planck oscillators as a formal way around a profound ignorance of the interaction of matter and the electromagnetic field.
4. No one other than Jeans had a proposal to get rid of the quanta, and Jeans's was not considered satisfactory by anyone but himself.

There is little if any evidence that these people felt themselves at the beginning of a new epoch.

Notes

1. *Planck's Original Papers in Quantum Physics* (New York: Wiley, 1972), p. 36.
2. M. Planck, *Annalen der Physik* **4**, 553 (1901).
3. H. A. Lorentz, *The Theory of Electrons* (New York: Dover, 1952), p. 90.
4. Lord Rayleigh, *Nature* **71**, 559 (1905); *Scientific Papers by John William Strutt, Lord Rayleigh* (Cambridge: Cambridge University Press, 1912), vol. 5, p. 248.
5. J. H. Jeans. *Philosophical Magazine* **11**, 604 (1906).
6. J. H. Jeans, *Philosophical Magazine* **18**, 209 (1909).
7. P. Ehrenfest, *Sitzungsberichte der Kaiserliche Akademie der Wissenschaften, Wien, Mathematische-wissenschaftliche Klasse* **114**, 1301 (1905).
8. P. Ehrenfest, *Physikalische Zeitschrift* **7**, 528 (1906).

The Radiation Quantum Appears, Is Rejected

For five years Planck's derivation of his distribution law for blackbody radiation stimulated almost no reaction. M. J. Klein discussed the reasons for this neglect, concluding that there were two of them.[1] There was the novelty and incomprehensibility of the energy quanta that Planck had associated with harmonic oscillators that made the subject a forbidding one, and there was so much else going on at the same time that seemed more accessible to understanding. Klein cited the discovery of X-rays in 1895, of radioactivity in 1896, of the electron in 1897, and of radium in 1898 as providing more than enough material for all the active workers in physics. Whatever the reason, no one made any constructive contribution to understanding Planck's work until 1905. As described in the last section, Rayleigh, Jeans, and Lorentz all considered the implications of the equipartition principle for blackbody radiation, but none of them contributed to understanding either the Planck law or the existence of a maximum in the blackbody spectrum.

In 1905 Einstein introduced the energy quantum associated with radiation itself in a paper entitled "On a Heuristic Viewpoint Concerning the Creation and Conversion of Light."[2] Wien, Planck, Rayleigh, and others all had started their investigations from some model of the radiation field and the matter with which it was supposed to be in equilibrium and worked toward the spectrum implied by that model. Planck had established the Wien law to his own satisfaction in 1899, but the experimental results of 1900 had forced him to modify his theory to make the low-frequency–high-temperature end agree with the observations. Thus his introduction of quantized energies for his oscillators was connected, from his point of view, with this end of the spectrum; the existence of a maximum and the exponential falloff at high frequencies required no further explanation. He did not associate his new low-frequency form with equipartition or recognize it as a quintessentially classical form.

In Einstein's view, what cried out for understanding was the existence of the maximum and the shape of the high-frequency–low-temperature end of the spectrum, the end at which the Wien law was confirmed experimentally. Einstein did not mention Planck's derivations of either the Wien or the Planck laws. Instead of deriving the spectrum of radiation to be expected on the basis of a theoretical model, he made a brilliant reversal of the argument and studied what conditions the model must meet in order to account for the observed spectrum in the region where the old ideas did not work.

In the first part of his "heuristic viewpoint" paper Einstein established that classical theory led inexorably to Rayleigh's expected result of 1900, namely, that in thermal equilibrium, energy should be shared equally by all modes of the radiation field. In this he redid what Rayleigh, Jeans, and Lorentz had already done, but in his own fashion. Einstein furnished the proportionality constant with which Rayleigh had not yet bothered.

Both the Wien and Planck laws were clearly contrary to classical theory. Yet the Wien law described the observations at high frequency or low temperature when λT is small. Einstein proceeded to investigate the consequences for thermodynamics of treating the Wien law as exact. It was this that led him to recognize the presence of energy quanta in the radiation field and not just in some auxiliary system such as Planck's oscillators. Planck's theory paper had been ignored; Einstein's was read and rejected out of hand. Not until the First Solway Conference in 1911 did it begin to achieve any recognition at all, and it was not until the discovery of the Compton effect in 1923 that it had to be accepted.

In this early paper Einstein recognized the basic importance of the idea that equipartition must apply to the radiation modes if classical statistical mechanics is correct. Einstein wrote the consequent distribution as

$$\rho_v = \frac{R}{N} \frac{8\pi v^2}{c^3} T.$$

The factor $8\pi v^2/c^3$ gave the number of radiation modes per unit volume per unit frequency. (Einstein did not mention Rayleigh or anyone else in this connection. Rayleigh had not yet bothered to obtain the coefficient of T, even though this was a calculation of a kind with which he was thoroughly familiar.) In this equation, R was the gas constant and N was Avogadro's number. Einstein remarked that

> this relation, which we found as the condition for dynamic equilibrium does not only lack agreement with experiment, but it also shows that in our picture there can be no question of a definite distribution of energy between aether and matter.

Next he stated the Planck law in the form

$$\rho_v = \frac{\alpha v^3}{e^{\beta v/T} - 1}$$

used by Planck in his first paper, not the form that he used later after he had derived it from statistics and had introduced in place of α and β the constant h—now known as Planck's constant—and the constant k that he called Boltzmann's constant. It might seem that Einstein was avoiding any association with that derivation. For small values of v/T, this can be approximated by

$$\rho_v = \frac{\alpha}{\beta} v^2 T,$$

which agrees with the equipartition expression, provided that

$$\frac{R}{N} \frac{8\pi}{c^3} = \frac{\alpha}{\beta}.$$

In this equation α/β can be found from the measured spectrum; R is known, and so N and Boltzmann's constant, $k = R/N$, can be determined from the low-frequency end of the spectrum. This was implied in what Rayleigh had done five years earlier, but Rayleigh did not appreciate this consequence of his work until 1905. Planck had evaluated k in his first quantum paper, but he had not made the connection with the equipartition of energy. Einstein pointed out that Planck's value for Boltzmann's constant (or Avogadro's number) is just that obtained by his equipartition argument and is therefore largely independent of his theory of the blackbody spectrum.

These preliminaries over, Einstein approached his main purpose:

> In the following, we shall consider "black-body radiation," basing ourselves on experience without using a picture of the creation and propagation of the radiation.

His approach was simple, using only the most basic principles of thermo-dynamics and kinetic theory. The starting point was his acceptance of the Wien law for large values of v/T as empirical fact for high frequencies and not-too-high temperatures. Einstein then calculated the entropy implied by this distribution in a completely unexceptionable way. Up to this point he was going back along the path that earlier had led Planck to the Wien law. But what he did next was sheer genius. In his own words:

> If we restrict ourselves to investigating the dependence of the entropy on the volume occupied by the radiation, and if we denote the entropy of the radiation by S_0 if it occupies a volume v_0, we get
>
> $$S - S_0 = \frac{E}{\beta v} \log \frac{v}{v_0}.$$
>
> This equation shows that the entropy of a monochromatic radiation of sufficiently small density varies with volume according to the same rules as the entropy of a perfect gas or of a dilute solution. The equation just found will in the following be interpreted on the basis of the principle introduced by Mr. Boltzmann into physics, according to which the entropy of a system is a function of the probability of its state.

It is important to realize that what Einstein had calculated was the difference in the entropies of radiation when it *happens to be* in volume v and when it *happens to be* in volume v_0, all external circumstances, including the energy E of the radiation, remaining unaltered. According to W. Pauli:

> In his first paper, Einstein computed for the region of validity of Wien's law, with the help of equation (1) [$S = k \log W + \text{const}$], the probability of the rare state in which the entire radiation energy is contained in a certain partial volume.[3]

This purely probabilistic approach embodies the very essence of Boltzmann's idea.* Einstein does the same thing later in the paper for an ideal gas. The change in volume occurring in the two cases of radiation and of a gas could be produced by a change in volume of the container, by an externally imposed compression, but this would have vastly different effects in the two cases. The total energy of an ideal gas is proportional to its temperature and is independent of its volume. In the case of radiation the Stefan–Boltzmann law tells us that the energy *density* of radiation depends only on the temperature, being proportional to T^4, so that the total energy is proportional to the volume and the fourth power of the temperature. The results of an imposed compression of the radiation and of the gas would not be comparable in any simple way. Einstein's calculation relates the change in entropy to the relative probability of *finding* the entire energy of the radiation or the entire sample of gas in a

---

* This point is not made clear in all descriptions of this development. In his scientific biography of Einstein, *Subtle Is the Lord* (New York: Oxford University Press, 1982), A. Pais stated that the calculation of the change in entropy of a gas is made where the "*n* gas molecules in the volume v_0 are confined to a subvolume *v*." This is ambiguous as to the ground for their confinement, compression, or fluctuation, but the word *confine* seems to favor an external influence.

subvolume v of the available volume v_0 as the result of a spontaneous fluctuation with the external circumstances remaining unchanged. His seeing that this was the thing to do must have resulted from this thorough appreciation of the statistical interpretation of entropy that he gained by reading Boltzmann and during his earlier research in statistical mechanics.

Einstein then made the Boltzmann-like calculation for an ideal gas. He found the change in entropy between two states of an ideal gas containing a fixed number n of molecules moving independently of one another when in one state all the molecules are within a volume v and in the other they all are within a volume v_0. As we just emphasized, the difference between the two volumes occupied by the gas is not caused by a compression of the gas but by fluctuations in the molecular distribution over the available volume, with the energy (and therefore the temperature) of the gas remaining unchanged. This is made explicit by Einstein's allowing for the presence of particles of another kind in the volume v_0 whose spatial distribution is not changed in correlation with the change in the spatial distribution of the molecules whose entropy he is examining. He simply compared the probabilities of finding all the molecules in the volume v with the probability of finding them all in the volume v_0. On average the molecules are uniformly distributed, and so the relative probability of finding any one molecule in the first or in the second volume, namely, the ratio of the relative times that a molecule spends in one volume or the other, is just the ratio of the volumes, v/v_0. At any one time the relative probability of finding all n molecules of the kind being studied in these two volumes is then the product of the probabilities of finding the individual molecules in them, namely $(v/v_0)^n$. Now, using Boltzmann's relation between entropy and probability,

$$S = R/N \log W,$$

Einstein obtained the difference in entropies for these two situations:

$$S - S_0 = \frac{R}{N} \log\left(\frac{v}{v_0}\right)^n$$

$$= R\frac{n}{N} \log \frac{v}{v_0}.$$

The formulas for the entropy differences for radiation and for a gas of n particles give the same answer if the coefficients of the logarithm are equal:

$$\frac{E}{\beta v} = R\frac{n}{N}.$$

Therefore a gas consisting of n molecules will have the same difference in entropy as will radiation of frequency v and energy E when both are found in a subvolume v of an original volume v_0, provided that the number of molecules is given by

$$n = \frac{N}{R} \frac{E}{\beta v}.$$

Einstein now drew this conclusion:

If monochromatic radiation of frequency v and energy E is enclosed (by reflecting walls) in a volume v_0, the probability that at an arbitrary time the total energy is in a part v of the volume v_0 will be

$$W = \left(\frac{v}{v_0}\right)^{NE/R\beta v}.$$

From this we can then conclude:

Monochromatic radiation of low density behaves—as long as Wien's radiation formula is valid—in a thermodynamic sense, as if it consisted of mutually independent energy quanta of magnitude $R\beta v/N$.

This extremely simple but vastly subtle argument led Einstein to the revolutionary conclusion that radiation itself has a discrete structure.* In the notation used by Planck, $R/N = k$ and $\beta = h/k$, so that Einstein's quanta were of magnitude hv.

In this argument Planck's formula played no role. The basis was simply the exponential falloff of the energy with increasing frequency, a falloff that directly contradicted the equipartition of energy and thus the classical theory that required it. Planck's constant appears in it only by writing the constants in the Wien law as Planck had done. It does not appear as a consequence of imposing the requirement that the energy of anything be quantized but, rather, as a consequence of requiring the nonclassical Wien law to be valid as a limiting case. Einstein himself did not use Planck's notation in his papers until several years later.

After the discussion of the behavior of blackbody radiation just described, Einstein turned to the creation and transformation of light:

If monochromatic radiation—of sufficiently low density—behaves, as far as the volume dependence of its entropy is concerned, as a discontinuous medium consisting of energy quanta of magnitude $(R/N)\beta v$ it is plausible to investigate whether the laws on creation and transformation of light are also such as if light consisted of such energy quanta. This question will be investigated in the following.

He considered three effects that were difficult to understand on the basis of classical electromagnetic theory, all quite distinct from questions of thermal equilibrium. We shall discuss them in the order that he did.

* In his *Statistical Theory and the Atomic Theory of Matter* (Princeton, NJ: Princeton University Press, 1983), S. G. Brush made the strange statement: "So Einstein's conclusion seems to be that if radiation in free space consists of quanta, it should obey Wien's rather than Planck's distribution law. He offers no explanation for this odd conclusion at this point" (p. 112). This has the direction of Einstein's argument backward.

On Stokes' Rule

When illuminated with monochromatic light, most materials simply reflect it to some degree without altering its color. The complete washing out of all apparent color from things illuminated by low-pressure sodium light shows this: No matter what color the thing is, the light it reflects is the color of the light from the sodium light. The only difference is between light and dark, but not between colors. There are photoluminescent materials that "reflect" light of a color different from that illuminating them. This is common when the illumination is with ultraviolet ("black") light. Stokes' rule is that the light so altered in frequency is changed toward lower frequency or longer wavelength. Concerning this Einstein wrote:

> Consider monochromatic light which is changed by photoluminescence of light of a different frequency; in accordance with the result we have just obtained, we assume that both the original and the changed light consist of energy quanta of magnitude $(R/N)\beta v$, where v is the corresponding frequency. We must then interpret the transformation process as follows. Each initial energy quantum of frequency v_1 is absorbed and is—at least when the distribution density of the initial energy quanta is sufficiently low—by itself responsible for the creation of a light quantum of frequency v_2; possibly in the absorption of the initial light quantum at the same time also light quanta of frequencies v_3, v_4, ..., as well as energy of a different kind (e.g. heat) may be generated. It is immaterial through what intermediate processes the final result is brought about. Unless we can consider the photoluminescence substance as a continuous source of energy, the energy of a final light quantum can, according to the energy conservation law, not be larger than that of an initial light quantum; we must thus have the condition
>
> $$\frac{R}{N}\beta v_2 \leq \frac{R}{N}\beta v_1, \qquad \text{or } v_2 \leq v_1.$$

This is the well-known Stokes' rule. Einstein then went on to discuss how exceptions to this rule could arise, as they do.

On the Production of Cathode Rays by the Illumination of Solids

Here came the photoelectric effect. According to Einstein,

> The usual idea that the energy of light is continuously distributed over the space through which it travels meets with especially great difficulties when one tries to explain photo-electric phenomena, as was shown in the pioneering paper by Lenard.
>
> According to the idea that the incident light consists of energy quanta with an energy $R\beta v/N$, one can picture the production of cathode rays by light as follows. Energy quanta penetrate into a surface layer of the body, and their energy is at least partly transformed into electron kinetic energy. The simplest picture is that a light quantum transfers all of its energy to a

single electron; we shall assume that that happens. We must, however, not exclude the possibility that electrons only receive part of the energy from the quanta. An electron obtaining kinetic energy inside the body will have lost part of its kinetic energy when it has reached the surface. Moreover, we must assume that each electron on leaving the body must produce work P, which is characteristic for the body. Electrons which are excited at the surface and at right angles to it will leave the body with the greatest normal velocity. The kinetic energy of such electrons is

$$\frac{R}{N} \beta n - P.$$

This last is the famous expression for the energy of photoelectrons. It was not accurately verified until 1916 when Millikan did definitive experiments confirming it, thereby contributing to his 1923 Nobel Prize.[4] Even while confirming the result, Millikan did not believe the theory yielding it and stated his conviction that even Einstein no longer held to the idea of radiation energy quanta.

On the Ionization of Gases by Ultraviolet Light

The ability of ultraviolet light and X-rays to make gases conductors of electricity had been known for some years. The conductivity produced increased with the intensity of the light and in all cases was rather small. When the molecular picture was developed—according to which this conductivity was due to the production of ionized molecules—the mechanism by which a uniform light wave ionized a few molecules and left the rest un-ionized posed a puzzle. In the classical picture, each molecule experienced the same perturbation by the light wave, and each molecule should therefore be affected in the same way. But this is obviously not what happens.

There were proposals that the internal motion of the molecules' constituents had different phases relative to that of the light wave, but none of these was satisfactory. In 1903 J. J. Thomson proposed that

> The difficulty in explaining the small ionization is removed if instead of supposing the front of the Röntgen ray to be uniform, we suppose that it consists of specks of great intensity separated by considerable intervals where the intensity is very small.[5]

Thomson's "needle radiation" was still classical. Although the wavefront was highly nonuniform, there was no hint of energy quantization in his suggestion. Nor was there any hint of how this needle radiation came about. Einstein's energy quanta accomplished the purpose of Thomson's needle radiation but went much further in giving the amount of energy in a quantum and thereby making it possible to estimate the number of ions formed. There would be none if the energy of a quantum were less than the energy needed to free an electron from the molecule. If the quantum had enough energy, Einstein proposed that there should be a relation between the absorbed light intensity and the number of ions produced:

> If each light energy quantum which is absorbed ionizes a molecule, the following relation should exist between the absorbed light intensity L and the number j of moles ionized by this light:
>
> $$j = \frac{L}{R\beta v}.$$
>
> This relation should, if our ideas correspond to reality, be valid for any gas which—for the corresponding frequency—does not show an appreciable absorption which is not accompanied by ionization.

These three applications of the energy quantum idea were in accordance with the little that was known about the corresponding phenomena. Of the three, the one most readily leading to a quantitative test was the photoelectric effect, and this test was not easy. Surfaces of materials that show the effect for visible or near-ultraviolet light are hard to prepare and keep clean during a prolonged experiment, and dirty surfaces have erratic photoelectric characteristics. In his definitive experiments Millikan had to prepare fresh surfaces of the alkali metals in a vacuum, a tricky and difficult process at the time, in order to obtain reproducible results.

The almost insurmountable obstacle preventing the acceptance of light energy quanta was that they were in violent contradiction with all that was known about the *propagation* of light. The theories of interference and diffraction, based on smooth classical waves, had given simple and accurate accounts of the behavior of electromagnetic waves under an enormous variety of circumstances, and these successes could not be given up for the seemingly small gains that Einstein had achieved in understanding three obscure effects involving the transformation and generation of light. As a result, the light energy quantum continued to be rejected long after the quantum had made itself indispensable in many other connections. Physicists were willing to have material oscillators quantized, but the Maxwell field was not to be tinkered with. It should be stressed that Einstein always made explicit the condition that light behaved like particles when "of sufficiently low density." His analysis had assumed the validity of the Wien law, but this law is not exact. Four years later he made a similar argument based on the Planck law.

The next year Einstein pursued the subject of thermal radiation in a paper entitled "Theory of Light Emission and Absorption"[6] in which he discussed the relation of Planck's theory to his work of the previous year. Planck retained the classically derived relation between the average energy of one of his oscillators and the energy density of the radiation of that frequency in equilibrium with it. Any question of whether this relation remained valid when the oscillator energy was quantized was not raised by Planck. Einstein proposed it as a postulate and then concluded that the existence of his energy quanta required the quantization of Planck's oscillators. The reversed direction of Einstein's argument, from the observed exponential drop-off of the blackbody spectrum with increasing frequency to the existence of energy quanta in the radiation field, and from that to the need to quantize Planck's material oscillators, made the urgency of the problem facing physicists much clearer than the original argument. Einstein started with observed behavior and, by applying very basic analysis, arrived at quanta. Planck started with a model involving quantized oscillators and arrived at the observed behavior. There could be and was hope that a modification of Planck's model was possible, avoiding quantization and still arriving at the desired conclusion. But Einstein dashed these hopes for those who read him and took him seriously. They were not many.

In 1909 Einstein was asked to give a talk at the Salzburg meeting of the German Physical Society. It was expected that he would talk mainly on relativity theory, a theory that had aroused intense interest and achieved wide acceptance. Instead, he devoted much of his attention to the radiation problem.[7] He considered it a major obstacle to the further understanding of physics, and although he had not achieved much progress since his work of 1905, he felt it important to discuss the problem.

This time Einstein started with the Planck law, not the Wien law which is only a limiting case of the Planck law. Instead of calculating the entropy as he had done before, he now calculated the mean square fluctuation of the energy per unit volume of radiation with frequency lying in the range between v and $v + dv$. Here ε is the value of this energy at any instant, and the mean square deviation from its mean value is

$$\langle (\varepsilon - \langle \varepsilon \rangle)^2 \rangle,$$

where the angular brackets denote the mean. Einstein showed that this mean square fluctuation contained two terms. One of these terms corresponded exactly to his earlier result obtained from the Wien law and was of the form to be expected from the fluctuation in the number of particles per unit volume in a gas or in any set of independently moving particles. The other was a new term and was shown to be of the form expected for the interference of waves of this frequency with random amplitudes and phases. Lorentz later demonstrated not only that the form was correct but also that the numerical coefficient was that arising from wave interference.[8]

This analysis leading to the simultaneous presence of wavelike and particlelike fluctuations in the energy density of blackbody radiation was the first clear statement of what became the waveparticle duality so characteristic of quantum mechanics. Still, no one was convinced of the necessity of regarding radiation in this dual way: It remained a pure wave phenomenon for essentially everybody but Einstein.

Notes

1. M. J. Klein, *Archive of the History of the Exact Sciences* 1, 32 (1961).
2. A. Einstein, *Annalen der Physik* 17, 132 (1905).

3. W. Pauli, in P. Schilpp, ed., *Albert Einstein, Philosopher-Scientist* (New York: Tudor, 1951), p. 152.

4. R. A. Millikan, *Physical Review* 7, 355 (1916); R. A. Millikan, *The Electron* (Chicago: University of Chicago Press, 1917).

5. J. J. Thomson, *Electricity and Matter* 1903, pp. 63–65.

6. A. Einstein, *Annalen der Physik* 20, 199 (1906).

7. A. Einstein, *Physikalische Zeitschrift* 10, 185 (1909).

8. H. A. Lorentz, *Les Théories statistique en thermodynamique* (Leipzig: Teubner, 1916), p. 59.

Amplification

Einstein's argument in the first part of his "heuristic viewpoint" paper was not novel; it was just clear and to the point. Here he was arguing from classical physics to the spectrum of blackbody radiation that it implied. For this purpose he could use Planck's system of charged oscillators interacting with the electromagnetic field in a cavity, their charge enabling them to emit and absorb radiation and so to establish equilibrium with the radiation in the cavity. To this he added a gas interacting with the oscillators. This gas acted as a gas thermometer and measured the temperature T of the entire system of radiation, oscillators, and gas. He stated the Maxwell–Boltzmann result that the average energy of each linear oscillator must be given by

$$\langle E \rangle = \frac{R}{N} T.$$

Einstein also used Planck's result on the connection between the average energy of an oscillator of frequency v and the energy density per unit frequency of radiation at this frequency, namely,

$$\langle E_v \rangle = \frac{c^3}{8\pi v^2} \rho_v.$$

Planck had derived this on the basis of classical mechanics and electromagnetic theory using an assumption of *natural radiation* that Einstein characterized as "the most random . . . imaginable." In a steady state these averages must be equal, so that

$$\frac{R}{N} T = \langle E \rangle = \langle E_v \rangle = \frac{c^3}{8\pi v^2} \rho_v$$

leading to

$$\rho_v = \frac{R}{N} \frac{8\pi v^2}{c^3} T.$$

Einstein's argument in the main portion of this paper was independent of any model for the radiation field. He did use Kirchhoff's result that the state of radiation in equilibrium at a given temperature is completely defined by its frequency distribution and that radiations of different frequencies can be separated from one another without doing work or supplying heat, as Newton had done with his prism. These assumptions allowed him to write the entropy S_R of radiation within a volume v in the form of an integral over frequency,

$$S_R = v \int_0^\infty \phi(\rho, v) \, dv.$$

The condition for equilibrium was that this entropy be a maximum with respect to variations of the frequency distribution keeping the total energy fixed. This led by straightforward mathematics to the equation

$$\frac{\partial \phi}{\partial \rho} = \frac{1}{T}.$$

"This is the blackbody radiation law," wrote Einstein.

For the frequency distribution Einstein took the Wien law

$$\rho = \alpha v^3 \, e^{-\beta v/T}$$

with full awareness of its limitations. Solving for $1/T$, he got

$$\frac{1}{T} = -\frac{1}{\beta v} \log \frac{\rho}{\alpha v^3}.$$

Equating the two expressions for $1/T$ led him to

$$\phi(\rho, v) = -\frac{\rho}{\beta v} \left[\log \frac{\rho}{\alpha v^3} - 1 \right].$$

The entropy per unit frequency of radiation at this frequency was thus

$$S = v\phi(\rho, v) = -\frac{E}{\beta v} \left[\log \frac{E}{v \alpha v^3} - 1 \right],$$

where E was substituted for ρv. The simple dependence on the volume v suggested comparing the entropies of the system when it occupied two different volumes, v and v_0, all other conditions being the same. The difference between the entropies of this radiation of this frequency v and this energy E in these two volumes is then given by

$$S - S_0 = \frac{E}{\beta v} \log \frac{v}{v_0}.$$

It is this expression that Einstein compared with the analogous one relating to a gas to get his startling result.

He would not have attained such a simple result had he used the Planck law instead of the Wien law. The reason for this remained obscure until 1924 when Einstein, following up on a suggestion by S. N. Bose, introduced a form of quantum statistics in place of the classical Boltzmann statistics. Quantum theory and classical theory treat identical particles differently. Classically interchanging two particles produces a new state even if they are identical in all respects, whereas in quantum electrodynamics the exchange of two light quanta does not produce a new state. This results in different ways of counting states. The resulting probabilities are affected if there are enough particles present to make it likely that two or more of them will be in the same state, but if there are so few present that the states are mostly empty and only a few have an occupant, then the results of classical and quantum counting are the same. The Wien limit is just the one in which this is the case. For lower frequencies the energy per quantum is reduced so the number of quanta corresponding to a given energy is increased. Then the probability that two quanta are in the same state can become appreciable, and the classical way of counting can no longer be used. Under these circumstances the thermodynamic behaviors of molecules and quanta are no longer the same. This confused experts on statistical physics such as Ehrenfest, who studied these questions before there was a quantum theory.

Other Energy Quanta Appear, Are Accepted

In 1905 Einstein had discovered the energy quantum in electrodynamics, basing his argument entirely on the high-frequency or Wien law end of the blackbody spectrum and in a way that was independent of Planck's quantization of the oscillators with which the radiation was in equilibrium. In 1906 he had understood Planck's derivation of the blackbody spectrum in a way different from Planck. Planck had retained the connection between the average energy of his oscillators and the average energy of the fields with the same frequency that he had derived classically even after abandoning the classical description of the behavior of the oscillators. Einstein now accepted this relation as a new postulate and could then regard the quantization of the oscillators as a consequence of the quantization of the field required by the Wien law. But Planck did not accept this interpretation. Later that year in a paper (published early the next year) entitled "Planck's Theory of Radiation and the Theory of Specific Heat," Einstein transferred the quantization idea from radiation to oscillators in an entirely different context.[1] In so doing he introduced a way of proceeding that later proved very fruitful in other hands.

In classical statistical mechanics, the value of any physical observable for a system in thermal equilibrium at temperature T is its average over all possible microscopic states of the system consistent with its macroscopic state, each state being given a statistical weight proportional to the Boltzmann factor $\exp[-E/kT]$, where the energy E is a continuous function of the variables such as positions and momenta specifying the microscopic state. Einstein modified this procedure by lumping together all the microscopic states whose energy lay within the range from E to $E + dE$, calling the resulting number $\omega(E)\,dE$. Then he expressed the value of an observable as the average of the product of the values of the observable, of the exponential Boltzmann factor, and of this *density of states* $\omega(E)$ over all the possible values of the system's energy. If the system is described by classical mechanics or classical electrodynamics, the new procedure involves an integration over the continuous range of possible energies and is just a rearrangement of the old one, giving the same result. Its advantage for Einstein lay in the way that new dynamical ideas could be introduced by using new expressions for the density of states $\omega(E)$ while leaving intact Boltzmann's statistical ideas represented by the Boltzmann factor.

Planck had found the average energy of his oscillators in thermal equilibrium from the statistically derived entropy, but had not considered the oscillators' specific heat that it implied. Einstein used his rearrangement to calculate this mean directly from the energy quantization. Instead of using the continuous classical form of the density of states, which in the particular case of the linear harmonic oscillator is simply $\omega(E) = 1$, Einstein assumed that only those states with an energy arbitrarily close to an integer multiple of $h\nu$ were to be included in the averaging procedure. This restriction of the energy to a discrete set of values reduced the classical integration over all energies to a summation of discrete terms and led immediately to an expression for the average energy of an oscillator quite different from the value kT coming from classical theory, namely,

$$\langle E \rangle = \frac{h\nu}{\exp[h\nu/kT] - 1},$$

which I have written using Planck's form of the constants. For high temperatures this is sufficiently approximated by the classical value kT.

Einstein applied this method to calculating the energy of vibration of the ions or atoms in a crystal. At first he thought his considerations would apply only to ions, but he soon realized that here the particles interacted through the forces that hold the crystal together and that therefore, unlike the case of radiation, the charge of the particles was irrelevant. For simplicity he assumed that all the atoms were able to vibrate in three directions about their equilibrium positions with the same frequency ν. The change in this energy per unit change of temperature gave the heat capacity of the crystal. The result was that at a high temperature the heat capacity of a crystal containing Avogadro's number of ions was just $3N_A k = 3R \approx 6$ cal/kelvin. This is the value found experimentally by Dulong and Petit early in the nineteenth century and confirmed for most crystalline solids at room temperature and above, with a few exceptions such as diamond, boron, and silicon. As early as 1872 H. F. Weber had started systematically measuring the specific heat of diamond at various temperatures and had found a marked rise with increasing temperature in a range near room temperature. By 1875 he had extended the range from $-100°C$ to $+1000°C$ and had included graphite, boron, and silicon in his list of substances. Over this range, the specific heat of diamond varied by more than a factor of ten, and it approached the Dulong–Petit value at the higher temperatures. These data were used by Einstein in comparison with his theoretical prediction. There was only one free parameter in his theory, the frequency v of the atomic vibration, and he could fit Weber's measurements quite well over their temperature range.

In 1876 Boltzmann had interpreted the law of Dulong and Petit as an example of the equipartition of energy. This interpretation met with difficulties in the exceptional cases just mentioned, and various possible explanations were suggested, but until Einstein's, none was persuasive. Boltzmann suggested that at low temperatures the atoms of the crystal tended to clump together, reducing the number of degrees of freedom. No mechanism for this clumping was known, but there was the analogy with the diatomic molecule in which the vibration of the two atoms relative to each other did not appear to contribute to the specific heat of diatomic gases, itself a mystery. After all, a crystal was a clump of atoms, and there the law of Dulong and Petit required them to vibrate, so why not the atoms in a molecule?

With a theory available purporting to explain the temperature dependence of the specific heat of crystalline materials, experimental work on the subject was stimulated. Another source of stimulation was Walter Nernst's heat theorem concerning the entropy of solids at very low temperatures. Classical thermodynamics offers no way of determining the absolute value of a system's entropy, only increases and decreases are specified by the heat added or removed and by the temperature. Starting in 1905 Nernst made a series of suggestions about the behavior of the entropy at temperatures approaching absolute zero that had implications for the specific heats of solids at low temperatures. These culminated in the *third law of thermodynamics*, a law whose basis can be understood only with the help of quantum theory. It soon emerged that the decline of the specific heat with falling temperatures that Einstein had predicted—an exponential one in fact—was a far too rapid one. This matter was completely cleared up in 1912 when P. Debye introduced his model in which Einstein's single vibration frequency was replaced by a continuous distribution of frequencies and when M. Born and T. von Karman made a detailed analysis of the spectrum of vibrations of a lattice.

Debye's model of a crystal of an elastic solid was similar to the air in a cavity as studied in acoustics, capable of supporting waves whose amplitudes had to vanish at the walls of the cavity. This is also like the model of black-body radiation used by Rayleigh in his 1900 and 1905 papers, in which the electromagnetic field replaced the sound field in air. Debye used the same counting procedure as Rayleigh had, except that the number of possible polarizations of waves in a crystal is different from that of light waves in a cavity. The crucial step in his calculation was that in a crystal made up of N atoms, there would be only $3N$ independent oscillations possible, and so only the lowest $3N$ modes should be kept.[2] This had the effect of excluding modes of vibration whose wavelengths were less than the interatomic spacing. Clearly, a wave with several oscillations between atoms is not helpful in describing the motions of the atoms. With this limitation on the number of modes to be used, Debye avoided the ultraviolet catastrophe of Rayleigh's theory of blackbody radiation, which had no such physical basis for a cutoff. Applying Einstein's result for the average energy of an oscillator of frequency v to these vibrational modes of the crystal led Debye to an expression for the specific heat that decreased as the third power of the temperature, not exponentially. This slow decline is due to the presence of some modes of long wavelength or low frequency for which quantum effects show up only at very low temperatures, so that these modes tend to act according to the classical Dulong–Petit law until the temperature approaches zero. Born and von Karman actually identified the vibrational modes in certain simple crystalline structures and applied Einstein's result to them. They obtained results more detailed than Debye's for the crystals they calculated that behaved in the same way at very low temperatures. The agreement with observation, though not exact, was very close.

It is perhaps strange that Einstein did not apply this idea to the molecular specific heat problem. A molecule is somewhat like a small piece of crystal, and if the constituent atoms vibrate with a high frequency about their positions of equilibrium in the molecule, at low temperatures the average energy of the vibrations will be very small and will not contribute appreciably to the specific heat. In any case, he did not, although he did attempt to account for the drop in the molar heat of hydrogen gas at very low temperatures (~ 60 K) from $5R$ to $3R$ by considering the suppression of the two rotations about axes perpendicular to the interatomic axis of the H_2 molecule.[3] His treatment of this problem was not the one that emerged as correct when quantum ideas were developed further.

The physical chemist Walter Nernst was much impressed with Einstein's theory of specific heats and became one of the first active supporters of the quantum as an essential physical idea rather than a formal calculational device. He was the prime organizer of the first Solvay Conference, held in 1911, where the role of the quantum in physics was discussed by the leading scientists of the time and where Poincaré, among others, was persuaded of the necessity of quantizing something to account for the blackbody spectrum having a maximum. By 1911, then, a widening circle of physicists and chemists was beginning to take Planck's constant h seriously even if they did not go as far as Einstein and believe that there were quantum effects in the radiation field itself.

Notes

1. A. Einstein, *Annalen der Physik* **22**, 180 (1907).
2. P. Debye, *Annalen der Physik* **39**, 789 (1912).
3. A. Einstein and O. Stern, *Annalen der Physik* **40**, 551 (1913).

Amplification

The effect of Einstein's introduction of an essentially discontinuous weight function in the expression for the average energy of an oscillator in equilibrium with its surroundings at temperature T was to convert the classical integral over all energies to a sum over a discrete set of energies. The classical expression for the average is

$$\langle E \rangle_{\text{classical}} = \frac{\int_0^\infty E \, e^{E/kt} \, dE}{\int_0^\infty e^{E/kt} \, dE} = \frac{(kT)^2}{kT} = kT.$$

With Einstein's modification this became

$$\langle E \rangle_{\text{quantum}} = \frac{\sum_0^\infty nh\nu \, e^{-nh\nu/kT}}{\sum_0^\infty e^{-nh\nu/kT}}.$$

The sum in the denominator was a geometric series with the sum given by

$$1 + x + x^2 + \cdots + x^n + \cdots = \frac{1}{1 - x} \qquad \text{for } |x| < 1$$

with x short for $e^{-h\nu/kT}$. The sum in the numerator was just the negative derivative of that in the denominator with respect to $1/kT$. Putting this all together yielded the result that

$$\langle E \rangle_{\text{quantum}} = \frac{h\nu}{e^{h\nu/kT} - 1}.$$

When $h\nu/kT$ was small, successive terms in the sum did not differ much from one another, and the sum was a good approximation to the classical integral, but when $h\nu/kT$ was large, successive terms did differ appreciably, and the sum and the integral had very different values.

h's Banishment to Obscure Places

Max Planck, as we saw, introduced his constant h in an obscure place. He had exploited to the full the freedom that thermodynamics gave him to choose the matter with which radiation interacts in order to reach thermal equilibrium, and he had picked harmonic oscillators. In order to find the entropy of these oscillators, he had had to imagine that they could have only a discrete set of values for their energy, just as Boltzmann had done with the kinetic energy of the molecules of a gas. This enabled Planck to count the number of ways of distributing a given amount of energy over a given set of oscillators, and so in the statistical spirit of Boltzmann he had arrived at an expression for their entropy. But here he had been forced to part ways with Boltzmann. The physical result that Boltzmann arrived at was achieved by going to the limit as the size of his molecular kinetic energy quanta approached zero. There was no impediment to taking this limit—the introduction of the quanta having been a purely formal device to enable him to count complexions—so in the end his quanta disappeared. Planck had to equate the average energy of his oscillators with the average energy of those modes of the electromagnetic field of the oscillator's frequency, because it was the energy of the field that he was after. The energy distribution over the possible frequencies of the field was restricted by Wien's displacement law, so the energy distribution over Planck's oscillators was similarly restricted. It was this restriction that forced Planck to make his energy quanta proportional to the frequency, and the proportionality constant h appeared in his final expression for the blackbody spectrum. It could not be set to zero if the expression was to yield values agreeing with observations. This was the indirect argument that led Planck to his condition that the energy element associated with an oscillator of frequency v was of size hv.

We also saw that Einstein introduced the energy quantum of the radiation field without mentioning Planck's h. He found that the radiation field behaves "in the region where the Wien law is valid," as though it consisted of independent energy quanta of size $R\beta v/N$, where β is one of the constants in the Wien law. Einstein's quanta were associated directly with the radiation field, not with an auxiliary set of oscillators. This association of quanta with the field itself was not taken seriously by the scientific community for almost twenty years.

Lord Rayleigh had, as early as 1900, recognized the failure of the "Boltzmann–Maxwell doctrine" of the equipartition of energy. When he pursued this subject in 1905, he declared himself unable to follow Planck's reasoning in the derivation of his radiation law, thereby in effect relegating to the realm of speculation the oscillator quantization included in that reasoning. J. H. Jeans, who joined Rayleigh in a discussion of the radiation law in 1906, insisted that Planck should have let the size of his quanta approach zero, even at the price of sacrificing the law that agreed so precisely with observation. Jeans regarded blackbody radiation as not being in thermal equilibrium with matter. By 1909 he had established that if Hamilton's equations described mechanical and electromagnetic systems, there was no escape from the conclusion that the spectrum of blackbody radiation must be that given in the Rayleigh–Jeans law, making more general and more rigorous the earlier work of Rayleigh, Lorentz, and himself.[1] The only way, to reconcile this with observation was, in Jean's view, to deny the existence of equilibrium. He asserted that the establishment of this equilibrium would be a remarkably slow process never to be completed in time measured on a human scale.

Paul Ehrenfest argued forcefully that Planck's oscillators were incapable of causing radiation to come to thermal equilibrium because of their unique quality of having exactly linear equations of motion.[2] The Maxwell equations were also exactly linear, and the equations of motion of the oscillators coupled to the radiation had this property. The linearity made the motion at one frequency completely independent of the motion at any other frequency. How then, Ehrenfest asked, could such a system have an entropy that approached the entropy of blackbody radiation if it did not have it to begin with? This indicated that Planck's derivation, which seemed to make the entropy do this, contained an arbitrary element, and this made questionable the need for quantization.

Planck himself tried to limit the role of his quantum hypothesis as much as possible. From the start he restricted it to the oscillators representing the matter with which the radiation was in equilibrium, and he did not apply it to the electromagnetic field modes. This was permissible because the connection he used between the oscillators and the field was a connection between average energies. Because the field had the same *average* energy as the oscillators, it did not follow that the possible energy values of the field were necessarily the same as the possible energy values of the oscillators. The field energy could take on a continuum of values, as Maxwell theory predicated. At first Planck quantized the energy values that an oscillator could have so that he could count "complexions." Later, the question arose of how an oscillator's energy changed as it absorbed and emitted radiation. In the second edition of his book *Heat Radiation* Planck formulated the theory, with absorption being continuous and only emission being abrupt:

> While the oscillator is absorbing it must also be emitting, for otherwise a stationary state would be impossible. Now, since in the law of absorption just assumed the hypothesis of quanta has as yet found no room, it follows that it must come into play in some way or other in the emission of the oscillator, and this is provided for by the introduction of the hypothesis of emission of quanta. That is to say, we shall assume that the emission does not take place continuously, as does the absorption, but that it occurs only at certain definite times, suddenly, in pulses, and in particular we assume that an oscillator can emit energy only at the moment when its energy of vibration, U, is an integral multiple n of the quantum of energy, $\varepsilon = h\nu$. Whether it then really emits or whether its energy of vibration increases further by absorption will be regarded as a matter of chance. This will not be regarded as implying that there is no causality for emission; but the processes which cause the emission will be assumed to be of such a concealed nature that for the present their laws cannot be obtained by any but statistical methods. Such an assumption is not at all foreign to physics; it is, *e.g.*, made in the atomistic theory of chemical reactions and the disintegration theory of radioactive substances.[3]

Still, later, just before the outbreak of World War I and just after the publication of Niels Bohr's paper on the hydrogen atom, Planck thought that he could bury the quantization process even more deeply than in the emission of radiation by his oscillators. In a presentation to the Prussian Academy of Science in 1914 he proposed that the quantized energy transfer from (and perhaps to) his oscillators that produced the thermal equilibrium energy distribution of the oscillators was not with the radiation field but with material particles such as electrons in the walls of the cavity.[4] This was done to preserve the Maxwell–Hertz theory of the propagation, absorption, and even emission of light completely unaffected by the quantum hypothesis. Planck was thoroughly aware of Einstein's considerations, of Thomson's proposal of "needle radiation" to explain the scarcity of ionization by X-rays, and of Johannes Stark's theory related to Einstein's work, and he ruled all these out because they were inconsistent with the Maxwell–Hertz theory. He was also aware of the success of the energy quantization in connection with the specific heat problem as treated by Einstein and of Poincaré's conversion to the quantum theory (at least of oscillator's) as a result of the first Solvay Conference.

In 1910 Nernst began to think of organizing a conference on quantum physics. He wrote to Planck concerning it and received a reply containing the following:

> But in my opinion another point argues even more strongly in favor of postponing such a conference for one year. The fundamental assumption of calling the conference is that the present state of theory, predicated on the radiation laws, specific heat, etc., is full of holes and utterly intolerable for any true theoretician and that this deplorable situation demands a joint effort toward a solution, as you rightly emphasized in your program. Now my experience has shown that this consciousness of an urgent necessity for reform is shared by fewer than half of the participants envisioned by you and that the others would hardly be motivated to attend such a conference. I need hardly mention the older ones (Rayleigh, Van der Waals, Schuster, Seeliger); I doubt that they will ever become enthusiastic about this matter. But even among the younger physicists the urgency and importance of these questions is still not fully recognized. I believe that out of the long list of those you named, only Einstein, Lorentz, Wien, and Larmor are seriously interested in the matter, besides ourselves.[5]

Despite Planck's reservations, the first Solvay Conference was held in Brussels in 1911. It was influential in spreading the quantum idea through the scientific world, but it did not convince any of the validity of Einstein's light quantum. We have encountered Einstein's introduction of quanta in the radiation field itself, an introduction that explained three puzzling phenomena occurring when light is emitted or absorbed by matter. Even though Einstein's argument was straightforward and rigorous, the conclusion to which it led was so surprising and so contradictory to everything that was understood about light that it was, and remained, simply unacceptable to the scientific community. This is well summed up in the famous letter by Planck, Nernst, Rubens, and Warburg proposing Einstein for membership in the Prussian Academy of Science in 1913, which ended:

> In sum, one can say that there is hardly one among the great problems in which modern physics is so rich to which Einstein has not made a remarkable contribution. That he may sometimes have missed the target in his speculations, as, for example, in his hypothesis of light-quanta, cannot be held too much against him, for it is not possible to introduce really new ideas even in the most exact sciences without sometimes taking a risk.

This was written the year in which Niels Bohr published his paper on the quantum theory of the hydrogen atom!

Notes

1. J. H. Jeans, *Philosophical Magazine* **18**, 209 (1909).

2. P. Ehrenfest, *Wien Berichte* **114**, 1301 (1905), included in his *Collected Papers*, ed. M. J. Klein (Amsterdam: North-Holland, 1959).

3. M. Planck, *The Theory of Heat Radiation*, trans. Morton Masius (Philadelphia: Blakiston, 1914), p. 153.

4. M. Planck, *Physikalische Abhandlungen und Vorträge* (Braunschweig: Vieweg, 1952), vol. 2, pp. 330–35.

5. Quoted in A. Hermann, *The Genesis of Quantum Theory*, trans. C. W. Nash (Cambridge, MA: MIT Press, 1971), p. 137.

CHAPTER 6

Relativity

The name Albert Einstein is associated with the theories of relativity, special relativity—which I discuss here—and general relativity or the relativistic theory of gravitation—which I do not. His enormous contribution to special relativity was not the immediate achievement of new results but the setting out of a new way of looking at phenomena. Nearly every equation that appeared in Einstein's first paper had previously been developed by Lorentz with the use of Maxwell's equations for the electromagnetic field, yet Lorentz himself credited Einstein with giving these equations a new and remarkable interpretation and a greater area of application than anyone else had proposed. This new interpretation swept away the concept of the ether as a tangible medium in which electromagnetic fields represented states of stress and thereby eliminated the problem of how motion through it was to be detected. It abolished distinctions that previously had been thought meaningful but that were then found to be concealed by strange quirks of electromagnetic theory. It reduced problems that had been classified as dynamical problems involving physical forces acting on bodies to kinematical problems that involved analysis of the means of determining where and when events took place rather than the physical causes of those events.

This replacement of dynamics with kinematics made the theory of special relativity especially hard to grasp. It was much easier to accept a new force or a new property of a material medium than it was to accept a new way of thinking about space and time. The results of the theory were much more readily incorporated into the content of physicists' statements than into the language in which these statements were made. Einstein's connection between energy and mass was used by people who continued to refer to the ether as the seat of electromagnetic happenings. In this chapter I shall describe the development of the ether-based theory of Lorentz and its interpretation by Einstein in this profoundly new way.

Conspiracy?

In Chapter 2, in the section entitled "The Elusiveness of the Ether," I described some of the problems that arose in accounting for the observed propagation of light. These were much older and quite distinct from the problems arising in the emission and absorption of radiation. To the mechanistically inclined physicists of the nineteenth century, the wave theory of light required a medium, the ether, in which the waves were disturbances that propagated like elastic waves in matter. The observation of stellar aberration showed that the earth's motion around the sun had to be taken into account when judging the position of a star in the heavens by determining the direction from which light from that star arrived at an earthly telescope, because this direction changed slightly as the direction of the earth's orbital motion changed. The only simple way to compound the velocity of light c with the velocity of the earth v was to think of the earth as moving through a stationary ether without disturbing it. We saw that this simple picture could not be correct because it would make the laws of optics on the surface of the earth depend on the angle between a beam of light in the laboratory and the direction of the earth's orbital motion. There was no such dependence, or at least none proportional to the ratio v/c. Before 1881 the existence of effects proportional to $(v/c)^2$ could not be excluded because of the small value of this quantity, ten parts per billion. Laboratory optics could be made immune to effects of this ether wind proportional to v/c by making Fresnel's assumption of a condensation of ether in matter and this excess ether being carried along by matter moving through the normal ether. We saw that other, later attempts—such as that by Stokes to reconcile the observation of aberration with an ether that did not move in the vicinity of the earth—though not utterly impossible, were highly unsatisfactory.

That section also discussed experiments that could observe effects proportional to $(v/c)^2$, the Michelson and the Michelson–Morley experiments. But they found no such effect. This null result could be explained by the FitzGerald–Lorentz contraction, a shortening of bodies moving through the ether in the direction of motion by the factor $\sqrt{(1 - v^2/c^2)}$. Another famous experiment, done much later than the Michelson ones, was performed by Trouton and Noble to detect a torque on a charged condenser moving through the ether, tending to align it so that the condenser plates were parallel to the direction of motion.[1] Such a torque was to be expected in the absence of the FitzGerald–Lorentz contraction, but it was not found. The contraction was a physical effect tailored just right to shield these second-order effects of the earth's motion through the ether from observation, just as the Fresnel drag had shielded first-order ones.

These explanations of the nonobservability of the earth's motion through the ether were quite *ad hoc* and had no basis other than success in arriving at the desired answer in these particular cases. The question naturally came up whether the Fresnel drag and the FitzGerald–Lorentz contraction were consequences of electromagnetic theory and whether still higher-order effects would also shield the earth's motion from detection by still more accurate experiments. By 1903 Lorentz had succeeded in deriving the first-order effects from electrodynamics, and by 1905 he could say that *if* intermolecular forces behaved in the same way as electromagnetic ones did in a moving system, then the FitzGerald–Lorentz contraction was also consistent with electromagnetism. He did it by detailed analyses of these effects, quite different for the two orders. He did not succeed in showing that still higher-order effects would also be shielded; there was then no experimental evidence that they were shielded, and no experiments had been proposed to find out whether they were.

This situation caused Henri Poincaré great dissatisfaction. As early as 1899 he had speculated that optical experiments would never be able to reveal more than the relative displacements of bodies so that whether or not they were moving relative to the ether could not be detected. In 1900 he pointed out that different explanations were given for the fact that effects proportional to v/c and effects proportional to $(v/c)^2$ could not be detected, thereby making the whole matter seem improvised. He suggested that a common explanation should be found, an explanation that would encompass the effects of all orders in v/c, ensuring that "the mutual destruction of the terms will be rigorous and absolute." He seemed to be envisioning a perfect *conspiracy* to shield the motion of bodies through the ether from detection by optical experiments. Poincaré called the principle ensuring the existence of this conspiracy the *principle of relativity*. He seemed to contemplate a theory in which dynamical effects would completely mask the effects of motion through the ether rather than a theory in which the motion was kinematically undefinable, just as motion through Newton's absolute space was kinematically undefinable in classical mechanics. It is this latter kind of theory that is now termed *relativistic*.

Poincaré repeated these expectations in 1904, stating:

> According to the Principle of Relativity the laws of physical phenomena must be the same for a 'fixed" observer as for an observer who has a uniform motion of translation relative to him: so that we have not, and cannot possibly have, any means of discerning whether we are, or are not, carried along in such a motion. . . . From these results there must arise an entirely new kind of dynamics, which will be characterized above all by the rule, that no velocity can exceed the velocity of light.[2]

It was Einstein who fulfilled Poincaré's expectations, but not in the dynamical way that Poincaré had foreseen. In the next section we will look at Lorentz's incorporation of Fresnel drag and the contraction hypothesis into electrodynamics. Later we will see how Einstein made this unnecessary by his analysis of space and time on the basis of two assumptions or postulates, the complete equivalence of reference systems moving with constant velocity with respect to each other and the independence of the velocity of light from the motion of its source.

Notes

1. F. T. Trouton and H. R. Noble, *Philosophical Transactions of the Royal Society*, A **202**, 165 (1903).
2. H. Poincaré, *Bulletin des sciences mathématique* **28**, 809 (1904).

The Ether Survives: Lorentz's Theory

Lorentz believed in the existence of an ether that was the seat of electromagnetic phenomena. He also believed that Maxwell's equations gave a complete and correct account of electromagnetic phenomena. His greatest work was in applying the Maxwell–Hertz methods to the discussion of systems containing electrons and other charged particles. His contributions in this area included the theory of electrical conduction in metals based on the Drude free-electron model, the theory of the dispersion of light in transparent materials, the emission of thermal radiation by thin metallic foils, and, above all for our concerns, the optics of moving media. In discussing his work I shall be guided by Lorentz's presentation of this last subject in his articles in volume 5 of the *Enzyklopädie der mathematischen Wissenschaften* written in 1903, in his 1904 report in the *Proceedings of the Academy of Sciences of Amsterdam*,[1] and in his *Theory of Electrons*[2] of 1909.

An early and unrecognized step in the direction of future success was made by Woldemar Voigt in 1887.[3] He showed that the *wave equation*—an equation satisfied by the individual components of the electric and magnetic fields in free space with no charges or currents present that described waves traveling with speed *c*—retained its form under a transformation of the coordinates and the time that was similar to the transformation from one Galilean system to another but that differed from that transformation in two essential ways. In order to appreciate this new transformation, we must first write the equations for the Galilean one. Consider a point that had coordinates x, y, z in one Galilean system and coordinates x', y', z' in another such system moving along the x-axis of the first with velocity v. Then the primed coordinates would be related to the unprimed ones by

$$x' = x - vt$$

$$y' = y$$

$$z' = z$$

provided that the two coordinate systems coincided at time $t = 0$. According to these equations, any point with a constant coordinate x' had a coordinate *x* that increased with time at the rate *v*. This means that as seen from the unprimed system, the primed coordinate system was moving in the direction of its *x*-axis with velocity *v* relative to the unprimed one. The measure of time was not affected by this transformation. Time was not on a common footing with space, and a transformation of time accompanying a transformation of space coordinates was simply beyond consideration. It is clear on inspecting these equations that the inverse of this transformation, the one that relates the unprimed set to the primed set, is the same except that the sign of *v* is reversed.

If one substituted the prime coordinates for the coordinates *x*, *y*, *z* occurring in Newton's equations of motion for a system of particles, the form of those equations would remain unchanged. If one did this for the coordinates appearing in the wave equation, leaving the time unchanged, the resulting equation would not have the form of the original wave equation. If Maxwell's equations were written in one Galilean frame, the equations that resulted from making this change of variables in them would not be Maxwell's equations. Maxwell's equations were thought to be valid in a coordinate system at rest in the ether, and so the fact that they were not valid in moving coordinate systems described by these transformations was no problem.

Voigt discovered that the wave equation did retain its mathematical form, including the value of the constant *c* giving the speed of waves, if the preceding transformation was replaced by a different one that also involved a change in the time variable. Voigt's transformation could be written

$$x' = \gamma l(x - vt)$$

$$y' = ly$$

$$z' = lz$$

$$t' = \gamma l\left(t - \frac{v}{c^2}x\right).$$

The first equation shows that a point with a fixed value of *x'* has a value of *x* that increases with the time at the rate *v* and so is moving in the positive *x*-direction with speed *v*, as seen from the unprimed coordinate system. This is similar to the Galilean case. Now, however, there is also a change in the length scale along the direction of motion by the factor γl. The second and third equations show a change in the length scale along the directions at right angles to the relative motion by the factor *l*. The fourth one replaces the time variable *t* with a new variable *t'* that depends on the spatial coordinate *x* as well as on the time *t*. This change in the variable that represents the time in the original coordinate system restores the transformed wave equation to its original form when expressed in the four primed variables, provided that the quantity γ is chosen so that

$$\gamma^2 = \frac{c^2}{c^2 - v^2}.$$

The quantity l can depend on the speed v. If the transformation is to have a vanishing effect on the coordinates when v becomes arbitrarily small, then $l(v = 0)$ must be set equal to unity. In the new coordinate system, waves travel with speed c when lengths are measured using the new, rather than the original, length scale and time is measured by the new time variable t', not the original time t. I have used present-day notation in writing these equations rather than one of the several that Lorentz used. Voigt did not give a detailed physical interpretation of these transformations but used the fact that the phase of a plane wave should be of the same form in both coordinate systems to derive the Doppler effect for light waves accurate to the first order in v/c. This work remained unknown to Lorentz until he prepared his 1906 Columbia University lectures for publication.

In studying optical phenomena in moving bodies, Lorentz started by looking only at first-order effects, effects proportional to the ratio of the speed, v, usually that of the earth in its orbit around the sun, to the speed of light, v/c. We have seen his dismissal of Stokes' theory of aberration. He showed how Fresnel's model of the earth moving through the ether worked for aberration. He then went on to discuss Fresnel's theory of ether drag and to see how it could be explained on the basis of electrodynamics.

Lorentz did this in two ways. First he considered a transparent material to have continuous distributions of positive and negative charge, one fixed relative to the "ponderable matter" and the other free to oscillate about its uniform distribution. In this connection he remarked,

> I shall make light of the difficulty that we are now obliged to imagine four different things, thoroughly penetrating each other, so that they can exist in the space, viz. 1. the ether, 2. the positive and the negative eletricity and 3. the ponderable matter.[4]

Lorentz wrote down the equations satisfied by the fields and the equations of motion for the movable charge and showed that they had solutions corresponding to waves traveling through the medium with a speed c/n corresponding to the medium having an index of refraction n specified by the values of the density of movable charge, the strength of the force tending to restore this charge to its equilibrium distribution and the frequency of the wave.

Next he repeated the calculation for a medium moving through the ether with velocity v as described by an observer stationary in the ether. These equations were different from those for the stationary medium, for two reasons. There was now an additional force acting on the movable charge in the medium because it was being carried along across the magnetic field of the wave. Also there was a change in the current associated with the motion of the movable charge because its motion was now the resultant of its motion relative to the medium and the motion of the medium itself relative to the ether. On taking these changes into account, keeping only those effects proportional to the first power of the ratio v/c, Lorentz found that the equations had the same form when written in coordinates fixed relative to the medium as did the equations for the wave in the stationary medium, except that the velocity of the wave relative to the medium was changed by an amount proportional to the speed v. On combining this velocity through the medium with the velocity of the medium itself, he obtained exactly Fresnel's expression. This derivation did not introduce any new coordinate systems; all the equations contained only coordinates referred to axes fixed in the ether.

Lorentz then redid the calculation using the language of the electron theory. Here he proceeded a little differently, considering a transformation from a system at "rest" to one in motion:

$$x' = x - v_x t, \qquad y' = y - v_y t, \qquad z' = z - v_z t.$$

These constitute a Galilean transformation for a relative velocity not necessarily along the x-axis. To these he added a transformation of the time,

$$t' = t - \frac{1}{c^2}(v_x x' + v_y y' + v_z z').$$

Concerning this Lorentz wrote:

We can regard t' as the time reckoned from the instant

$$\frac{1}{c^2}(v_x x' + v_y y' + v_z z')$$

which changes from one point to the other. This variable is therefore properly called the *local* time in order to distinguish it from the *universal* time t.[5]

Lorentz also introduced transformed field variables. The new electric field $\mathbf{E}'$ was the old $\mathbf{E}$ plus a part of the old magnetic field $\mathbf{B}$ proportional to the ratio v/c, and the new magnetic field $\mathbf{B}'$ was the old $\mathbf{B}$ plus a part of the old electric field $\mathbf{E}$ proportional to the ratio v/c. If terms of the magnitude $(v/c)^2$ were neglected, the equations satisfied by these primed fields using the four primed variables would be in nearly the same form as those satisfied by the unprimed fields in the four unprimed variables, the Maxwell equations. From this Lorentz concluded that

for each state in which $\mathbf{E}$, $\mathbf{B}$, $\mathbf{D}$ are certain functions of x, y, z, t, there is a corresponding state in the moving system, characterized by values of $\mathbf{E}'$, $\mathbf{B}'$, $\mathbf{D}'$ which depend in the same way on x', y', z', t'.[6]

This was an early form of his theorem of *corresponding states*, established here only to the first order in the velocity of the moving system. Fresnel's result followed as a consequence of this theorem.

By these arguments Lorentz demonstrated that the observation of stellar aberration could be reconciled with the observed lack of effect of the earth's orbital motion on the behavior of light in the laboratory, at least to the first order in the ratio v/c, on the basis of Maxwell's electrodynamics, to the extent that all the forces acting in the systems studied behaved in the same way as electromagnetic forces did.

The irrelevance of the earth's orbital motion to the behavior of light in the laboratory remained to be explained when experiments sensitive enough to detect effects proportional to $(v/c)^2$ were done. The contraction hypothesis of FitzGerald and Lorentz provided a place to put the blame for the null rests of Michelson and Morley and of Trouton and Noble, but the contraction itself called for an explanation on some fundamental basis..

In his encyclopedia article (finished in December 1903) Lorentz considered in some detail the electrodynamics of moving media, but mainly only in the first approximation. What he did do exactly was to study the fields produced by a system of charges having only a common translational motion with velocity v parallel to the x-axis and therefore at rest relative to one another. For this case he showed that with such a moving system of charges, one could associate a system in which the charges were at rest in the ether but were distributed in space with all dimensions parallel to the x-axis *elongated* by the factor

$$1/\sqrt{(1 - v^2/c^2)}$$

Here there was no thought that the two coordinate systems involved were equivalent to each other. The one in which the charges were moving was *the* physically relevant system. If the moving charges included an electron, that electron was spherical in the system in which it was moving because that was the *real* system and electrons were presumed really to be spheres. The other coordinate system in which the charges *appeared* to be at rest was a mathematical artifact that was useful for calculation, and the elongated image of the spherical electron that appeared in this system had no physical reality. Later this was modified: A stationary electron was taken to be spherical, and when it was put in motion, it was *contracted* in the direction of motion.

At the end of this important article Lorentz discussed the Michelson–Morley experiment and the hypothetical contraction to which he was driven in order to explain it. He noted that if the forces that hold a solid material together were transmitted through the ether, if they were to experience the same change in strength and direction that electrostatic forces do when the matter moves, and if the matter were really in static equilibrium under these forces, then the contraction would occur. He cited Poincaré's reproach that each new null experiment called for a new hypothesis and observed that things would be better if it could be shown on the basis of some fundamental assumptions that electromagnetic phenomena are independent of the translational motion of the earth. He even contemplated having to accept the Stokes–Planck picture of a compressible ether moving slowly enough near the surface of the earth to make second-order fringe shifts sought in the Michelson–Morley experiment undetectable if the difficulties arising from analyzing second-order effects could not be overcome.

By 1904 Lorentz had made considerable progress. In pursuing the study of second-order effects, he rediscovered the transformations that Voigt had introduced some seventeen years earlier. Using them he discussed anew the problem of explaining the FitzGerald–Lorentz contraction as a consequence of electrodynamics. Extending his calculation to include possible molecular motion, he found it necessary to introduce the idea that the mass of a particle depends on its velocity and on the direction of the acceleration relative to the particle's velocity. Considering two reference systems S and S_0—the former moving with velocity v and the latter at rest in the ether—he found the following:

If m_0 is the mass of a molecule in the absence of a translation, the formulae

$$\text{Longitudinal mass} = \frac{m_0}{(1 - v^2/c^2)^{3/2}}$$

$$\text{Transverse mass} = \frac{m_0}{(1 - v^2/c^2)^{1/2}}$$

contain the assumptions required for the establishment of the theorem, that the systems S and S_0 can be the seat of molecular motions of such a kind that in both the effective coordinates of the molecules are the same functions of the effective time.[7]

The word *effective* is used here because to Lorentz the two systems were not equivalent, the "effective" coordinates and time in the moving system S being only useful auxiliary quantities. This amounted, however, to stating the conditions under which the mechanics of particles transforms in a way consistent with the way that electromagnetic theory transforms, and so this was an introduction of relativistic mechanics.

With this groundwork laid, Lorentz refined his theory of corresponding states so that it was valid to orders higher than the first in v/c:

The theorem amounts to this, and in two systems S and S_0, the one moving and the other stationary, there can be motions of such a kind, that not only the effective coordinates which determine the positions of the molecules are in both the same functions of the effective time (so that the translation is attended with the change of dimensions which we have discussed) but that the same rule holds for the effective coordinates of the separate electrons. Moreover, the components of the vectors **E'** and **B'** will be found to be identically determined by x', y', z', t', both in S and in S_0.

Here Lorentz used the primed notation for "effective" coordinates in both systems. In the rest system, the "effective" coordinates were the "true" coordinates.

He went on to draw some remarkable conclusions from this theorem:

It is remarkable that, when the source of light forms part of the system, so that it shares the translation v, . . . we may assert that in the source of light too, the effective coordinates of the electrons can be the same functions of the effective time, whether the source move or not. If the vibrations are represented in both cases by formulae containing the factor nt', the frequency will be n when the source is at rest, and

$\dfrac{l}{k} n\left[\dfrac{1}{\gamma} - n\right]$ when it moves. This shows that in all experiments made with a
terrestrial source of light, the phenomena will correspond quite accurately to
those which one would observe, using the same source on a stationary planet;
the course of the relative rays, the position of interference fringes, and, in
general, the distribution of light and darkness will be unaltered.[8]

Further on in this discussion Lorentz also pointed out the existence of what
is now called the *transverse Doppler effect*, the reduction in the frequency of
light from a source moving *across* the line of sight. He considered it
undetectable because the frequency shift was proportional to $(v/c)^2$. This is
now a well-known phenomenon and is interpreted as demonstrating the
slowing down of moving clocks.

Pursuing still further the consequences of the theorem on corresponding
states, Lorentz found expressions for the longitudinal and transverse mass of
a charged particle arising from the electromagnetic fields that they generate.
They agreed with those obtained earlier only if the parameter l appearing in
the transformation equations was chosen to be unity. From then on Lorentz
used this value. The transformation equations then took on the form that we
now call the *Lorentz transformations*, so named by Poincaré.

Lorentz's expressions for the mass of a moving particle were based on a
picture in which a moving electron becomes flattened into an ellipsoid if it is
moving. M. Abraham calculated the transverse and longitudinal masses of an
electron, assuming that the electron was spherical both when at rest and when
moving with any velocity less than that of light, thereby contradicting the
contraction hypothesis used by Lorentz.[9] He obtained rather complicated
formulas. Kaufmann measured the transverse mass of the electron over a range
of velocities, and his data agreed better with Abraham's result than with
Lorentz's.[10] Kaufmann's experiments were done with β-rays. In the measurement a β-ray beam was deflected sideways by an electric or a magnetic field.
There were good reasons to believe that the size of the electron charge was
independent of the velocity of the electron, so any change in its charge-to-mass
ratio was interpreted as produced by a change in the mass. As late as the 1909
publication date of *The Theory of Electrons* Lorentz had doubts about
whether Abraham's rigid electron or his "deformable" electron was correct,
remarking:

> Kaufmann, who, as early as 1901, had deduced from his researches on this
> subject that the value of e/m increases [*sic*] most markedly, so that the mass
> of an electron may be considered as wholly electromagnetic, has repeated his
> experiments with the utmost care and for the express purpose of testing my
> assumption. His new numbers agree within the limits of experimental errors
> with the formulae given by Abraham, but not so with the second of the
> equations (313) [Lorentz's transverse mass formula], so that they are
> decidedly unfavorable to the idea of a contraction, such as I have attempted
> to work out.[11]

This was an example of experiments that misled. By 1915 Lorentz could add note 86 to the second edition of *The Theory of Electrons*:

> Later experiments by Bucherer, Hupka, Schaefer and Neumann, and lastly Guye and Lavanchy[12] have confirmed the formula (313) for the transverse electromagnetic mass, so that, in all probability the only objection that could be raised against the hypothesis of the deformable electron and the principle of relativity has now been removed.[13]

This account shows how far Lorentz had gone in developing electromagnetic theory as it applied to moving bodies. He had derived results that now are regarded as results of relativity theory, results that have stood the test of experimental observations. His conclusions were drawn from a model in which the ether played a fundamental role. He applied the Lorentz transformations as a calculational device to derive results in complicated situations involving matter in motion, and not as a means of connecting coordinate systems equally valid for the description of physical phenomena so that a coordinate system especially appropriate to the specification of a given problem could be used. Even though Lorentz had shown that no experiment, no matter how precise, made with terrestrial sources of light would ever show an influence of the earth's motion, he did not draw the conclusion that the ether was a superfluous concept. As late as 1915, when he added notes to the second edition of the *Theory of Electrons*, Lorentz still expressed a preference for the ether picture over Einstein's relativistic picture. The two pictures differed in fundamental attitudes toward the description of nature, but they did not differ in their predictions of the outcome of observations that had been made.

Notes

1. H. A. Lorentz, reprinted in *The Principle of Relativity* (New York: Dover, 1952).

2. H. A. Lorentz, *The Theory of Electrons* (New York: Dover, 1952). This is a reprint of the second edition, published in 1915. It contains the text of the first edition of 1909 modified only by notes and an appendix added later.

3. W. Voigt, *Göttingen Nachrichtungen* (1887), p. 41.

4. Lorentz, *The Theory of Electrons*, p. 182.

5. Ibid., p. 57.

6. Ibid., p. 189

7. Ibid., p. 205.

8. Ibid., p. 208.

9. M. Abraham, *Annalen der Physik* 10, 105 (1903).

10. W. Kaufmann, *Annalen der Physik* 19, 487 (1904).

11. Lorentz, *The Theory of Electrons*, p. 212.

12. A. H. Bucherer, *Physikalische Zeitschrift* 9, 755 (1908); E. Hupka, *Annalen der Physik* 31, 169 (1910); C. Schaefer and G. Neumann, *Physikalische Zeitschrift* 14, 1117 (1913); C. E. Guye and C. Lavanchy, *Comptes rendus* 161, 52 (1915).

13. Lorentz, *The Theory of Electrons*, p. 339.

Amplification

In his 1904 report Lorentz introduced the transformation that became the Lorentz transformation in three steps. First he made the usual Galilean transformation, leaving the length scales unchanged. He did not distinguish these new variables in his notation; here I denote them by ^. Then,

$$\hat{x} = x - vt$$

$$\hat{y} = y$$

$$\hat{z} = z.$$

He next changed the scale along the x-axis by the factor γl and the scales along the other two axes by the factor l that was to differ from unity by at most a quantity of the second order in v/c:

$$x' = \gamma l \hat{x}$$

$$y' = l \hat{y}$$

$$z' = l \hat{z}.$$

Finally, he introduced the new "local" time variable t' by

$$t' = \frac{l}{\gamma} t - \gamma l \frac{v}{c^2} \hat{x}.$$

The net effect of all these changes was to obtain the transformation given in the main text.

The new, primed fields that Lorentz introduced were defined by

$$E'_x = \frac{1}{l^2} E_x \qquad\qquad B'_x = \frac{1}{l^2} B_x$$

$$E'_y = \frac{\gamma}{l^2}\left(E_y - \frac{v}{c} B_z \right) \qquad B'_y = \frac{\gamma}{l^2}\left(B_y + \frac{v}{c} E_z \right)$$

$$E'_z = \frac{\gamma}{l^2}\left(E_z + \frac{v}{c} B_y \right) \qquad B'_z = \frac{\gamma}{l^2}\left(B_z - \frac{v}{c} E_y \right).$$

Together with these fields he defined a transformed charge density ρ' defined by

$$\rho' = \frac{\rho}{\gamma l^3}.$$

This related the effective charge density to the true charge density, so that

$$\rho \, dx \, dy \, dz = \rho' \, dx' \, dy' \, dz'$$

and the charge contained in a volume element specified by the coordinate differentials retained its value on carrying out the transformation, provided that the primed differentials were evaluated at a constant value of the universal time t and not of the local time. This reflected the fact that to Lorentz simultaneity was not a relative concept but an absolute one. The boundaries of the volume element $dx' \, dy' \, dz'$ had to have their locations specified at a definite value of the universal time t, not of the local time t'. The mathematical simplification that resulted from using a common value of the local time was recognized by Poincaré in 1905. The physical necessity for insisting on the use of the local time only appeared in Einstein's work of that same year.

If a point was moving with a velocity **u** in addition to the motion of the moving coordinate system, then Lorentz wrote its velocity in the stationary system, using the classical vector addition of velocities, as

$$v_x = v + u_x, \qquad v_y = u_y, \qquad v_z = u_z.$$

He then introduced a new vector **u'** defined by

$$u'_x = \gamma^2 u_x, \qquad u'_y = \gamma u_y, \qquad u'_z = \gamma u_z$$

so that in combination with the effective charge density just defined the Maxwell equation involving the current took the form

$$\nabla' \times \mathbf{B}' = \frac{1}{c}\left(\frac{\partial \mathbf{D}'}{\partial t'} + \rho' \mathbf{u}'\right).$$

This is just like the corresponding equation in the stationary system except that all the variables are primed. The two Maxwell equations not involving charges or currents also retained their form when written in the primed variables. The one equation to change was that giving rise to Coulomb's law,

$$\nabla \cdot \mathbf{E} = \rho \Rightarrow \nabla' \cdot \mathbf{E}' = \left(1 - \frac{v u'_x}{c^2}\right)\rho'.$$

The parenthetical expression is a factor that appears in Einstein's formula for the addition of two velocities, here the velocity v of the coordinate system and the velocity u of the charge. Its presence here is a result of Lorentz's insistence on using the universal time when describing processes taking place in the moving coordinate system, in particular the motion of the charge in that system. With a slight change in his definitions, this factor could have been moved to the Maxwell equation involving the current, where its connection with the velocity addition would be more apparent. Lorentz had no reason to think this preferable, with Einstein's velocity addition formula not appearing for another year.

The Ether Retires; Einstein's Theory

Einstein's approach to the electrodynamics of moving bodies, leading to his theory of relativity, was altogether different in spirit from Lorentz's.* It may be compared with his approach to the problem of blackbody radiation in the "Heuristic Viewpoint" paper that appeared earlier in the same year as his "On the Electrodynamics of Moving Bodies," 1905. In the earlier paper we saw that instead of trying to calculate the blackbody spectrum from some fundamental laws of radiation and statistical physics, Einstein sought the consequences for those laws of the observed rapid falloff of the spectrum at high frequency. Here Einstein did not start out from Maxwell theory in the manner of Lorentz but, instead, sought the consequences of the apparent impossibility of detecting any effects of the earth's motion on electrodynamic observations. One might say that Einstein applied Occam's razor to the problem of the ether and, in the words of the *Oxford English Dictionary*, concluded that "for purposes of explanation things not known to exist should not, unless it is absolutely necessary, be postulated as existing." Not knowing of Lorentz's work of 1904, Einstein wrote in the second paragraph of his relativity paper:

> Examples of this sort [the asymmetry between the description of faraday induction with a moving magnet and stationary wire and with a stationary magnet and moving wire], together with the unsuccessful attempts to discover any motion of the earth relatively to the "light medium," suggest that the phenomena of electrodynamics as well as of mechanics possess no properties corresponding to the idea of absolute rest. They suggest rather that, as has already been shown to the first order in small quantities, the same laws of electrodynamics and optics will be valid for all frames of reference for which the equations of mechanics hold good. We will raise this conjecture (the purport of which will hereafter be called the "Principle of Relativity") to the status of a postulate, and also introduce another postulate, which is only apparently irreconcilable with the former, namely, that light is always propagated in empty space with a definite velocity c which is independent of the state of motion of the emitting body. These two postulates suffice for the attainment of a simple and consistent theory of the electrodynamics of moving bodies based on Maxwell's theory for stationary bodies.

Einstein's concept of relativity, the one now universally used, requires the form of the laws of physics to be unchanged when the reference frame in

---

* A dissenter from this opinion is E. T. Whittaker. In the 1951 edition of his monumental *A History of the Theories of Aether & Electricity*, he ascribed the entire theory to Lorentz and Poincaré. Fourteen pages into his chapter "The Relativity Theory of Poincaré and Lorentz," he wrote: "In the autumn of the same year, in the same volume of the *Annalen der Physik* as his paper on the Brownian motion, Einstein published a paper which set forth the relativity theory of Poincaré and Lorentz with some amplifications, and which attracted much attention. He asserted as a fundamental principle the *constancy of the velocity of light*, i.e. that the velocity of light in vacuo is the same in all systems of reference which are moving relatively to each other, an assertion which at the time was widely accepted, but has been severely criticized by later writers."

which they are stated is changed in certain ways. This is a more stringent demand than merely requiring physical phenomena to be describable equally accurately when viewed from any appropriate reference frame. It demands that the way the laws describing these phenomena are stated, the recipes used to predict their outcomes, be given in a way equally applicable in any appropriate frame. In *Galilean* relativity the form of Newton's equations of motion, and those arising from them such as Lagrange's and Hamilton's, is independent of the inertial reference frame used in writing them. Relativity in this sense is somewhat difficult to grasp at first. It appeared in an implicit form when Galileo identified *acceleration* as the quantity characterizing the effect of gravity on bodies and so demoted velocity from its primary importance. Acceleration is a quantity that has the same value in all Galilean reference frames, while neither position nor velocity is. Therefore, Galileo's statement that the effect of gravity on a body is to give it a definite downward acceleration is applicable in all reference frames moving relative to one another with constant velocity. Galileo described the dropping of an object in a moving ship: The dropped object falls straight down as judged from inside the ship no matter what the ship's uniform motion may be. This illustrate that his description of the effect of gravity on bodies is independent of any uniform motion of the environment in which this effect is observed.

Two quotations from Aristotle's *Physics* illustrate some of the difficulties faced by anyone trying to formulate the concept of relativity.[2] The first concerns the idea of *motion* and the second the idea of *place*.

> Now no motion exists apart from things; for that which changes always does so either with respect to substance or with respect to quantity or with respect to quality or with respect to place, and there can be no thing common to these which is not, as is our manner of speaking, a *this* or a quantity or some one of the other categories. Thus neither motion nor a change can exist apart from these [categories] if nothing else exists but these.

This defines motion, or more specifically *locomotion*, as relative to *place* and so does not conflict with our concept of relativity. Aristotle devoted a great deal of space to the discussion of *place*, and his conclusion was one that was denied by Galilean relativity and again later by Einstein's more extensive form of relativity.

> A place is thought to be an object of importance and difficult to grasp, both because it appears to be present with matter and the shape [of a thing] and because the displacement of a body in locomotion occurs in a container which is at rest; . . . Now just as the vessel is a place which can be in locomotion, so a place is an immovable vessel. So when something inside is moving or changing in something else which is in motion, as a boat in a river, it uses the containing vessel as a vessel rather than as a place. Now a place is meant to be something motionless; so in the example it is rather the whole river [basin] which is a place, since the whole river is motionless. A place, then, is this, namely, the primary motionless boundary of that which contains. And it is in view of this that the Center of the heaven and the last [inner surface]

of the rotating part of heaven toward us are thought to be, most of all and principally, *down* and *up* respectively.

Here the center of the earth or heavens is given the privileged position of Center; the directions *down* and *up*, and the laws governing the behavior of bodies—such as that heavy bodies go down to seek their natural place near the Center—refer to this Center explicitly. Also, terrestrial and celestial realms are distinguished. The statement of Aristotle's laws of nature is not independent of the reference frame in which that statement is made, and therefore in our terms his laws are nonrelativistic.

Galileo had started the trend toward formulating the laws of nature in a way that is independent of the reference frame in which they are to be applied. As we saw in Chapter 1, in the section "From Galileo and Newton: Classical Mechanics," Newton introduced an absolute space and an absolute time, but he did not specify how they were to be identified. His laws of motion took the same form in any reference system that moved uniformly, as measured using absolute time, through absolute space, that is, in any Galilean reference frame. The acceleration of a body was the same in all reference frames moving uniformly relative to one another, and so were the relative positions of bodies the same in them all. Absolute time was completely independent of space and so was the same in all reference frames. Any forces that depended on the relative positions of bodies, such as Newton's gravity, were also the same. Therefore Newton's second law of motion—force equals mass times acceleration—took the same form in all these reference frames.

When the wave optics of Young and Fresnel displaced the corpuscular theory of light, it was taken for granted that light waves had to be elastic waves propagating through a material medium, just as sound waves are compressional waves propagating through air. The electromagnetic theory of light inherited this assumption, and the ether of optics became the ether in which electromagnetic fields represented states of stress. If there were an ether, the light from a small source at rest in it should propagate outward, as seen by an observer at the source, in spherical waves with the source and the observer as center. If the ether and source were moving past the observer, the waves emitted by the source just as it passes him should not appear to be spherical, with him as center, because those parts traveling in the same direction as the ether would have a higher speed relative to him than would those parts emitted in the opposite direction. This lack of symmetry about the observer, if it arose from the cause described, should have been detectable when the motion was taken to be the orbital motion of the earth. It was looked for in the experiments of Michelson[3] and of Michelson and Morley[4] discussed in Chapter 1 in the section "An Elusive Ether" and in others such as that of Trouton and Noble involving expected forces on moving charged condensers but it was not found.[5]

In his paper "The Electrodynamics of Moving Bodies," Einstein gave a clear statement of the two principles outlined in his introductory remarks:

(1) The laws by which the states of physical systems undergo change are not affected, whether these changes of state be referred to the one or the other of two systems of co-ordinates in uniform translatory motion.

(2) Any ray of light moves in the "stationary" system of co-ordinates with the determined velocity c, whether the ray be emitted by a stationary or by a moving body.

These principles do not involve electrodynamics directly, but only indirectly through the introduction of a means of sending signals with a defined speed independent of the coordinate system in which it is measured. This speed c appears in Maxwell's equations which, according to item (1) have the same form and hence contain the same value of c in all the reference frames that Einstein is considering. Nowadays a neutrino beam might replace the light ray without changing anything except the technical difficulty of producing and detecting signals.

In the first part of his paper subtitled "Kinematical Part," Einstein presented his discussion of the meaning of the statement that two events at different places are simultaneous. Here he defined a way to synchronize clocks located at rest in different places in some coordinate system, by means of sending light signals between them. (Earlier, Poincaré had raised the issue of what the simultaneity of events at different places means, and he proposed almost simultaneously with Einstein in 1905 the synchronization of clocks by means of light signals. He did not, however, recognize the implications of this for the measurement of lengths but kept the FitzGerald–Lorentz contraction as a separate hypothesis.) In any one coordinate system this leads to nothing surprising. Doing the same thing with clocks at rest in another system moving uniformly with respect to the first one also leads to nothing new, as the speed of the light signals used as measured in this system is by hypothesis the same as that of those used in the previous system as measured in that system. The postulate that the speed of light in any coordinate system was c no matter what the motion of its source did, however, make the reconciliation of the results of these two synchronization processes bring in a startling new feature. Clocks that had been synchronized by Einstein's procedure in one system would not appear to be synchronized as viewed from another system moving with respect to the first, because each observer would think the other had used the wrong speed of light to carry out his synchronization. Newton's absolute time no longer served its purpose but had to be replaced by a relative time! The relation between the times in the two systems was exactly the relation between Lorentz's universal time and local time, except that Lorentz and Poincaré thought they knew which was universal and which was local, whereas Einstein put them on exactly the same footing.

Why pick out light to synchronize clocks? Why not use sound? On a still day the synchronization of clocks at rest would be the same whether one used light or sound. With a steady wind blowing, however, the results would not agree. This is because of the observable fact that the velocity of sound measured in a coordinate system moving through air is not the same as the velocity of that sound as measured in a coordinate system at rest in that air. Only the velocity of light has the characteristic of having the same velocity in "all frames of reference in which the equations of mechanics hold good."

His new operational definition of simultaneity led Einstein directly to the FitzGerald–Lorentz contraction. If a rod is at rest along the *x*-axis of a coordinate system, it makes no difference when you look at the two ends to read off the values of *x* there in order to determine its length. If, however, the rod is moving relative to you, the situation is quite different, and you must be sure to look at the two ends at the same time as determined by clocks synchronized in your reference frame. An observer moving along with the rod will judge your two acts of looking not to be simultaneous, and so because he can see the rod move between your two observations, he will not be surprised that you get an answer different from his. When you measure the length of the moving rod, what you get is the contracted length, while he gets the "true" length. If the rod were at rest in what he called "your" system, you would get the "true" length, and he would get the contracted length. The epithet "true" is completely inappropriate. We can better refer to the length of a rod at rest as its *rest* length. If the rod is not parallel to the *x*-axis but is moving parallel to that axis, the situation is slightly more complicated. The values of the *y*-coordinate and the *z*-coordinate of the ends of the rod do not change as a result of a motion parallel to the *x*-axis, and so it makes no difference when these coordinates of the ends of the rod are observed, and no question concerning simultaneity arises. The only contraction is in the dimension parallel to the direction of motion.

It is important to realize that the contraction arrived at in this way has nothing to do with the forces holding the atoms of a material body together. It is a purely kinematic effect arising from the definition of the length of a rod as the difference between the coordinates of its ends *at the same time* as judged by the one making the measurement. Lorentz and Poincaré had not arrived at this view of it, and it is not clear when, if ever, they fully accepted it.

Another surprising result of Einstein's analysis is the way in which velocities were to be compounded. Consider motion along the *x*-axis only. Suppose that as seen from O, point A was moving away at speed v_{OA} and that as seen from A, point B was moving away at speed v_{AB} (see Fig. 6.1). At what speed as seen from O was B moving away? Einstein showed that the speed

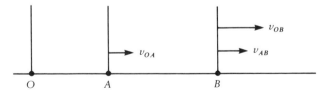

Fig. 6.1 The velocity of A relative to O cannot simply be added to the velocity of B relative to A to obtain the velocity of B relative to O.

v_{OB} is not the sum of the separate speeds v_{OA} and v_{AB}, but because of the effects of the FitzGerald–Lorentz contraction and the slowing of clocks, it is given by the formula

$$v_{OB} = \frac{v_{OA} + v_{AB}}{1 + \dfrac{v_{OA}v_{AB}}{c^2}}.$$

If both the speeds v_{OA} and v_{AB} are very small compared with the speed of light, the denominator can be considered to be unity, and the usual rule for adding speeds holds, again giving a result much smaller than the speed of light. There was no conflict with everyday experience on this formula. If one of the speeds, say v_{AB}, was the speed of light, then no matter what the other was, the resulting speed would again be the speed of light. This showed that Einstein's analysis of space and time led to a picture in which his postulate that the speed of light is independent of the motion of its source is satisfied. A pulse of light was seen to move away from O with speed c whether it was emitted from a source fixed at O or from a source fixed at A moving away from O at speed v_{OA}. If the velocity v_{AB} is comparable to that of light and v_{OA} in the same direction is smaller, the resulting velocity will be appreciably less than that expected from the nonrelativistic way of combining speeds and will always be less than the speed of light. (A wag once proposed that a high-energy accelerator that produced particles moving within one meter per second of the speed of light be mounted on a platform that moved two meters per second, thus producing particles with speeds one meter per second greater than that of light!)

All this was done without using Maxwell's equations, although the speed of light played an essential part in the argument. This is important because it meant that Einstein's relativity theory was not just a theory of electrodynamics but one that provided a framework into which any physical theory incorporating the measurement of space and time would have to fit as long as light signals were the means of synchronizing clocks. If another form of signal should be discovered that had a speed s independent of its source's state of motion and, further, if this speed differed from that of light, then his theory would be untenable because there would be no way to decide whether to use one signal or the other in synchronizing clocks or whether to use c or s in the transformation equations. So far no signal of this kind has been found, but other signals such as those transmitted by neutrinos that do travel with (or perhaps just very close to and less than) the speed of light are now known. This is evidence for the limiting character of the speed of light for all means of transmitting signals.

By carefully analyzing the consequences of the postulate that light spreads out at speed c in spherical waves as seen by two observers moving uniformly with respect to each other and the postulate that the form of the mathematical description of this spreading is the same for both observers, Einstein derived the equations connecting the coordinates and time in one frame to those in another. They were the Lorentz transformations. Recall that Voigt's and Lorentz's original equations contained a parameter *l* that allowed for a scale change in the directions at right angles to the relative motion of the two coordinate systems. A value of this parameter other than unity would destroy any symmetry between the two coordinate systems because the effect of one transformation would not be undone by a second transformation with reversed velocity. The net effect of these two transformations would be a scale change of l^2. This did not disturb them because they had no need for such a symmetry, as the system at rest in the ether was a distinguished one, no others being equivalent to it. For Einstein, the effect of one transformation had to be undone by applying another one just like it with reversed velocity, because neither system was preferred over the other. Lorentz found it advantageous to choose *l = 1* for dynamical reasons arising from electrodynamics, not for reasons of symmetry.

In the electrodynamic part of his 1905 paper introducing his theory of relativity, Einstein used the principle of relativity to find the transformed fields that obey Maxwell's equations in the new coordinate system in terms of the fields that obey them in the original coordinate system. To do this he wrote out the equations for empty space in component form for one coordinate system. He then replaced the coordinates and time for that system with those for a moving system, which produced a set of equations for the original field components that was not Maxwellian in form. The principle of relativity, however, required the equations for the field components as seen in the new system to be of the same form as the original equations for the original field components. Einstein showed that this was indeed the case if new field components that were linear combinations of the original ones were introduced. The relation between these new field components and the original ones was just that used by Lorentz in his theory of corresponding states. The difference between the two theories lay in the significance attributed to their results, not in their form. Einstein regarded the transformed coordinates, time, and electromagnetic field in one system as fully equivalent to those in the other, neither being preferred in any way (except that in one system the equations describing a particular situation might be easier to work with than in the other), while Lorentz regarded the set of equations written in ether-at-rest frame as fundamental and all the others as useful auxiliary constructs involving auxiliary quantities such as the *local* time and the *effective* fields.

The difference between his view and Lorentz's was illustrated by Einstein's discussion of the expression for the force on a charged particle moving through an electromagnetic field following his derivation of the transformation of the electric and magnetic fields. Lorentz had needed to postulate this force independently of Maxwell's equations, writing it as

$$\mathbf{F} = e\left(\mathbf{E} + \frac{1}{c}\mathbf{v} \times \mathbf{B}\right).$$

In discussing the meaning of the transformation between coordinate systems moving relative to each other Einstein noted:

> The . . . equations themselves to be clothed in words in the two following ways:—
>
>   1. If a unit electric point charge is in motion in an electromagnetic field, there acts upon it, in addition to the electric force, an "electromotive force" which, if we neglect the terms multiplied by the second and higher powers of v/c, is equal to the vector-product of the velocity of the charge and the magnetic force, divided by the velocity of light. (Old manner of expression.)
>
>   2. If a unit electric point charge is in motion in an electromagnetic field, the force acting upon it is equal to the electric force which is present at the locality of the charge, and which we ascertain by transformation of the field to a system of coordinates at rest relatively to the electrical charge. (New manner of expression.)

In the "old manner of expression" the force on a charge was the Lorentz force just written. In the "new manner of expression" a coordinate system in which the charge was instantaneously at rest was selected, in that system the force was that due to the electric field, by definition of the electric field. The Lorentz force appeared on transforming back to the original coordinate system: What had been just an electric field became a combination of electric and magnetic fields. Now the Lorentz force was not a new postulate needed to account for the force on a particle moving in the ether-at-rest system but followed from the definition of the electric field as the force on a unit test charge placed at rest in the field.

Einstein considered how light from a distant source, say a star, would appear to an observer moving with a given velocity as seen from a coordinate system in which the source is at rest. He found that according to his theory, the color of light would be altered as predicted by the Doppler effect and changed in apparent direction as in the observed stellar aberration, just as Lorentz had discovered before him. Einstein also considered how the electromagnetic energy contained in a volume changes on transforming from one coordinate system to another moving with constant velocity relative to it. He remarked about the result that

> it is remarkable that the energy and the frequency of a light complex vary with the state of motion of the observer in accordance with the same law.

He gave no hint here of a connection with his photoelectric equation, $E = hv$, that he had published just a few months earlier!

Einstein also considered the "dynamics of the slowly accelerated electron" in this paper. In it he gave definitions of the longitudinal and transverse masses that did not turn out to be useful. He took as the measure of the acceleration the acceleration in the frame in which the electron has velocity v, and as the measure of the force as it would be described in the frame in which the electron is at rest. Equating this force with a mass times this acceleration, he obtained the following as expressions for the longitudinal and transverse masses:

$$\text{Longitudinal mass} = \frac{m}{(\sqrt{1 - v^2/c^2})^3}$$

$$\text{Transverse mass} = \frac{m}{1 - v^2/c^2}.$$

He noted here that

> with a different definition of force and acceleration we should naturally obtain other values for the masses. This shows us that in comparing different theories of the motion of the electron we must proceed very cautiously.[8]

This may have been Einstein's only comment on the interpretation of Kaufmann's experiments. His transverse mass defined in this peculiar way differed from both Abraham's and Lorentz's.

At the end of his paper Einstein calculated the radius of curvature of the path of an electron moving perpendicularly to a uniform magnetic field. Here the force on the electron is perpendicular to its velocity, so interpreting the result of an observation of this radius leads to something that could be called a *transverse mass*. His result here is that the electron acts as though its mass were increased by the factor

$$1/\sqrt{1 - v^2/c^2}.$$

If he had taken this as defining the transverse mass, he would have obtained Lorentz's result. A measurement of this radius was involved in Kaufmann's experiments.

No explanation was needed for the null result of the Michelson–Morley experiment. What Lorentz had explained was why light traveling in different directions relative to the earth's motion about the sun—and hence in his view traveling with different speeds—took equal times to go back and forth across the interferometer. He explained it by having the distance traveled also depend on the direction of the light path relative to the motion of the earth around the sun because of the FitzGerald–Lorentz contraction. According to Einstein's postulates, the speed of light is the same in all inertial reference frames, and so naturally the times—as measured by clocks at rest relative to the interferometer—required for the two beams to cover the two equal arms of the interferometer were equal no matter what the motion of the interferometer itself happened to be.

In a short note published later that same year, Einstein raised the question "Does the inertia of a body depend upon its energy content?" By considering the emission of light energy in two opposite directions by a moving body, he arrived at the following conclusion:

> *If a body gives off the energy L in the form of radiation, its mass diminishes by L/c². The fact that the energy withdrawn from the body becomes energy*

of radiation evidently makes no difference, so that we are led to the more general conclusion that

The mass of a body is a measure of its energy-content; if the energy changes by L, the mass changes in the same sense by $L/(9 \times 10^{20})$, the energy being measured in ergs, and the mass in grammes.[6] (italics in original)

This led to perhaps the most famous equation in physics,

$$E = mc^2.$$

The contribution that Einstein made to this subject that went well beyond the contributions of Lorentz and Poincaré was this: whereas they had conceived of the existence of an ether that provided a preferred reference frame and had then proposed that the detection of this frame was precluded by a conspiracy between distinct effects depending on this frame that made them cancel each other out, Einstein proposed that there is complete symmetry between reference frames moving uniformly with respect to each other, one being completely equivalent to any other and none being privileged over any other. The ether in the sense of an exotic material medium that provided the means for the propagation of an electromagnetic field in space and time—simply had no place in this picture. The electromagnetic field was viewed as a basic element of the world not requiring explanation in terms of a still more basic element such as ether. The consistent development of this view required some modification of Newtonian mechanics to include such things as the equivalence of mass and energy and the existence of a limiting speed, c, but it left the theory's essential structure unchanged. It left Maxwell's electromagnetic theory untouched. After all, the covariance of the Maxwell equations under the Lorentz transformations had inspired the recognition of the symmetry of all reference frames in translational motion with respect to one another.

Notes

1. A. Einstein, *Annalen der Physik* 17, 891 (1905). An English translation, *On the Electrodynamics of Moving Bodies*, by W. Perrett and G. B. Jeffery, appears in *The Principle of Relativity* (New York: Dover, 1952), from which the quotations have been taken.

2. The quotations are taken from the translation of Aristotle's *Physics* by H. G. Apostle (Bloomington: Indiana University Press, 1969). The first is from Book Γ, p. 42, and the second is from Book Δ, p. 67.

3. A. A. Michelson, *American Journal of Science* 22, 20 (1881).

4. A. A. Michelson and E. W. Morley, *American Journal of Science* 34, 333 (1887).

5. F. T. Trouton and H. R. Noble, *Philosophical Magazine* 202, 165 (1904).

6. A. Einstein, *Annalen der Physik* 18, 639 (1905). An English translation, *Does the Inertia of a Body Depend upon Its Energy Content?* by W. Perrett and G. B. Jeffery, appears in *The Principle of Relativity*.

The Acceptance: A Special Status for c

As we have seen, the phrase "the theory of relativity" meant different things to different people. To Lorentz and Poincaré and many others it meant a dynamical theory based on the existence of an ether that was the seat of electrodynamic phenomena, a theory that rendered it impossible to detect a uniform translation with respect to that ether. To Einstein it meant the existence of a fundamental symmetry that made it meaningless rather than just impossible to distinguish between reference frames in uniform translational motion with respect to each other. Poincaré came very close to viewing the theory as a symmetry. He discussed the difficulty of the concept of simultaneity for separated events, and he recognized that the transformations considered by Lorentz with the transverse scale change parameter l put equal to unity constituted a group. This means that the result of making two successive transformations can be achieved by making a single transformation of the same kind with an appropriate choice of the velocity that characterizes these transformations. This mathematical property immediately makes it impossible to select a particular reference frame as an identifiable starting point. If I am in a frame S thinking of it as moving, how can I tell whether I should consider frame S_0 to be the rest frame with respect to which it is moving with velocity v_0 or whether I should consider S_1 to be the rest frame with respect to which it is moving with velocity v_1, since the transformations from these two frames to S have exactly the same form and differ only in the velocities v_0 and v_1 characterizing them? With the mathematical ground thoroughly prepared, Poincaré did not make the step back to the physical world and announce that the existence of this new symmetry made the identification of a unique rest frame impossible and a hypothesis of an ether completely redundant. Lorentz had always started from a frame in which his ether was at rest, and he never had to require the result of successive transformations to be another transformation of the same form, which it was not unless l was unity. He put l equal to unity to make the equation of motion of an electron come out right, not to give the transformations the group property.

In his 1909 *Theory of Electrons* Lorentz discussed Einstein's theory in the following way, using the terminology that he had developed for use in his own theory:

> The denominations "effective coordinates" [i.e., coordinates determined with the help of a measuring rod subject to the FitzGerald–Lorentz contraction], "effective time" [i.e., time as determined by a moving clock giving the "local time"] etc. of which we have availed ourselves for the sake of facilitating our mode of expression, have prepared us for a very interesting interpretation of the above results, for which we are indebted to Einstein. Let us imagine an observer, whom we shall call A_0 and to whom we shall assign a fixed position in the ether, to be engaged in the study of the phenomena going on in the stationary system S_0. We shall suppose him to be provided with a measuring rod and a clock, even, for his convenience, let us say, with a certain

number of clocks placed at various points of S_0, and adjusted to each other with perfect accuracy. By these means he will be able to determine the coordinates x, y, z for any point, and the time t for any instant, and by studying the electromagnetic field as it manifests itself at different places and times, he will be led to the equation (321, 323) [Maxwell's equations written in the stationary system].

Let A be a second observer, whose task is to examine the phenomena in the system S [moving through the ether with velocity w], and who himself also moves through the ether with the velocity w, without being aware either of this motion or of that of the system S.

Let this observer use the same measuring rod (or an exact copy of it) that has served A_0, the rod having acquired in one way or another the velocity w before coming into his hands. Then, by our assumption concerning the dimensions of moving bodies, the divisions of the scale will in general have a length that differs from the original one, and will even change whenever the rod is turned round, the law of these changes being, that, in corresponding positions in S_0 and S, the rod has equal projections on the plane YOZ [i.e., the lateral dimensions are unchanged], but projections on OX whose ratio is as k to 1 [i.e., the dimension in the direction of motion is contracted]. It is clear that, since the observer is unconscious of these changes, he will be unable to measure the true relative coordinates x_r of the points of the system. His readings will give him only the values of the effective coordinates x' and, of course, those of y', z' which, for $l = 1$, are equal to y_r, z_r. Hence, relying on his rod, he will not find the true shape of bodies. He will take for a sphere what really is an ellipsoid, and his cubic centimetre will be, not a true cubic centimetre, but a parallelepiped k times smaller. This, however, contains a quantity of matter, which, in the absence of the translation, would occupy a cubic centimetre, so that, if A counts the molecules in *his* cubic centimetre, he will find the same number N as A_0. Moreover, his unit of mass will be the same as that of the stationary observer, if each of them chooses as unit the mass of the water occupying a volume equal to *his* cubic centimetre.

With the clocks of A the case is the same as with his measuring rod. . . . Therefore, a clock in the system S will indicate the progress of the effective time, and without his knowing anything about it, A's clocks will go k times slower than those of A_0. . . .

It is of importance not to forget that, in doing all that has been said, the observer would remain entirely unconscious of his moving (with himself) through the ether, and of the errors of his rod and his clocks. . . .

Attention must now be drawn to a remarkable reciprocity that has been pointed out by Einstein. Thus far it has been the task of the observer A_0 to examine the phenomena in the stationary system, whereas A has had to confine himself to the [moving] system S. Let us now imagine that each observer is able to see the system to which the other belongs, and to study the phenomena going on in it. Then, A_0 will be in the position in which we have all along imagined ourselves to be (though strictly speaking, on account of the earth's motion we are in the position of A); in studying the electromagnetic field in S, he will be led to introduce the new variables x', y', z', . . . etc. and so he will establish the equation (326) and (322, 324) [Maxwell's equations for a moving body]. The reciprocity consists in this

that, if the observer A describes in exactly the same manner the field in the
stationary system, he will describe it accurately. . . .

It will be clear by what has been said that the impressions received by
the two observers A_0 and A would be alike in all respects. It would be
impossible to decide which of them moves or stands still with respect to the
ether and there would be no reason for preferring the times and lengths
measured by the one to those determined by the other, nor for saying that
either of them is in possession of the "true" lengths. This is a point which
Einstein has laid particular stress on, in a theory in which he starts from what
he calls the principle of relativity, i.e. the principle that the equations by
means of which physical phenomena may be described are not altered in form
when we change the axes of coordinates for others having a uniform motion
of translation relatively to the original system.[1]

Here Lorentz described accurately the content of Einstein's theory as
applied to the problems that had occupied his own attention for years. Yet
the symmetry that he acknowledged as existing in the mathematical description
did not persuade him that the physics really possesses that symmetry so that
the ether has no existence. He went on to write:

I cannot speak here of the many interesting applications which Einstein
has made of this principle. His results concerning electromagnetic and optical
phenomena agree in the main with those which we have obtained in the
preceding pages, the chief difference being that Einstein simply postulates
what we have deduced with some difficulty and not altogether satisfactorily,
from the fundamental equation of the electromagnetic field. By doing so, he
may certainly take credit for making us see that in the negative result of
experiments like those of Michelson, Rayleigh and Brace, not a fortuitous
compensation of opposing effect, but the manifestation of a general and
fundamental principle.

Yet, I think, something may also be claimed in favour of the form in
which I have presented the theory. I cannot but regard the ether, which can
be the seat of an electromagnetic field with its energy and its vibrations, as
endowed with a certain degree of substantiality, however different it may be
from all ordinary matter. In this line of thought, it seems natural not to
assume at starting that it can never make any difference whether a body
moves through the ether or not and to measure distances and lengths of time
by means of rods and clocks having a fixed position relatively to the ether.

It would be unjust not to add that, besides the fascinating boldness of
its starting point, Einstein's theory has another marked advantage over mine.
Whereas I have not been able to obtain for the equations referred to moving
axes *exactly* the same form as for those which apply to a stationary system,
Einstein has accomplished this by means of a system of new variables slightly
different from those which I have introduced. I have not availed myself of
his substitutions, only because the formulae are rather complicated and look
somewhat artificial, unless one deduces them from the principle of relativity
itself.[2]

I have included these extensive quotations from Lorentz because he was
the outstanding contributor to the subject of the electrodynamics of moving

bodies before Einstein. He had derived most of the results bearing on this subject, but he had done so on the assumption that an ether existed and that its escape from direct observation was due to the cancellation of dynamic effects arising from the theory formulated in a frame at rest in the ether. His statements that Einstein had simply postulated what he had deduced contained something of a put-down of Einstein's achievement. Lorentz still regarded a dynamical derivation as superior to the assertion of a symmetry principle. He differed from many of his contemporaries in understanding thoroughly and correctly the difference between them. By the end of the century ushered in by these developments, the imposition of symmetry requirements on the structure of theories had become an essential ingredient in theoretical physics in areas extending from the smallest quark to the whole of the cosmos, but at its beginning the exploitation of symmetries other than the familiar geometrical ones was an unfamiliar tool whose power had yet to be demonstrated.

One of the obstacles to the immediate acceptance of relativity theory in either Lorentz's or Einstein's formulation was the work by Kaufmann on the mass of the electron. Both of these formulations gave an expression for the transverse mass of the electron as measured by the magnetic and electric deflection of a particle beam that differed from one derived by Abraham on the basis of a rigid spherical electron not affected by the FitzGerald–Lorentz contraction, and Kaufmann's results agreed more closely with Abraham's theory than with the relativistic prediction. Kaufmann used a beam of β-rays that had a continuous spectrum of velocities. He deflected the beam by passing it through parallel electric and magnetic fields at right angles to the beam direction. The electric field produced a deflection of the electrons in a direction opposite to the field direction (because the electron charge was negative), and the magnetic field produced a deflection at right angles to the direction of the fields and to the beam direction. Kaufmann then let the beam strike a photographic plate that registered the deflections suffered by electrons of differing velocities.

To see how this gave information about the way that the electron mass depended on the velocity, imagine that it did not, so that all the electrons in the beam had the same mass. The electric force on an electron was the same for all, but the time for which the force acted was inversely proportional to the velocity of that electron, and so the transverse velocity it acquired would be inversely proportional to its speed. The angle of deflection would be the ratio of the transverse velocity acquired to the original velocity and so would be inversely proportional to the square of the original velocity. The magnetic force on an electron of the beam was proportional to its velocity, and so the angle of deflection produced by the magnetic field would be inversely proportional to the original velocity, not to its square. The very fast electrons would go through the fields with almost no deflection, while slower ones would be deflected more, with the electric deflection increasing more than the magnetic deflection for electrons of lower and lower original velocity. The

resulting trace on the photographic plate would be a parabola. If the faster
electrons had greater masses than the slower ones did, the transverse velocities
acquired by the electrons would depend differently on their velocity, and the
shape of this trace would be altered. It was the deviation of this trace from a
parabola that Kaufmann used to determine how the mass depended on the
velocity.

As was mentioned in this chapter in the section "The Ether Survives;
Lorentz's Theory," Kaufmann's results favoring Abraham's rigid sphere model
disturbed Lorentz greatly and made him doubt the validity of the contraction
concept. Einstein seemed to be much less bothered by them, having confidence
in the relativity principle because it was so successful in relating "wider
complexes of phenomena." Eventually, as stated in the section "The Ether
Retires; Einstein's Theory," experiments confirmed the relativistic prediction
of the variation of mass with velocity.

Planck was the first to use Einstein's approach in discussing the fundamental equations of mechanics.[3] In March 1906 he gave an address to the
German Physical Society at the beginning of which he remarked on the doubt
cast on the Lorentz–Einstein theories by Kaufmann's results but expressed the
opinion that the experiment was a very delicate one and that the relativity
theory might still be consistent with the observations. Planck then proceeded
to derive in a simple and straightforward way, explicitly using the principle
of relativity, the relativistic equations of motion for a particle in Newtonian
form, in Lagrangian form, and in Hamiltonian form. The Newtonian form
differed from the traditional one only in substituting the rate of change of
momentum for the product of mass and acceleration, a difference that allowed
for the dependence of the mass on the velocity:

$$\frac{d}{dt}\left(\frac{mv_x}{\sqrt{1 - v^2/c^2}}\right) = F_x$$

and similarly for the y and z equations. The presence of the variable mass

$$m/\sqrt{1 - v^2/c^2}$$

in place of the constant mass m was the only effect of relativity that showed
here. If the force F was perpendicular to the velocity v, then v^2 was constant,
and the left side of the equation was of the usual Newtonian form, with the
transverse mass in place of the constant Newtonian mass. If the force was
parallel to the velocity, then v^2 was not constant and had to be differentiated.
An equation containing the longitudinal mass resulted.

In September of that same year Planck gave a lecture on Kaufmann's
experiments as summed up in a paper published that year.[4] In this lecture he
described his analysis of Kaufmann's data. The difference between the results
of the "Sphere-Theory," as he called Abraham's, and the "Relativity-Theory"
of Lorentz and Einstein was less than the difference of either from the
experimental values, although the former was closer. Planck did not regard

this as a definitive test. He pointed out also using the experimental parameters published by Kaufmann, some of the electron velocities came out higher than the velocity of light, thereby contradicting both theories and demonstrating the large probability that there were flaws in the experiment that needed to be corrected before it could be regarded as persuasive evidence in choosing between the two theories. Planck did not commit himself to the relativity theory this early but considered it an important contribution and worthy of detailed study.

A. H. Bucherer had been concerned about the velocity dependence of the electron mass, and his theory was one of those favored by Kaufmann's results. In 1907 he published a paper entitled "A New Principle of Relativity in Electromagnetism."[5] It contained no reference to Einstein and claimed to establish the validity of Newton's third law of motion for the interaction of charged particles, the one leading to the conservation of momentum, without the introduction of electromagnetic momentum. Much earlier, Lorentz had used this momentum to salvage the law of momentum conservation. My reason for mentioning this paper is that Bucherer was among those whose later experiments confirmed the relativistic variation of mass with velocity. Apparently there were those deeply interested in these problems who did not consider Einstein's or Lorentz's formulation to be persuasive or even important during the first few years after Einstein's paper.

In his paper of 1905 Einstein had laid out the complete theory of special relativity. The qualification *special* refers to the restriction to coordinate systems that are in uniform translational motion with respect to one another. By no means were all the implications of this theory developed, and a clear way to visualize the effect of the combined transformation of space coordinates and time was not presented. This latter was done by Minkowski in 1908.[6]

Einstein had emphasized the need to describe events by giving their location in space relative to some frame of reference and giving the time at which they occurred as read on a clock at their location. Minkowski seized on this need and proposed combining the space coordinates giving the location of an event with the clock reading giving the time of that event into a set of four coordinates for the event, and representing it as a point in a four-dimensional space. He called this space the *world* and the point representing an event a *world point*. Now the term *space-time* is generally used rather than *world*. He then compared the space-time corresponding to Newtonian mechanics and that corresponding to relativity.

Minkowski pointed out that in both theories the uniformity of space made arbitrary the directions in which the spatial axes are drawn, as long as they are at right angles to each other. Any given set of spatial axes can be rotated in an arbitrary way to form a new set, and the statement of any law of physics will be the same in the rotated axes as in the original axes, although, of course, the description of any given situation will not be the same. In Newtonian mechanics the impossibility of identifying absolute space introduces another arbitrary element: Any set of axes may be replaced by a set having a uniform

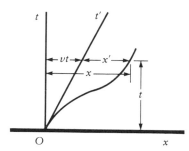

Fig. 6.2 Minkowski's Galilean space-time. The motion of a reference frame is represented by the tipping of the time axis, all space axes remaining unchanged.

translational motion with respect to the original set, leaving the absolute time unchanged, without affecting the statement of the laws of motion. Minkowski described this graphically, for reasons of simplicity restricting himself to systems moving parallel to the x-axis. (This is not really a restriction; he could always rotate the coordinate axes so that the x-axis lay in the direction of the translatory motion.) He chose to start with a system in which he drew the t-axis at right angles to the x-axis, as in Fig. 6.2. This angle has no meaning in the sense of the angle between two spatial directions.

The path or *world line* of a particle can be drawn giving the value of x at every value of t, also as shown in Fig. 6.2. At any time t specified by the height above the origin along the t-axis, the x-coordinate to a particle is given by the distance from the t-axis to the event, measured parallel to the x-axis.

If a primed coordinate system moving with speed v along the x-axis is introduced by the equation

$$x' = x - vt,$$

the x'-coordinate of the particle is reduced by an amount proportional to the time. This is represented on the graph by drawing a new time axis labeled t', and the value of the x'-coordinate at any time is read off as the distance from this new axis to the event, again measured parallel to the x-axis. This new time axis can make any angle between $0°$ and $180°$ with the x-axis, depending on the velocity v. The time is determined by the distance above the x-axis measured parallel to the original t-axis in all cases, the measure of time being the same in all systems, according to Newton. The x-axis is drawn in heavy type to show its fixity under these transformations. It was this picture that Minkowski adapted to the Lorentz transformation.

Einstein's postulate that the speed of light is the same in all reference frames moving with uniform translational motion with respect to one another implied the following: If a set of observers moving uniformly with respect to one another all were at the same point at the instant that a source of light there emitted a pulse, then every one of these observers would see the light

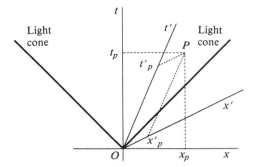

Fig. 6.3 Minkowski's relativistic space-time. The motion of a reference frame is represented by the tipping of both the time axis and the coordinate axis in the direction of the motion.

wave as a shell expanding at the rate c with that observer as center.* Einstein explained how this was possible by virtue of the way that the various observers disagreed on when events at distant places were simultaneous. Unlike Newton's absolute time, the time that Einstein used was affected by the transformation from one reference frame to another. Minkowski showed how to represent this on a space-time diagram.

To an observer using an unprimed coordinate system, a pulse of light sent out from the origin along the x-axis at time $t = 0$ will travel with speed c and will have a world line, shown as the right-hand *light cone* in Fig. 6.3. The scale of the figure has been chosen so that this light cone has unit slope. This same light pulse observed by another observer moving in the x-direction will also be seen to move with speed c in the same direction. This is clearly inconsistent with the situation represented by Fig. 6.2. According to that figure, the primed observer should see the light move with speed $c - v$. Minkowski made a change in the drawing to make it consistent with the new situation, as illustrated in Fig. 6.3. Instead of keeping the x-axis unchanged in the transformation, Minkowski kept the light cone unchanged, as indicated by the heavy line. To do this, he had to tilt the x'-axis up toward the light cone as well as to tilt the t'-axis down toward it. Now the time of an event P in the primed coordinate system was its distance from the x'-axis measured parallel to the t'-axis, and the x'-coordinate of the event was its distance from the t'-axis measured parallel to the x'-axis. The larger the velocity of the primed coordinate system relative to the unprimed was, the greater would be the tilt of these two axes, but clearly there is a limit to how much they can tilt; they draw closer and closer to meeting at the light cone as the velocity

---

* Ehrenfest used this example in a dramatic way in his inaugural lecture as Lorentz's successor at Leyden in 1912. See M. J. Klein's *Paul Ehrenfest* (Amsterdam: North-Holland, 1979) for a vivid description of this lecture and its implications for the acceptance of relativity at that time.

increases. The velocity of light is the limit of the velocity that can be reached by making a Lorentz transformation.

Having introduced this picture of space-time and the way Lorentz transformations affect it, Minkowski introduced some new terminology that has become the standard language of the subject. An event lying within the light cone shown has a time coordinate greater than zero as determined in all space-time coordinate systems sharing the same origin. This cone contains the future for the origin of space-time, that is, for the origin of the spatial coordinate system when the clock there reads zero, in the sense that these future events could be influenced by the event of the origin. There is another light cone extending down from the origin; all events within it have negative times in all these coordinate systems, and this cone constitutes the past as seen from the origin. The event at the origin could be influenced by these past events. The space-time vector connecting the origin with an event in either the past or future light cone Minkowski called a *timelike* vector. The space-time vector connecting the origin with events outside both light cones he called a *spacelike* vector. Events lying outside both these light cones cannot influence the event at the origin, nor can it influence any of them. The temporal order of the event at the origin and an event between the two light cones depends on the reference frame. They can be simultaneous or either can be earlier than the other.

At first Einstein rather pooh-poohed this geometrization of special relativity, considering it an unnecessarily arcane way of treating it. But later he changed his mind. He found it an indispensable aid in formulating his theory of general relativity, a theory of gravitation in which the geometry of space-time is determined by its content. I shall not discuss the general theory.

An indication of the lack of appreciation of the difference between the dynamic formulation of the theory by Lorentz and the kinematic formulation by Einstein is given in an article written by Wien in 1909 on the electromagnetic theory of light. Wien devoted Section 27 to "the relativity theory," introducing the subject in the following way:

> Lorentz has shown that Maxwell's equations for stationary bodies retain their form when one refers the quantities defining the state to a coordinate system moving with constant velocity and introduces new variables.
>
> Einstein has introduced the same transformation from a somewhat different standpoint namely in that he postulated that by a coordinate transformation to a second, relatively moving system Maxwell's equations remain invariant, and that the [electromagnetic] phenomena are independent of whether one refers them to one or the other system.[7]

The restriction of Einstein's considerations to the invariance of the form of Maxwell's equations completely obscures the essential difference between these two approaches. Einstein postulated the invariance of the velocity of light and the invariance of the form of *any* set of laws describing natural phenomena, not just electrodynamics. For him the invariance was a purely kinematic matter, completely independent of the dynamics resulting from any particular theory. Thus to Lorentz the FitzGerald–Lorentz contraction was a consequence of the way that electromagnetic forces transform and the assumption that other forces transform in the same way. The change in these forces changes the dimensions of bodies. To Einstein the contraction was a result of the way in which one carries out the process of measurement and is independent of the nature of the body measured. This confusion makes it hard to decide who had and who had not accepted the full theory of relativity, as distinct from the dynamical or "conspiracy" theory. Either way, the *results* of the theory were accepted nearly universally after 1915 when the matter of the velocity dependence of the electron mass had been settled beyond dispute.

Notes

1. H. A. Lorentz, *The Theory of Electrons* (New York: Dover, 1952), pp. 223–26.

2. Ibid., p. 229.

3. M. Planck, *Physikalische Abhandlungen und Vorträge* (Braunschweig: Vieweg & Sohn, 1958), band II, pp. 115–20.

4. Ibid., pp. 121–35.

5. A. H. Bucherer, *Philosophical Magazine* **13**, 413 (1907).

6. H. Minkowski, an address given in Cologne. An English translation by W. Perrett and G. B. Jeffery is included in *The Principle of Relativity* (New York: Dover, 1952). Minkowski died before he could prepare the material for publication.

7. W. Wien, *Encyklopädie der Mathematischen Wissenschaften* (Leipzig: Teubner, 1909), vol. 22.

CHAPTER 7

Models of Atoms

O nce the existence of the chemical atom had been accepted, the question of its structure arose. This was inevitable because there were so many distinct kinds of atoms, at least one for every chemical element, and yet these atoms had many common features and similar types of behavior. They emitted spectra when excited; they formed molecules with one another; they collided with one another; and they fell into the pattern provided by Mendeleev's periodic table of the elements. Each separate kind of atom could not be treated as an independent ingredient in a satisfactory picture of matter. The discovery of the electron did not raise the question of what the electron is made of in nearly such an acute form. There was only one kind of electron, and it seemed to be completely specified by its position, its mass, and its charge. Even today, when the proton, the neutron, and the pion are considered to be made of quarks and gluons, the electron is a structureless particle. The discovery of the electron and its identification as a universal constituent of atoms gave the quest for a satisfactory atomic model a good starting point.

All the models I shall describe in this chapter are built around electrons. There had been pre-electron models, the best known of which was Kelvin's ether vortex model, based on Helmholtz's theorem concerning the permanence of vortex motion in an ideal fluid. If the ether were such a fluid, any vorticity in it would be a permanent feature of the ether, impossible to destroy and impossible to create. A variety of kinds of atoms could be provided by having these vortex rings knotted or braided with themselves so that they could not be manipulated into the form of simple rings without cutting themselves, which was excluded by the Helmholtz theory. There were attempts to explain spectra as resulting from oscillations of these rings, but these attempts never succeeded in reducing the number of arbitrary elements needed to account for what was seen. They had the virtue of giving the ether something else to do besides escaping detection by electromagnetic means, but they became unimportant before the ether disappeared. There was no place in a vortex atom for electrons.

Early Models of Atoms

Newton's picture of atoms as "solid, massy, hard, impenetrable, moveable particles" had served those who thought of such things at all up through the early days of the kinetic theory of gases, but by the second half of the nineteenth century it was inadequate to account for what had been learned.[1] As early as 1857 Robert Wilhelm Bunsen and Gustav Kirchhoff concluded that metal vapor in a sufficiently hot flame emits light of one or more well-defined colors that uniquely identify the metal responsible for the emission.[2] The yellow color of a flame into which a sodium compound such as table salt has been introduced is a familiar example. Yellow light with a wavelength of 589.3 nanometers is characteristic of sodium. The emission of these spectral lines by the heated vapor and the ability of that vapor to absorb light from a brighter source of that same wavelength were early ascribed to the ability of atoms or molecules to vibrate internally with characteristic frequencies, an ability indicating that atoms and molecules must have some internal structure.[3] The electromagnetic theory of light suggested that these vibrations were executed by charged particles within the atoms or molecules. These oscillating particles would constitute oscillating electric currents, and according to Maxwell and Hertz they would radiate light with the frequency of their vibrations. After the discovery of the electron, these currents were pictured as consisting of moving electrons.

The possibility of motions with an atom caused difficulty for the kinetic theory of gases. We have seen that linear and some rotational motions of gas molecules could by themselves account for all the energy supplied to a gas when it was heated. There would be no energy left over to excite even vibrations of the atoms making up the molecule relative to one another, to say nothing of still more vibrations within the atoms. The atoms constituting the molecule seemed not to acquire the energy that the equipartition theorem would have led one to expect. This failure of equipartition was totally unaccounted for. As we pointed out earlier, Maxwell regarded these anomalies as the most serious problem facing the atomic theory of matter.

Putting aside the molar heat difficulty, the question of how many electrons there are in any given kind of atom became urgent. At first it was estimated that this number was large, hundreds or thousands. Electrons are much lighter than atoms. If electrons are to make up, say, half of the mass of a hydrogen atom, that atom would have to contain roughly nine hundred electrons, and heavier atoms would have to contain proportionately more. This was not in conflict with what was expected from the great number of lines to be seen in the spectra of the elements if, as the Maxwell–Hertz theory implied, there had to be a separate electronic motion for each frequency radiated. The nature of the positive charge needed to give the atom its electrical neutrality was also unknown. No positively charged particles lighter than the positive hydrogen ion had been observed in gas discharges, the place where they should have been most conspicuous if they existed. The positive rays called *canal rays*, the counterpart of the cathode rays, that traveled toward the cathode of a gas discharge were massive, and their nature depended on the gas used in the discharge, so they were not universal constituents of all atoms as the electron was.

Here we shall discuss two atomic models that were developed in the first few years of this century, one by Hantaro Nagaoka, the other by Lord Kelvin (William Thomson) and J. J. Thomson.

Nagaoka's model was based on a planetary analog, the rings of Saturn.[4] Maxwell had won the University of Cambridge Adams Prize for 1856 for his essay on the stability of Saturn's rings. In this essay Maxwell had analyzed the motion of a ring of small masses under the gravitational attraction of a planet and of one another, and he concluded that

> a ring of satellites can always be rendered stable by increasing the mass of the central body and the angular velocity of the ring.[5]

The coulomb force between two charged particles depends on their separation, just as the force of gravity depends on the separation of the gravitating bodies, as the inverse square of the separation. This suggested to Nagaoka that the atom might consist of a positively charged body around which circulated rings of electrons. One difference between the electrical and the gravitational cases was that while the particles making up Saturn's rings all attract one another as well as the central body, the electrons in Nagaoka's model would be attracted to the central body but would repel one another. The analysis of the effect of this is not easy, but since by far the largest single force on an electron was that due to the central body, it was assumed that the difference it would make would be small. Eventually it turned out that this made Nagaoka's rings unstable, that slight perturbations of the motion of the ring particles would have a large and probably catastrophic effect on their motion.

There was another serious difference between the electrical and gravitational systems. Newton's theory of gravitation was an action-at-a-distance theory in which the force between two gravitating bodies was given directly by their masses and their separation. No medium was contemplated through which the gravitational influence of one body was transmitted to another, and so there was no possibility of gravitational radiation in the theory. Maxwell–Hertz theory dealt with the electromagnetic fields in space, and it was these fields that transmit electromagnetic forces between charge bodies. It most certainly did include the processes of emission and absorption of electromagnetic radiation, of light, by accelerated charges such as electrons moving in circles. The particles circulating around the central body of one of Nagaoka's atoms should emit light continuously and should lose energy continuously in the process. This difficulty could be partly overcome by assuming the presence of a great many electrons. If each ring contained many uniformly spaced electrons, the total current caused by all the electrons of a ring moving together around the center would approximate a current not changing with time, that is, a direct current, and a direct current does not radiate. How the oscillations giving rise to the spectrum of the element composed of these atoms were constituted never became clear. The central charge of the atom needed to be large to neutralize the charges of the many electrons and to give hope that the system was stable.

Kelvin introduced a static model, elaborated by J. J. Thomson, and so avoided the radiation trouble of Nagaoka.[6] There was no stable equilibrium for any set of discrete charges acted on only by the coulomb forces between them. If all the charges were separated from one another, each one would be effectively at the electric potential produced by the others at its location and would not be affected by its own charge. The effective potential was just the potential that would be produced in free space at that point by all the other charges. The coulomb potential at any point in free space had neither a maximum nor a minimum value. If it decreased along one direction away from the point, it necessarily increased along some other direction, so there was no place where there was a position of stable equilibrium for a charged particle. If, however, there was a fixed continuous distribution of charge present, stable equilibrium points were not excluded. Of course, forces other than electromagnetic ones would be needed to keep the fixed positive charge fixed, forces that did not affect the electrons. Neither Kelvin nor J. J. Thomson said what these forces were nor how they arose. They assumed a uniform spherical distribution of positive charge with a radius equal to the radius of the atom. A number of electrons was considered to be embedded in this sphere so that the whole was electrically neutral. Thomson calculated possible equilibrium configurations of this system.[7] (He also considered rings that rotated.) Each electron felt a force toward the center of the sphere of strength proportional to its distance from the center due to the smeared-out positive charge and a Coulomb's law repulsion from each other electron. Thomson showed that there are equilibrium configurations in this system, and a simple one is given in Fig. 7.1.

Two electrons are at a distance r from the center along a diameter of a sphere of radius a and of charge $2e$. The attractive force exerted by the uniformly distributed positive charge on an electron a distance r from the center is given according to Gauss's law by Coulomb's law with an attractive charge equal to that contained inside the sphere of radius r. (The proof of the analogous theorem for gravitation puzzled Newton for a considerable time and delayed his conviction that the force of gravity acting on the moon was simply related to that acting on a body at the surface of the earth.) If the distance of one electron from the center is slightly increased, then the attractive

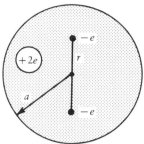

Fig. 7.1   A two-electron Thomson atom showing a possible stable configuration of the two electrons within the fixed sphere of uniform positive charge density.

force on that electron is slightly increased because of the increase in the contained charge, and the repulsive force is decreased because of the increased separation from the other electron, so this electron is urged back to its original position, and the system is stable against radial motion of the electrons relative to the center of the atom. The same is true for three, four, or five electrons arranged in regular plane polygons, but not for six. In order to stabilize a ring of six electrons, Thomson had to introduce one electron inside the ring. For outermost rings with large numbers of electrons, he had to introduce many electrons inside, and these in turn had to be arranged in rings in order to be stable. In this way he could build up a sequence of atoms with increasing numbers of electrons. His calculations dealt mainly with electrons arranged in a single plane, although he did consider the case of four electrons at the corners of a regular tetrahedron. He even analyzed the Zeeman effect for such a system. Lorentz extended this model to one with a density of positive charge that decreased with increasing distance from the center of the atom.[8] With three-dimensional arrangements, different stable configurations would exist, so that the sequence of planar rings he studied did not necessarily correspond to the sequence of atoms in nature. In order to account for spectra, this model would require a large number of electrons in even the lightest elements. Rayleigh went so far as to extend Thomson's picture to an infinite number of electrons that could be approximated by a charge fluid.[9]

Thomson's model is reminiscent of Benjamin Franklin's picture of "electrical matter" in "common matter":

> If a piece of common matter be supposed entirely free from electrical matter, and a single particle of the latter be brought nigh, it will be attracted, and enter the body, and take its place at the center, or where the attraction is every way equal. If more particles enter, they take their places where the balance is equal between the attraction of the common matter, and their own mutual repulsion. It is supposed they form triangles, whose sides shorten as their number increases; till the common matter has drawn in so many that its whole power of compressing those triangles by attraction, is equal to their whole power of expanding themselves by repulsion; and then will such piece of matter receive no more.[10]

Franklin was, of course, thinking of bulk matter rather than atoms. His force of attraction was the polarization force that attracts uncharged light objects to charged ones and not the force between two charged bodies, but the concept was remarkably similar to the Kelvin–Thomson one and was advanced for its date of 1749, some thirty-five years before the appearance of Coulomb's papers on the law of force between charges.

These two models were not successful in the form they were proposed. Nagaoka's was clearly a precursor of Rutherford's nuclear model. Both contradicted classical physics, as we shall see in detail, and in the end Rutherford's model could be salvaged only by the application of quantum concepts by Niels Bohr. Thomson's model did provide for some properties of atoms that were known at the time, but not in a satisfactory way. The reason that the sphere of positive charge in which the electrons were embedded did not blow apart was not explained, and the model required a separate electron for each spectral line so that the number of electrons needed was enormous and any regularities exhibited by these lines would appear to be accidental. The models represented serious attempts to understand the existence of atoms with the chemical and physical attributes required, but they became less and less acceptable as further knowledge about atoms accumulated.

Notes

1. I. Newton, Query 31 in his *Opticks* (New York: Dover, 1952).

2. R. W. Bunsen and G. Kirchhoff, *Berlin Monatsberichte*, October 1859, p. 662; *Annalen der Physik* **109**, 193 (1860); *Philosophical Magazine* **21**, 240 (1861).

3. See, for example, J. Norman Lockyer in Mary Jo Nye, ed., *The Question of the Atom* (San Francisco: Tomash, 1984).

4. H. Nagaoka, *Philosophical Magazine* **7**, 445 (1904).

5. J. C. Maxwell, *Collected Works* (New York: Dover, 1965), vol. 1, p. 318.

6. Lord Kelvin (William Thomson), *Nature* **67**, 45, 103 (1902).

7. J. J. Thomson, *Philosophical Magazine* **7**, 237 (1904).

8. H. A. Lorentz, *The Theory of Electrons* (New York: Dover, 1952), sec. 100.

9. Lord Rayleigh, *Philosophical Magazine* **11**, 117 (1906).

10. *The Works of Benjamin Franklin, L.L.D.* (London: J. Johnson, St. Paul's Churchyard; and Longman, Hurst, Rees and Orme, Paternoster Row, 1806), vol. 1, p. 219; cover letter dated July 29, 1750, to Peter Collinson, Esq. F.R.S.

How Many Electrons?

The discovery of the electron as a universal constituent of matter immediately led to the question, How many electrons are there in an atom? The early cathode ray experiments did not give any information about this number except that there were some in the atoms of all the elements tested. The fact that the charge-to-mass ratio for the electrons was so much larger than that for any ion occurring in electrolysis meant that the mass of the electron was much less than that of a chemical atom, and this in turn made it possible for there to be a large number of electrons in an atom, in fact thousands of them. However many there might be, the absence of highly charged positive ions in electrolysis indicated that only a few electrons could be easily extracted from a neutral atom, but this put no limit on the number that were there but were, for some reason, difficult to remove. The atomic ions observed in the canal rays carried a positive charge corresponding to the absence of one or, at most, a few electrons. This was consistent with electrolysis and also led to no limit on the number of electrons present in the neutral atom.

The emission and absorption of spectral lines of frequencies characteristic of the emitting or absorbing material, together with the discovery of electrons in atoms, led naturally to the assumption that these optical processes were connected with the presence of electrons within the material. According to Maxwell–Hertz theory, the only explanation for this emission and absorption was the existence of oscillating electric currents within the material. Since the line spectra were emitted and absorbed by material in the form of vapor—in which the individual atoms or molecules acted independently of one another— the currents had to be within the atoms or molecules, and the obvious candidate for the charge carrier was the electron. The presence of electrons able to oscillate about positions of equilibrium in an atom seemed to be required.

The theory of vibrating systems was well advanced by the end of the nineteenth century, largely in connection with acoustics. Hermann Helmholtz had published his monumental work *On the Sensations of Tone* in 1862. The fourth edition was published in 1877 and was available in an English translation in 1885:

> The whole sensation excited in the ear by a periodic vibration of the air we have called a *musical tone*. We now find that this is *compound*, containing a series of different tones, which we distinguish as the *constituent* or *partial tones* of the *compound*. The first of these constituents is the *prime partial tone* of the compound, and the rest its *harmonic upper partial tones*. The *number* which shews the *order* of any partial tone in the series shews how many times its vibrational number exceeds that of the prime tone.

This is just an application of Fourier's general theorem, which Helmholtz stated informally as

> The multiplicity of vibrational forms which can be thus produced by the composition of simple pendular [harmonic] vibrations is not merely extra-ordinarily great: it is so great that it cannot be greater.[1]

The first edition of Rayleigh's *Theory of Sound* appeared in 1877 and was followed by a second edition in 1894. This work carried the subject further than Helmholtz had but shared the starting point based on Fourier's theorem regarding the decomposition of any periodic motion into a series of harmonics.

According to the classical theory of vibration, one could expect a source of sound to emit sounds of any frequency, but only those sounds composed of integer multiples of a "prime partial tone" or fundamental could come from a source vibrating periodically. Applying the same theory to the emission of light, if in an atom there were an electron moving in some periodic way other than pure simple harmonic motion, it would emit light of the frequency corresponding to that period and some harmonics of that frequency. The absence of any harmonics would indicate that the electron was moving with simple harmonic motion and could therefore account for the emission of light of only one frequency. It seemed that the atoms of a given chemical element had to contain electrons bound to positions of equilibrium about which they could oscillate. Because in the spectra observed there were no lines harmonically related to one another, each spectral line would have to be emitted by a separate electron that oscillated with the frequency of that line. Zeeman's observation of the broadening of neither spectral lines emitted in a magnetic field and Lorentz's interpretation of this broadening as a splitting into three components due to the effect of the magnetic field on the vibrating electron strongly reinforced this picture. Thus the number of electrons in an atom was expected to be large. Both the Nagaoka and the Kelvin–Thomson models of the atom were based on the supposed existence of many electrons in the atoms of all the chemical elements.

It was J. J. Thomson who opened the way to determining the number of electrons in atoms.[2] He suggested three separate lines along which the problem could be attacked. One method was based on the slowing down and dispersion of light in passing through matter. The slowing down is measured by the index of refraction n defined as the ratio of the speed of light in vacuum to that in the material and is the cause of the refraction of light by a prism. Dispersion is the variation of the index of refraction with frequency and is the cause of the separation of the colors when white light is refracted by a prism as observed by Newton. According to the electromagnetic theory of light, the electric field of the light wave causes charged particles in transparent materials to oscillate with the frequency of the wave. These oscillating charges emit radiation, and if they are uniformly distributed throughout the material, the radiation from them combines to interfere with the incident light wave, in effect slowing it down. The extent of this slowing is measured by the material's index of refraction. The response of an individual particle to this field depends on its binding to its equilibrium position in the material, and in particular on the natural frequency with which it oscillates about that position. The variation of the index of refraction with frequency offers information about the natural frequencies of the electrons in the material. The theory of this was described by P. Drude.[3] The total slowing effect is proportional to the number density of electrons in the material, and if the resonance frequencies of the electrons in an atom are known, the proportionality constant can be calculated. In this way a measurement of the index of refraction yields the number density of electrons in the material. If Avogadro's number is known, the number density of atoms in the material can be found, and combining this with the number density of electrons determined optically yields the number of electrons per atom. Thomson worked this out for monatomic gases. Monatomic gases other than helium were not well known, but by extrapolating the results to diatomic gases, of which many were well known, he found that the number of electrons per atom was somewhere near one per unit atomic weight of the material. This was enormously lower than had been expected.

The second method that Thomson suggested was based on a somewhat similar idea, using X-rays rather than visible light. The electrons in a material traversed by a beam of X-rays are made to oscillate at the frequency of the X-rays, just as before they oscillated at the optical frequency. The difference between the two cases is that X-ray frequencies are much higher than those of atomic spectral lines. This has two consequences. First, the binding of the electrons in the atoms does not appreciably affect their response to fields of frequency as high as that of X-rays; the X-ray field reverses itself before an electron has time to move far enough to notice the change in the atomic force acting on it, so that the electrons can be treated as free, in contrast with the optical case in which the binding plays an essential role. This free-electron approximation was recognized as good only for light elements because the heavy elements were known to emit characteristic X-rays of the frequency of the X-rays from the source used, and so the atoms of the heavy elements had to contain electrons bound tightly enough to emit at these high frequencies. Second, the wavelength of the radiation that these rapidly oscillating electrons emit was so short that there were not many electrons within a wavelength. This made the emissions by the separate electrons mutually independent so that they did not systematically interfere with one another to cancel radiation in directions other than that of the exciting beam, as was the case at optical frequencies. Instead of the X-ray beam just being slowed down like a beam of light, it was spread out. The radiation emitted sideways had an appreciable intensity and could be observed. Its intensity was expected to be proportional to the number of electrons in the atoms that scatter the X-rays. Thomson calculated the power scattered by each electron and in this way showed how to interpret data on X-ray scattering in terms of the number of electrons per atom.

A series of measurements of the power scattered from samples of various light elements was made by C. G. Barkla.[4] The interpretation of his results also depended on the value of Avogadro's number, as he had to know the number of atoms in his sample. He arrived at an estimate of one hundred to two hundred electrons per molecule of air, or about four to eight electrons per unit atomic weight of nitrogen. Several years later, by using improved values of the constants, Barkla reduced this to about one electron per two units of atomic weight, its present value.[5]

Thomson's third method involved measuring the ability of matter to absorb β-rays. The mechanism proposed for this process was much more speculative than the other two, and so I shall not describe it here. The result was similar to that of the first two.

This small number of electrons per atom posed a difficulty. How could a system of so few charges emit so many spectral lines of definite frequency? Thomson proposed that under the circumstances present in a flame or an electric discharge when spectral lines are emitted, the atom also interacts with other electrons and ions present in the vicinity. This would imply that when the atom was not in a flame or discharge, it would not absorb light of all the frequencies it emitted when it was under excitation. It was not then known whether or not this was true. It turned out to be true, but the explanation was not what Thomson suggested it might be. I shall discuss this later in connection with the more successful quantum models of the atom. In the richness of its spectrum, in its paucity of electrons, and in the seemingly perfect identity of all the atoms of a given element, the atom was presenting its investigators with a contradictory array of properties that defied explanation on the basis of the physics accepted at that period.

Notes

1. Hermann Helmholtz, *Die Tonempfindungen*, 4th ed. (1877). The English translation is *The Sensations of Tone*, trans. A. J. Ellis (New York: Dover, 1954), pp. 22, 34.
2. J. J. Thomson, *Philosophical Magazine* 11, 769 (1906).
3. P. Drude, *Lehrbuch der Optik* (Leipzig, 1900); English translation by C. R. Mann and R. A. Millikan, *The Theory of Optics* (1902) (New York: Dover, 1959).
4. C. G. Barkla, *Philosophical Magazine* 7, 543 (1904).
5. C. G. Barkla, *Philosophical Magazine* 21, 648 (1911).

Amplification

The method proposed by J. J. Thomson to count the electrons in an atom by measuring the scattering of X-rays by atoms was a seminal one. The beams produced by the great particle accelerators of high-energy physics provide the means of probing much smaller entities than the atoms that the early X-ray beams probed, but the manner of their use is the same as that proposed by Thomson.

The central idea is to calculate the amount of power scattered from a collimated beam of X-rays by a single free electron. A measurement of the power scattered by a sample can then be interpreted as yielding the number of electrons in the sample. If Avogadro's number is known, weighing the sample will give the number of atoms it contains, and so one can arrive at the number of electrons per atom. Thomson did the calculation assuming that the electrons were free. In the presence of the X-ray beam, an electron obeys the equation of motion

$$eE_0 \cos 2\pi vt - kx = m\frac{d^2x}{dt^2}.$$

The resulting periodic motion has the amplitude

$$A = \frac{eE_0}{4\pi^2 m(v_0^2 - v^2)},$$

where v_0 is the natural frequency of the bound electron. If $v \gg v_0$, the amplitude of motion hardly depends on the natural frequency. This requirement restricts the application of the method to the lighter elements because the heavier elements are known to have characteristic X-ray lines of high frequency.

The method of calculation uses the assumption that the fraction of the incident X-ray power scattered by an electron is very small. This permitted Thomson to neglect the effect of radiation when calculating the motion of the electron produced by the incident X-ray beam. Once he found this motion, he could evaluate the radiation that the electron undergoing this motion would emit.

The electric field of the X-ray beam at the location of an electron varies with time according to

$$E(t) = E_0 \cos \omega t, \qquad \omega = 2\pi v = 2\pi\frac{c}{\lambda}.$$

The equation of motion for the "free" electron is

$$m\frac{d^2x}{dt^2} = -eE.$$

The radiation emitted by the electron is directly proportional to the square of its acceleration and is given according to electromagnetic theory by the formula

$$S = \frac{2}{3c^3}\left(e\frac{d^2x}{dt^2}\right)^2 = \frac{2}{3}\frac{e^4E^2}{mc^3},$$

The intensity S_0 of the original beam is $(c/4\pi)E^2$, so the fraction of the incident energy per unit area scattered by the electron is given by

$$S = \frac{8}{3}\left(\frac{e^2}{mc^2}\right)^2 S_0.$$

The quantity in parentheses has the dimensions of a length and is called the classical radius r_0 of the electron. Its current value is 2.818×10^{-13} cm, so the scattering cross section is of the order of magnitude of the area presented to the incident beam by a sphere of radius r_0. The value of r_0 accepted by Thomson's group at this time was quite inaccurate but was of the right order of magnitude.

Electromagnetic waves are transverse, and so the electric field is perpendicular to the beam direction. This implies that even if the original X-ray beam is unpolarized, as it usually is, the X-rays scattered in a definite direction at right angles to the original beam direction will be polarized perpendicular to the scattering plane because they will arise from incident X-rays with their electric vector perpendicular to the scattering direction where $\sin^2 \theta$ is unity. (If this once-scattered beam is now scattered from a second target, the intensity of the rescattered X-rays will not be uniform in angle around the direction of the once-scattered beam because of its polarization. Barkla used this effect to confirm the electromagnetic wave nature of X-rays.) Thomson's proposal was not concerned with the polarization, but only with the intensity of the scattered radiation.

Assuming that the electrons in an atom scatter X-rays independently, the total power scattered from an atom containing n electrons is n times that amount. As soon as the number of atoms in the target was known, the number of electrons per atom could be found.

Obscure and Complicated Patterns in Spectra

Patterns in the measured wavelengths of some spectral lines began to emerge late in the nineteenth century. The process began in 1885 with Johan Jakob Balmer's discovery of a formula for the wavelengths of visible lines in the spectrum of hydrogen. This was mentioned in Chapter 2, "The Connection Between Matter and Radiation." At first Balmer fitted just four lines by his equation, but other lines in the hydrogen spectrum soon followed, and all fitted the formula with truly remarkable precision. The equation has a simpler form when written in terms of the reciprocal of the wavelength, the number k of waves per centimeter called the *wave number*. Written in terms of wave numbers Balmer's formula is

$$k_m = 109721 \left(\frac{1}{2^2} - \frac{1}{m^2} \right), \qquad m = 3, 4, \cdots.$$

Five years later it was generalized by Johannes Robert Rydberg to apply also to some of the lines in the spectra of the alkalis, lithium, sodium, potassium, and the like. Rydberg kept the essential feature of expressing the wave number of a line as the difference of two numbers, one of which depends on an integer variable m, so that successive lines in a series have wave numbers corresponding to successive integers:

$$k_m = n_0 - \frac{R}{(m + p)^2}.$$

The numerator R of the fraction depending on the variable integer m is the same as the numerator of the variable fraction in Balmer's formula and is now known as the *Rydberg*, but a constant p now called the *quantum defect* was added to m. The constant term n_0 is not written in terms of an integer. Because as $m \to \infty$ the wave number approaches n_0, this quantity is called the *series limit*. The lines of a series obviously arose from a common source, the atoms of an element, but lines of frequencies given by this kind of formula cannot be emitted by any one periodic motion of particles within the atom. These formulas suggest no general physical principle governing the relation between matter and radiation.

The analysis of spectroscopic data led to more detailed relations among the wave numbers of spectral lines of the elements, particularly the alkalis, which had relatively simple spectra. There is no need to go into detail about these developments, but a sample of the kind of numerology that was done illustrates the hard and persistent efforts needed to get anywhere in this field rich in data and lacking any clear theoretical guideposts. I take as my sample the examples selected by Lorentz in his *Theory of Electrons* and from the work of Rydberg and of H. Kayser and C. Runge cited there.[1]

Three series were identified in the alkalis, called the *principal series*, the *diffuse series*, and the *sharp series*, all of whose lines were double, like the famous pair of yellow D-lines in the spectrum of sodium. Rydberg showed that the wave numbers of the two components of the principal series can be given by formulas

$$k_m = n_0 - \frac{R}{(m + p_1)^2} \quad \text{and} \quad k_m = n_0 - \frac{R}{(m + p_2)^2},$$

differing only in the values of the quantum defect added to the index m. The series limit is the same for both the upper and the lower lines of the doublets. The formulas for the two components of the sharp series were found to be

$$k_m = n_0' - \frac{R}{(m + s)^2} \quad \text{and} \quad k_m = n_0'' - \frac{R}{(m + s)^2},$$

and those for the diffuse series were

$$k_m = n_0' - \frac{R}{(m + d)^2} \quad \text{and} \quad k_m = n_0'' - \frac{R}{(m + d)^2}.$$

These look much like the formula for the principal series but differ in an essential way. In the sharp and diffuse series, the two members of a doublet share the same quantum defect but have different series limits, and these series limits in turn are shared between the two series!

After the discovery of these obscure patterns Rydberg and A. Schuster noticed independently that the difference between n_0' and n_0 was the wave number of one component of the first line in the principal series, that the difference between n_0'' and n_0 was the wave number of the other component of this first line, and further that the series limit of the principal series had the value $R/(1+s)^2$. This permitted the principal series to be written in the form

$$k_m = \frac{R}{(1+s)^2} - \frac{R}{(m+p_{1,2})^2} \qquad m = 2, 3, 4, \cdots$$

with similar expressions for the other series.

The discovery of these regularities in the hydrogen and in the alkali spectra was made possible, though not easy, by their relative simplicity. In this connection Lorentz remarked:

> It is only for a comparatively small number of chemical elements, that one has been able to resolve the system of their spectral lines, or at least the larger part of them, into series of the kind we have been considering. In the spectra of such elements as gold, copper and iron, some isolated series have been discovered, but the majority of their lines have not yet been disentangled. Nevertheless, it cannot be denied that we have made a fair start towards the understanding of line spectra, which at first sight present a bewildering confusion. There can be no doubt that the lines of a series really belong together, originating in some common cause, and that even different series must be produced by motions between which there is a great resemblance.
>
> The similarity of structure in the spectra of elements that resemble each other in their chemical properties, is also very striking. The metals in whose spectra the lines are combined in pairs are all monovalent, whereas the above series of triplets belong to divalent elements. Perhaps the most remarkable of all is the fact that Rydberg was able to represent all series, whatever be the element to which they belong, by means of formulae containing the same number N_0 [denoted above by R]. This equality, rigorous or approximate, of a constant occurring in the formulae of the different elements, must of course be due to some corresponding equality in the properties of the ultimate particles of which these elements consist, but at present we are wholly unable to form an idea of the nature of this similarity, or of the physical meaning of the length of time corresponding to $1/N_0$.

In the second edition of these lectures issued in 1915 Lorentz added a footnote referring to Bohr's 1913 paper "On the Constitution of Atoms and Molecules," without, however, describing it as the desired explanation of these regularities.

Walther Ritz made the generalization that provided the framework for analyzing these more complex spectra, the *Ritz Combination principle*. He stated that the wave numbers of the spectral lines of the elements are always the differences between two *spectral terms*, each of which depends on an integer. This simplified things a great deal because the number of terms required is much smaller than the number of spectral lines to be accounted for. For example, if there are ten terms, there will be forty-five ways of selecting them two at a time, and so ten terms could account for forty-five spectral lines. Not all pairings of terms correspond to lines of detectable intensity, but even so it was found that the number of terms needed to specify a spectrum was much smaller than the number of lines present in that spectrum. The physical meaning of this regularity was totally obscure. No classical model led to a dependence like this on integers, and there was no way seen to account for the emission of more than one spectral line by one electron. These regularities did nothing to help reconcile the wealth of spectra with the small number of electrons in an atom.

Even the Zeeman effect that had provided such welcome support for Thomson's identification of the electron as a universal constituent of atoms turned out to be incomprehensible. Lorentz had shown how a spectral line would be expected to split into three components when the emitting atom was in a magnetic field, even correctly predicting some of the polarization properties of these components. As long as the effect was observed as a broadening of the line, the polarization predictions of Lorentz's theory agreed with what was seen. When, however, the resolving power of spectroscopes became greater, it was found that the D-lines of sodium did not split into triplets. Rather, the lower-frequency component split into four lines and the higher-frequency one into six. This required more terms for its description than would have been needed if the expected triplet had been there. Other elements whose spectra showed a Zeeman effect had even more complicated splittings. The terms accounting for these lines were given labels, but what these labels signified about the physics of the atom remained a mystery until Bohr's proposed model of the hydrogen atom began to make some sense of it, although even it did not explain the Zeeman effect.

The first decade of the twentieth century was awash with extraordinarily precise data on spectra. Henry A. Rowland in Baltimore and John A. Brashear in Pittsburgh had provided the spectroscopists with diffraction gratings of marvelous precision leading to wavelength determinations good to six or seven figures. Johannes Stark had shown that atoms radiating while in a strong electric field also radiated lines that were split into characteristic patterns distinct from those produced by magnetic fields. Yet none of these wonderful data led to a picture of what was going on within atoms to account for the emission of spectral lines with these characteristics.

Notes

1. J. R. Rydberg, *Svenska Vetensk. Akad. Handl.* **23** (1889); and *Rapports présenté au Congrès de physique* **2**, 385 (1900); H. Kayser and C. Runge, *Annalen der Physikalische Chemie* **41**, 302 (1890); **43**, 385 (1891).

2. H. A. Lorentz, *The Theory of Electrons* (New York: Dover, 1952), p. 106.

The Atom Acquires New Aspects: Radioactivity

The discovery of radioactivity by Henri Becquerel in 1896 was a direct outgrowth of the discovery of X-rays by Röntgen in the previous year. X-rays were produced in gas discharge tubes such as those used in cathode ray experiments. Their outstanding characteristic was their ability to penetrate matter and to register on a photographic plate. Their production mechanism was unknown, but it was thought to be possibly associated with the fluorescence of the glass envelope of the discharge tube when struck by the cathode rays. This led to investigations of other fluorescent and phosphorescent materials. While pursuing an investigation of this kind, Becquerel exposed a uranium-containing phosphor lying on a carefully wrapped photographic plate to the sun and found on developing it that the plate had been exposed. It showed the shape of the phosphor and also the shape of a small piece of copper that had been placed between the phosphor and the plate. This positive effect was duly reported. The production of the radiation that exposed the plate was taken to be a consequence of the exposure of the phosphor to the sun. Another carefully wrapped photographic plate and associated uranium containing phosphor were prepared for exposure to the sun, but because of cloudy weather lasting some days, it was not exposed. For some reason the plate was developed anyway and, to Becquerel's surprise, showed the shape of both the phosphor and the copper that shielded the photographic plate from the phosphor just as clearly as the earlier one had. The effect could have been due to an extraordinarily prolonged phosphorescence, but Becquerel doubted this. Further experiments soon showed that the effect had nothing to do with phosphorescence but did have to do with a variety of compounds of uranium. Finally an experiment with metallic uranium definitely showed that the effect was caused by that element.

In his first studies of the radiation emitted by uranium, Becquerel showed that the radiation would cause the discharge of a charged electrometer, as do X-rays. He also believed that he had shown the rays to be both refracted by glass prisms and polarized by passing them through polarizing crystals such as tourmaline. This distinguished them from X-rays that had been neither refracted nor polarized at that time. The truly remarkable feature of this *Becquerel radiation* was its lack of a need for excitation. X-rays require a gas discharge for their formation, and their production ceases the moment the gas discharge is turned off. Phosphorescence does persist after the source of excitation, usually exposure to light or to cathode rays, is turned off, but only for a brief time. (We all are familiar with the lingering dim glow of a television tube or a computer screen after the instrument has been switched off.) Becquerel's radiation persisted for months (and years, but this could not be verified quickly) without any perceptible decrease in its intensity. This caused consternation because of the apparent violation of the first law of thermodynamics, the conservation of energy.

Soon after the discovery of the activity of uranium, a similar effect was found to be associated with thorium by Marie Curie and independently by G. Schmidt in Germany. They found this by noting the ability of a thorium compound to cause the discharge of an electroscope. M. Curie* also established the proportionality between activity as measured by the ability to discharge an electroscope and the amount of uranium present, independent of the chemical compound containing the uranium. Having established this, she was in a position to appreciate the observation that certain minerals containing uranium were more active than could be accounted for by their uranium content. This led eventually to the discovery of polonium by Marie and Pierre Curie and then of radium by the Curies and Gustav Bémont. These materials are much more radioactive than uranium.

Shortly after the discovery of radioactivity, Rutherford began the experimental career that formed the main line of progress in the subject. He had prepared for this as a student of J. J. Thomson at Cambridge, where he worked on the effect of ultraviolet light in producing the discharge of an electrometer. Their picture was that these radiations produced ions in the gas and these ions moved under the influence of the electric field of the charge on the electrometer, constituting a current that discharged it. A hint of things to come first appeared in a paper Rutherford wrote while still a student at Cambridge:

> In order to obtain a discharge with ultra-violet light, the light must fall on a negatively electrified surface. There is no discharge produced by allowing the light to fall between two plates without impinging on either. In this respect the action of ultra-violet light is very different to Röntgen and uranium radiation, which produce a volume ionization of the gas through which they pass. . . . The result of Wilson that a cloud is formed in moist air with strong ultra-violet light renders it possible that there is a slight volume ionization of the gas through which the light passes, but the effect appears to be too small to be determined by electrical means.[1]

In this short introduction to a paper "The Discharge of Electrification by Ultra-violet Light" he brought in the photoelectric effect, the ability of the newly discovered uranium rays to create ions in a gas, and the effect that soon led to the invention of the cloud chamber, all in connection with his problem of how ionized gases conducted current.

Two years later, as the McDonald Professor of Physics at McGill University, Rutherford published a paper "Uranium Radiation and the Electrical Conduction Produced by It."[2] Early in this paper Rutherford described his repetition of Becquerel's experiments that purported to show that uranium rays could be refracted and polarized just as light could. Rutherford could find no evidence for either effect, and Becquerel acknowledged his mistake. As a preliminary to discussing the details of the conduction produced in gases by this radiation, Rutherford investigated the nature of the radiation itself. To do this he determined the rate of leak of charge from a

---

* In the case of the Curies, one should note that a citation of M. Curie in the French literature probably means Monsieur Curie and not Marie Curie. Here it means Marie Curie.

pair of zinc plates charged to a potential difference of fifty volts when a uniform layer of powdered metallic uranium or compound of uranium was placed on the lower plate. He then measured the change in the rate of leak when successive thin metal films were put over the uranium, reducing the ionization of the gas and therefore the current. The first set of foils produced the kind of diminution one gets when passing light through tinted glass; each thickness of the foil reduced the leakage current to a certain fraction of its previous value. Then suddenly the addition of more foils produced very little effect. Whatever radiation got through the first foils got through the rest with hardly any loss!

Rutherford's conclusions from this series of measurements were as follows:

> These experiments show that the uranium radiation is complex, and that there are present at least two distinct types of radiation—one that is very readily absorbed, which will be termed for convenience the α radiation, and the other of a more penetrative character, which will be termed the β radiation.

His "terms of convenience" have stuck. The γ radiation came along the next year, a very penetrating and magnetically undeviable radiation identified by P. Villard.[3] Already in this first announcement of the existence of these two distinct types of uranium radiation, Rutherford discovered:

> The amount of the α radiation depends chiefly on the surface of the uranium compound while the β radiation depends also on the thickness of the layer. The increase of the rate of leak due to the β radiation with the thickness of the layer indicates that the β radiation can pass through a considerable thickness of the uranium compound. Experiments showed that the leak due to the α radiation did not increase much with the thickness of the layer. I did not, however, have enough uranium salt to test the variation of the rate of leak due to the β radiation for thick layers.
>
> The rate of leak from a given weight of uranium or uranium compound depends largely on the amount of surface. The greater the surface, the greater the rate of leak. A small crystal of uranium nitrate was dissolved in water, and the water then evaporated so as to deposit a thin layer of the salt over the bottom of the dish. This gave quite a large leakage. The leakage in such a case is due chiefly to the α radiation.

Similar experiments were done at about the same time by M. and P. Curie, Becquerel, S. Meyer and E. von Schweidler, F. Giesel, and P. Villard.

In "Uranium Radiation" Rutherford speculated that the "apparently very powerful radiation obtained from pitchblende by Curie may be partly due to the very fine state of division of the substance rather than to the presence of a new and powerful radiating substance." He did not consider this a possibility for long. Shortly afterward, P. Curie showed that the radiation from radium was also complex and consisted of an easily absorbed part and a more penetrating part.

Hearing of the radiation of thorium while doing the experiments on uranium radiation, Rutherford compared the radiations of the two elements. Both radiations were complex, and the α-radiation from thorium was much more penetrating than that from uranium, requiring between three and four foils to give the same reduction in leakage that one foil gave for uranium.

In the balance of this long paper Rutherford studied primarily the properties of the ionized gas, the rate of recombination of the ions, their drift velocity in electric fields, and other topics of importance to the study of gas discharges. In the last section, entitled "General Remarks," he wrote:

> The cause and origin of the radiation continuously emitted by uranium and its salts still remain a mystery. All the results that have been obtained point to the conclusion that uranium gives out types of radiation which, as regards their effect on gases, are similar to Röntgen rays and the secondary radiation emitted by metals when Röntgen rays fall upon them. If there is no polarization or refraction the similarity is complete.

This was the situation some three years after the discovery of radioactivity by Becquerel. The nature of the X-rays was as yet uncertain, and the electron had been identified as a constituent of atoms for only about two years. An enormous new area of research had opened up, its various aspects being connected by the means available to study them, principally the production of electrical conductivity in gases and secondarily the ability of the radiations to expose photographic plates. These means were exploited to the full over the next few years.

The next great advance was hinted at when Rutherford and R. B. Owens did some comparisons between uranium and thorium radiation.[4] The radiations were of the same general kind, the thorium α-radiation being more penetrating, as noted earlier. What was strikingly different was the variability of the strength of the radiation from thorium, especially thorium oxide, in contrast with the constancy of the strength of the radiation from uranium. They reported:

> It was found that if the substance was inclosed in a lead box with a door, the rate of leak was much slower with the door open than closed. The addition of a slight draught of air caused by opening or shutting the door to the room diminished the rate of leak still more. Under similar conditions the rate of leak due to the sulphate and nitrate of thorium and the uranium compounds is not appreciably affected. . . . With the air quite still, the substance in a few minutes regained its normal activity. The recovery was quite gradual. . . . It appears as if in the pores of the thick layer of thorium oxide some changes takes place with time, which increases the intensity of the radiation, and if the result of the action is continually removed, the intensity of the radiation is diminished. This would explain why the action is shown chiefly in thick layers, and depends on the current of air.

This preliminary note was followed a few months later by the announcement of "A Radioactive Substance Emitted from Thorium Compounds." The paper begins:

> It has been shown by Schmidt that thorium compounds give out a type of radiation similar in its photographic and electrical actions to uranium and Röntgen radiation. In addition to this ordinary radiation, I have found that thorium compounds continuously emit radioactive particles of some kind, which retain their radioactive powers for several minutes. This "emanation," as it will be termed for shortness, has the power of ionizing the gas in its neighbourhood and of passing through thin layers of metals, and, with great ease, through considerable thicknesses of paper.... The fact that the effect of air currents is only observed to a slight extent with thin layers of thorium oxide is due to the preponderance, in that case, of the rate of leak due to the ordinary radiation over that due to the emanation. With a thick layer of thorium oxide, the rate of leak due to the ordinary radiation is practically that due to a thin surface layer, as the radiation can only penetrate a short distance through the salt. On the other hand, the "emanation" is able to diffuse from a distance of several millimetres below the surface of the compound, and the rate of leak due to it becomes much greater than that due to the radiation alone.[5]

The idea of the transmutation of the elements was not yet being advanced, but we can see the groundwork for it being laid in these careful and imaginative investigations of what might have been taken for irrelevant nuisances. Further on in the paper Rutherford enumerated the properties of the emanation:

> The emanation passes through a plug of cotton-wool without any loss of its radioactive powers. It is also unaffected by bubbling though hot or cold water, weak or strong sulphuric acid. In this respect it acts like an ordinary gas. An ion, on the other hand, is not able to pass through a plug of cotton wool, or to bubble through water, without losing its charge.

That is, the emanation retained its ability to produce ions in a gas after passing through the plug, but the ions produced could not get through. The idea of a radioactive gas produced by radium was not accepted right away by everyone working in the area. In her thesis of 1903 Marie Curie wrote:

> Mr. Rutherford suggests that radioactive bodies generate an *emanation* or gaseous material which carries the radio-activity. In the opinion of M. Curie and myself, the generation of a gas by radium is a supposition which is not so far justified. We consider the emanation as radio-active energy stored up in the gas in a form hitherto unknown.[6]

In his paper Rutherford reported that the rate of discharge due to the emanation varied with the time in a way that was simple to describe mathematically:

> When a thick layer of thorium oxide, covered over with several thicknesses of paper, is placed inside a closed vessel, the rate of discharge due to the emanation is small at first, but gradually increases, until after a few minutes a steady state is reached.
>
> These results are to be expected, for the emanation can only slowly

diffuse through the paper and the surrounding air. A steady state is reached when the rate of loss of intensity due to the gradual decay of the radioactivity of the emanation is recompensated by the number of new radioactive centres supplied from the thorium compound.

Let n = number of ions produced per second by the radioactive particles between the plates.

Let q = number of ions supplied per second by the emanation diffusing from the thorium.

The rate of variation of the number of ions at any time t is given by

$$\frac{dn}{dt} = q - \lambda n$$

where λ is a constant.

The results . . . show that the rate of diminution of the number of ions is proportional to the number present.

This was the first appearance of the law of radioactive decay. It was somewhat disguised by being applied to the number of ions between the plates of his electrical detector rather than to the number of atoms or molecules of emanation present there, but the direct proportion between these had already been noted.

At the very end of this paper Rutherford wrote that he had found "that the positive ion produced in a gas by the emanation possesses the power of producing radioactivity in all substances on which it falls." This induced radioactivity lasted several days, although the activity of the emanation itself decayed to half in about a minute and was more penetrating than either thorium or uranium rays.

Rutherford and his associates at McGill University spent the years 1901–2 pursuing the study of radioactive substances and X-rays using what Rutherford called the *electrical method*, the production of electrical conductivity in gases. In particular, Rutherford and F. Soddy investigated the radioactivity of thorium and the nature of the thorium emanation. They published important results of this work in two papers, "Radioactivity of Thorium Compounds I and II."[7] In the first of these they considered the following possibilities for the nature of the emanation:

1. It was a radioactive gas.
2. It was an atmospheric gas with induced radioactivity produced by the thorium.

Noting that the emanation survived passage over or through an extremely wide variety of chemical reagents, the authors remarked:

> It will be noticed that the only known gases capable of passing in unchanged amount through all the reagents employed are the recently discovered gases of the argon family.

The induced radioactivity alternative was ruled out in a series of experiments in which carbon dioxide was substituted for air in blowing over thoria, or thorium oxide. The carbon dioxide was then mixed with air and removed from the mixture by passing it over soda lime. They found:

> The amount of emanation found was quite unchanged, whether carbon dioxide was sent over thoria in the manner described, or whether an equally rapid current of air was substituted for it, keeping the other arrangements as before. The theory that the emanation may consist of the surrounding medium rendered radioactive is thus excluded, and the interpretation of the experiments must be that the emanation is a chemically inert gas analogous in nature to the members of the argon family.

Another important result was the establishment of the existence of ThX, a substance chemically distinct from thorium, which was responsible for most of the radioactivity of thorium and its compounds. Similar work had been done with uranium using the photographic method by Sir William Crookes, who showed that photographically inactive uranium could be obtained by a single chemical separation of what he called *Uranium X*. After a year the uranium was seen to have recovered its activity. What Rutherford and Soddy could do using the electrical method was to measure the changes in the activity of the radioactive material with time. They showed that these changes could be accurately described by assuming a constant rate of production of ThX from thorium, and a decay of the ThX at a rate proportional to the amount of it present. This was a refinement of Rutherford's discussion of the rate of diminution of the number of ions produced by thorium emanation that was described earlier. The mechanism of this production of ThX was not clear; the thorium left after the separation of ThX was still somewhat active, but this activity was not identified as the source of the ThX. The ThX was the source of the emanation that had been observed earlier.

In the second of the two papers on the "Radioactivity of Thorium Compounds" Rutherford and Soddy announced their breathtaking conclusion that radioactivity involved the spontaneous change of an atom of one element into an atom of another element. Here they pointed out the complications arising from energy conservation:

> Energy considerations require that the intensity of radiation from any source should die down with time unless there is a constant supply of energy to replace that dissipated. This has been found to hold true in the case of all known types of radioactivity with the exception of the "naturally" radioactive elements, to take the best established cases, thorium, uranium, and radium. . . . In the case of the three naturally occurring radioactive elements, however, it is obvious that there must be a continuous replacement of the dissipated energy, and no satisfactory explanation has yet been put forward to account for this.

> The nature of the process becomes clear in the light of the foregoing results. The material constituent responsible for the radioactivity, when it is separated from the thorium which produces it, behaves in the same way as the other typically radioactive substances cited. Its activity decays geometrically

with the time, and the rate of decay is independent of the molecular conditions. The normal radioactivity is, however, maintained at a constant value by a chemical change which produces fresh radioactive material at a rate also independent of the conditions. The energy required to maintain the radiations will be accounted for if we suppose that the energy of the system after the change has occurred is less than it was before.

Later in the paper they came to the crux of the matter:

The present researches had as their starting point the facts that had come to light with regard to the emanation produced by thorium compounds and the property it possesses of exciting radioactivity on surrounding objects. In each case, the radioactivity appeared as the manifestation of *a special kind of matter* in minute amount. The emanation behaved in all respects like a gas, and the excited radioactivity it produces as an invisible deposit of intensely active material independent of the nature of the substance on which it was deposited, and capable of being removed by rubbing or the action of acids.

The position is thus reached that radioactivity is at once an atomic phenomenon and the accompaniment of a chemical change in which new kinds of matter are produced. The two considerations force us to the conclusion that radioactivity is a manifestation of sub-atomic chemical change.

Here was the end of the atom as an immutable smallest piece of a chemical element. The discovery of the existence of electrons in atoms had already shown that atoms have some structure and therefore can be changed in some way, but this had not broken the unique association of an atom with a chemical element. Rutherford and Soddy broke this association; atoms could change spontaneously from being atoms of one element into being atoms of another element!

In a pair of papers, "The Cause and Nature of Radioactivity," published later that same year of 1902, they reviewed the work that pertained to the radioactivity not only of thorium but also other radioactive elements, such as uranium and radium, coming to the same conclusions they had reached for thorium.[8]

This conclusion was restated emphatically in a paper they published the next year:

The law of radioactive change, that the rate of change is proportional to the quantity of changing substance, is also the law of monomolecular chemical reaction. Radioactive change, therefore, must be of such a kind as to involve one system only, for if it were anything of the nature of a combination, where the mutual action of two systems was involved, the rate of change would be dependent on the concentration, and the law would involve a volume-factor. This is not the case. Since radioactivity is a specific property of the element, the changing system must be the chemical atom, and since only one system is involved in the production of a new system and, in addition, heavy charged particles, in radioactive change the chemical atom must suffer disintegration.

The radio-elements possess of all elements the heaviest atomic weight.

This is indeed their sole common chemical characteristic. The disintegration of the atom and the expulsion of heavy charged particles of the same order of mass as the hydrogen atom leaves behind a new system lighter than before, and possessing chemical and physical properties quite different from those of the original element. The disintegration process, once started, proceeds from stage to stage with definite measurable velocities in each case.[9]

Rutherford and Soddy pointed out the similarity of the radioactive decay law, the proportionality of the number of decays per unit time to the number of undecayed atoms present, to the rate of a "monomolecular chemical reaction." The rates of chemical reactions are governed by the probability of the necessary encounters between the reacting molecules, and these encounters are random, their frequency depending on the concentrations of the various kinds of atoms. The rate of a monomolecular chemical reaction is governed by the interaction of the reacting molecules with an environment such as a solvent that is not affected by the reaction and so is simply proportional to the concentration of the reacting matter. In the chemical case there is a visualizable random element present, such as the thermal motion of the molecules of the solvent. The radioactive rate was seen, however, to be independent of the chemical environment of the decaying atom, so here no external source of random influences on the reaction was present. This heralded the appearance of "chance" at a new level in science. Rutherford's law of radioactive decay was later used by Einstein in connection with the emission of light by atoms, as described in Chapter 8. Only with the development of quantum mechanics did it fit into a consistent theoretical framework.

It was still not clear that the first step in the chain of processes suggested by Rutherford and Soddy was also a radioactive transformation, one with a very long half-life. There was speculation in Europe that the energy came from an as yet unnoticed radiation from the sun. Julius Elster and Hans Geitel compared the activity of uranium at the surface of the earth and at the bottom of an 850-meter mine shaft, finding them the same. Evidently 850 meters of earth did not shield against this radiation if it existed. The Curies compared the activity at midday and at midnight to see whether the diameter of the earth would be sufficient to shield against it. They found no difference.

The identification of the radiations that signaled the transformation of atoms of one kind to another was incomplete. The β-rays had been identified with the cathode rays quite early. Their ability to be deviated by a magnetic field had been observed by several investigators, starting with Giesel and including Meyer and von Schweidler, Becquerel, and P. Curie. Becquerel estimated the value of e/m by observing the deviation of a pencil of rays in electric and magnetic fields. The β-rays have a wide range of velocities, so that a pencil of rays is spread out in either kind of field, making precise measurements difficult, but it was clear that the value of e/m was of the same general size as that of the cathode rays and not that of even the least massive atom, the hydrogen atom, as determined by electrolysis. That the β-rays from radium carried a negative charge away from the host material was shown by the Curies.[10] This established that the charge on the rays was there at their expulsion from the atom and was not acquired during their flight.

The α-rays were harder to study. Because they were easily absorbed, they would travel only a few centimeters through air. They were thought not to be deviated by a magnetic field; their deviation, if any, was much less than that of the β-rays. Sources of α-rays had to be thin to prevent most of the radiation from being absorbed in the source, and so it was hard to get a strong beam of the particles from a radioactive source. Rutherford overcame these difficulties in an ingenious experiment.[11]

The α-rays from a radium source placed underneath the set of metal plates (shown in Fig. 7.2) passed up between the plates and ionized the gas in the test chamber containing a goldleaf electroscope. The gas was hydrogen, a stream of which was kept flowing down through the plates to carry away the emanation emitted by the radium so that its radiation would not mask that due to the α-rays from the radium itself. By applying the strongest magnetic field available in his laboratory Rutherford was able to cut the ionization in the electroscope by 30 percent, which he interpreted as indicating that 30 percent of the α-particles struck the plates as a consequence of following a curved path between the plates. When he used a generator magnet in the electrical engineering laboratory, he was able to deviate all the rays.

This experiment showed that the rays were deviated by a magnetic field, but it did not determine the sign of their charge. Rutherford then put a lip on one side of the top edge of the plates. If the rays were deviated toward the lip, fewer of them would be able to enter the test chamber than when they were deviated away from the lip. Reversing the direction of the magnetic field then told him which sign of charge they carried. It was positive.

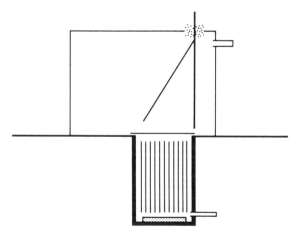

Fig. 7.2 Rutherford's device for detecting the very small deviation of the α-rays of radium produced by the then available magnetic field.

Next Rutherford insulated the plates from each other and applied a potential difference between adjacent plates. In this way he achieved an electrostatic deviation of the rays. Combining these results he obtained an estimate of their speed of 2.5×10^9 centimeters per second, a tenth of the speed of light, and a rough value for the charge to mass ratio of the α-particles,

$$\frac{e}{m} = 6 \times 10^3 \text{ emu per gram.}$$

This was much closer to the values found for canal rays than to the value for the cathode rays. The α-particles were seen to be of atomic mass.

The e/m experiments were refined and repeated for the α-rays from a variety of sources. The velocities were found to vary from one radioactive material to another, but the values of e/m were always the same within the experimental uncertainty, namely, about half the value for the hydrogen atom, or 5070 emu per gram. This made it seem that the α-particles from all sources were alike and that each was either a helium atom with two positive charges or a hydrogen molecule with a single positive charge, that is, a molecular ion, assuming that all the charges were positive or negative multiples of the electron charge. The fact that Curie had found helium associated with radium made helium the more likely possibility. Up to this point no effects produced by single α-particles had been identified. The particles were detected by observing the deflection of an electroscope as the ionization produced by the particles allowed it to discharge.

Two detectors were soon devised that were sensitive to individual α-particles. The first was the Geiger counter, the second the spinthariscope. The Geiger counter, developed by Hans Geiger in Rutherford's laboratory, took advantage of the ability of ions formed in a gas in a strong electric field to acquire enough energy from the field to produce more ions.[12] In this way a big enough avalanche of ions could be produced by a single α-particle passing through a gas in the neighborhood of a highly charged wire where the electric field was high, to be detected by the deflection of an electroscope. In its original form the Geiger counter could be used only with very feeble sources, as the response time of the electroscope was relatively long, and to distinguish the effect of individual particles they had to be produced at intervals longer than the response time. The counter was bulky, the chamber being about 20 cm long and almost 2 cm in diameter. The central wire was charged from a bank of batteries providing 1320 V. According to Rutherford:

> If a strong electric field acts on a gas at a low pressure, the number of ions produced in the gas by an external source is greatly increased by the movement of the ions in the electric field. If the voltage applied is near the sparking value, the ionisation current may in this way be increased more than a thousand times. In the experiments of Rutherford and Geiger to detect a single α particle, it was arranged that the α particles were fired through a gas at low pressure, exposed to an electric field somewhat below the sparking value. In this way, the small ionisation produced by one α particle in passing

along the gas could be magnified several thousand times. The sudden current through the gas due to the entrance of a single α particle in the detecting vessel, was by this method increased sufficiently to give an easily measurable deflection to the needle of an ordinary electrometer.... The intensity of the source and its distance from the aperture were adjusted so that from 3 to 5 α-particles entered the detecting vessel per minute. The entrance of an α-particle in to the detecting vessel was signified by a sudden *ballistic* throw of the electrometer needle. By adjusting the potential, it was not difficult to obtain a throw of 50 to 100 mms. on the electrometer scale for a single α-particle.[13]

This was an epoch-making development because it formed the basis of counting methods in particle physics. The size and slowness of the detector limited its usefulness in its original form, but it really did show the discrete nature of the α-radiation, still questioned by a few at the time of this work, 1908. Another important consequence of the invention of the Geiger counter was the confirmation that the spinthariscope also counted individual α-particles.

Crookes had shown that the illumination produced by α-radiation on certain phosphors, in particular phosphorescent zinc sulfide, consists of a number of scintillating points of light distributed over the entire surface, each point emitting for only a short time interval. This effect had also been observed by Elster and Geitel. The mechanism was not understood, and it was not clear whether each flash of light represented the effect of one or of several α-particles. Crookes developed the spinthariscope to show this effect. It consisted of a small tube at one end of which he placed a small trace of radium a few millimeters away from a zinc sulfide screen and observed the scintillations through a lens at the other end. In Rutherford's words:

> In a dark room the screen is seen as a dark background dotted with brilliant points of light, which come and go with great rapidity. The experiment is extremely beautiful, and brings vividly before the observer the idea that the radium is shooting out a stream of particles, the impact of each of which on the screen is marked by a flash of light.

By comparing the number of scintillations produced by a given source under controlled conditions with the number of counts in a Geiger counter, Rutherford and Geiger showed that the number of scintillations produced corresponded exactly to the number of counts on the counter; one α-particle, one scintillation. This now gave a way of counting α-particles with good spatial resolution because a scintillator could be made very small in comparison with a Geiger counter. Rutherford used this with memorable results to begin a series of experiments on the scattering of α-particles.

Even before the development of counting methods, von Schweidler had pointed out an interesting consequence of the law of radioactive decay, stating that the rate of decrease in the number of radioactive atoms of a given kind is proportional to the number of them present.[14] He recognized that this was

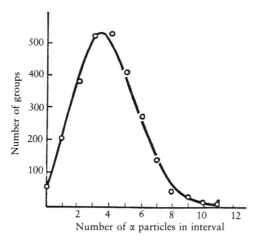

Fig. 7.3 Rutherford and Geiger's distribution of the number of α-particles detected in equal successive time intervals, showing the fluctuations to be expected from their random times of emission.

a statement of a probability law; each separate radioactive atom had a definite probability of decaying in unit time. There should, therefore, be fluctuations in the number of α-particles emitted per unit time from a radioactive material. As in the case of Brownian motion, discreteness of events leads to fluctuations in the results of many successive individual trials. These fluctuations were observed by Friedrich Kohlrausch. He balanced the ionization currents from two electrical detectors activated by two distinct radioactive samples and measured the fluctuations from exact balance. The possibility that these fluctuations were instrumental and not due to the discrete nature of the decay process was eliminated by Geiger, who showed that the fluctuations in balance of the ionization currents from two detectors were much smaller when the detectors were excited by the same beam of α-particles passing through both than when separate α-particle beams activated them.[15]

When individual particles could be counted, it became possible to do a much more detailed study of these fluctuations. Such a study was done by Rutherford and Geiger,[16] who based their analysis on Harry Bateman's working out of the relevant probability theory. The number of particles observed in successive equal time intervals should have a Poisson distribution. The theoretical expectation and the experimental results are shown in Fig. 7.3. By this time the atomic picture of matter was so well established that these experiments were no longer needed to confirm it. They were important, however, to confirm the transformation theory of radioactivity.

Notes

1. E. Rutherford, *Proceedings of the Cambridge Philosophical Society* 9, pt. 8, p. 401 (1898). *Collected Papers of Lord Rutherford of Nelson*, ed. J. Chadwick (New York: Interscience, 1962), vol. 1, p. 149.

2. E. Rutherford, *Philosophical Magazine* 47, 109–63 (1899); *Collected Papers*, vol. 1, p. 169.

3. P. Villard, *Comptes rendus* 130, 1010, 1178 (1900).

4. E. Rutherford and R. B. Owens, *Transactions of the Royal Society of Canada* (1899); *Collected Papers*, vol. 1, p. 216.

5. E. Rutherford, *Philosophical Magazine* 49, 1 (1900); *Collected Papers*, vol. 1, p. 220.

6. Marie Curie, "Radio-active substances," *Chemical News* (London: Chemical News Office, 1904).

7. E. Rutherford and F. Soddy, *Transactions of the Chemical Society* 81, 321; 837 (1902); *Collected Papers*, vol. 1, pp. 376, 435.

8. E. Rutherford and F. Soddy, *Philosophical Magazine* (6) 4, 370, 569 (1902); *Collected Papers*, vol. 1, pp. 472, 495.

9. E. Rutherford and F. Soddy, *Philosophical Magazine* 5, 576 (1903); *Collected Papers*, vol. 1, p. 376.

10. M. and P. Curie, *Comptes rendus* 130, 647 (1900).

11. E. Rutherford, *Philosophical Magazine* 5, 177 (1903); *Collected Papers*, vol. 1, p. 549.

12. E. Rutherford and H. Geiger, *Proceedings of the Royal Society* A 81, 141 (1908); *Collected Papers*, vol. 2, p. 89.

13. E. Rutherford, *Radioactive Substances and Their Radiations*, 2nd ed. (Cambridge: Cambridge University Press, 1913), pp. 129–31.

14. E. von Schweidler, *Congrès international pour l'étude de la radiologie et de l'ionisation* (Liège, 1905).

15. H. Geiger, *Philosophical Magazine* 15, 539 (1908).

16. E. Rutherford and H. Geiger, *Philosophical Magazine* 20, 698 (1910); *Collected Papers*, vol. 2, p. 203.

The Atom Acquires a Nucleus; the Rutherford Model

One of the continuing motifs in Rutherford's research career was the α-particle. In 1899 he had named it. By 1903 he had shown the magnetic and electric deviation of the α-particles from radium and had arrived at an estimate of their charge-to-mass ratio, with a value of about 6×10^3 emu per gram. Assuming the charge to be the ionic charge measured by Townsend and Thomson, this led to an estimated mass of the α-particle of about twice that of the hydrogen atom. If the charge were $2e$ instead of e, the mass would be close to that of helium. Later that year, after Ramsay and Soddy showed that helium is produced from radium bromide and radium emanation, Rutherford stated:

> The determination of the mass of the α body, taken in conjunction with the

experiments on the production of helium by the emanation, supports the view that the α particle is in reality helium. In addition, the remarkable experiment of Sir William [Ramsay] and Lady Huggins in which they found that the spectrum of the phosphorescent light of radium consisted of bright lines, some of which within the limit of error were coincident with the lines of helium in the ultra-violet, strongly supports such a view. For as a consequence of the violet expulsion of the α particle it is to be expected that it would be set into powerful vibration and give its characteristic spectrum.[1]

Until 1905 no direct demonstration that the α-particles carry a positive charge had been successful, although attempts had been made by J. J. Thomson, Rayleigh, and Rutherford. The idea of the experiments was simple. A thin layer of α-emitting material was placed on the lower plate of a capacitor, and the space between the plates was evacuated to prevent ionization of the gas by the radiation. It was expected that a positive charge would accumulate on the upper plate from the α-particles crossing the gap between the plates. It proved impossible to be sure this positive charge did accumulate. The trouble was the presence of β-particles, presumably knocked out of the plates by the α-particles. Finally, Rutherford introduced a strong magnetic field parallel to the plates. This bent the paths of the β-particles into circles smaller than the interplate separation so they could not get from one plate to the other but returned to the plate they came from, therefore carrying no current across the gap. The α-particles were hardly affected. He then could see the accumulation of positive charge on the upper plate and could estimate the charge carried in unit time from a source of known strength, assuming that half the emitted particles went in the direction of the upper plate. He got a crude measure of the number of α-particles emitted by one gram of radium bromide per second, namely, 1.4×10^{11}, which agreed more or less with other estimates based on the value of e/m he had obtained.

By 1906 Rutherford had obtained a better measure of e/m for the α-particles from radium using electric and magnetic deflection, between 5.0 and 5.2×10^3 emu per gram.[2] This made it more definite that its value was half that for hydrogen, but he still considered three possible explanations of this. The α-particle might be

1. A *molecule* of hydrogen carrying the ionic charge of hydrogen.
2. A helium atom carrying *twice* the ionic charge of hydrogen.
3. *One half* of the helium atom carrying a single ionic charge.

Of these he concluded that the second was the most probable, but the others were not completely excluded by the available evidence. Taking it to be a helium atom, Rutherford was not clear whether it was expelled with a double ionic charge or whether it acquired it in its passage through matter. The knowledge of the structure of atoms was in no state to make the answer to this question obvious.

In this paper, Rutherford used the tentative identification of the α-particle with a helium atom, together with the amount of helium found in uranium-bearing rocks by Ramsay and Travers, to estimate the age of the mineral. He arrived at a minimum age of 400 million years, pointing out that any escape of helium from the material would make the estimate too low.

In 1907 Rutherford moved from McGill University to the University of Manchester. By the end of 1908 he had laid to rest any doubts he might have had in the past concerning the nature of the α-particle. With Geiger he measured the charge carried from a source consisting of radium C to a receiver attached to an electroscope. A magnetic field in the space between the source and the receiver served to prevent any β-particles from influencing the charge accumulated on the receiver. The number of α-particles emitted per unit weight of radium C had been determined earlier, and the number reaching the receiver was found, assuming that they were emitted uniformly in direction. Then the amount of charge arriving at the receiver per unit time from a known amount of radium was measured. The result arrived at was a charge per particle $q_\alpha = 9.3 \times 10^{-10}$ esu. This was almost three times the ionic charge measured by Thomson's group and over twice Millikan's value of 4.06×10^{-10}. Believing for various reasons that two was a more probable multiple than three, Rutherford went on to suggest strongly that the measurements of e involving cloud formation, as in the experiments of Townsend, Thomson, Wilson, and Millikan, were bound to give too small a value because of the evaporation of the drops. If the drops lost mass by evaporation, they would be lighter when weighted at the end of an experiment than earlier when their radius was determined by their rate of fall, as described in Chapter 3. This would lead to an overestimate of the number of drops in a cloud and so to too small a value for the charge. In this connection Rutherford pointed out that Planck had given a value of 4.69×10^{-10} esu from his blackbody radiation theory and that 4.69 is much closer than even 4.07 to $9.3/2 = 4.65$. This is the only citation of Planck's value up to this time that I have found. I rather doubt that Rutherford cared particularly about the reasoning behind Planck's value. He might have been aware of its existence by having been in Cambridge when the Townsend and Thomson measurements were being carried out, when it must surely have been discussed, if not appreciated.

The conclusion that the charge of the α-particle was $2e$ made its mass 3.84 on the scale of atomic weights, while that of helium was 3.96. This was a strong argument, but not conclusive.

The final proof came in the work of Rutherford and T. D. Royds. The glassblower at Manchester, Baumbach, succeeded in blowing glass tubes that had walls thin enough, about 1/100 millimeter, to permit the passage of fast α-particles but thick enough to contain helium under the experimental conditions. A small amount of radium emanation was introduced into the tube, which was surrounded by a thick-walled glass tube. Most of the α-particles from the emanation and its α-emitting decay products, radium A and, after a while, radium C, were able to go through the thin-walled tube but were stopped by the thick-walled one. Any helium appearing in the space between the two tubes could be detected spectroscopically. They reported:

> At intervals after the introduction of the emanation the mercury was raised
> [to compress any gas present] and the gases in the outer tube spectroscopically
> examined. After 24 hours no trace of the helium yellow line was seen; after
> 2 days the helium yellow was faintly visible; after 4 days the helium yellow
> and green lines were bright; and after 6 days all the stronger lines of the
> helium spectrum were observed. The absence of the neon spectrum showed
> that the helium present was not due to a leakage of air into the apparatus.[3]

The long times needed for any helium to be seen were due to the fact that it was formed in a thin layer on the inside of the outer tube's thick wall and had to diffuse out of the glass into the space where it could be compressed and its spectrum could be excited.

Several other possible sources of error were discussed and ruled out by the appropriate experimental checks. Rutherford and Royds concluded:

> These experiments thus show conclusively that the helium could not have
> diffused through the glass walls, but must have been derived from the α
> particles which were fired through them. In other words, the experiments
> give a decisive proof that the α particle after losing its charge is an atom
> of helium

That α-particles are scattered on passing through matter was well known. The sharpness of the edges of the image formed on a photographic plate by a collimated beam of α-particles was noticeably greater when the beam went through a vacuum than when it went through air. This scattering was studied by Geiger using the scintillation method.[4] A narrow beam of radiation from a radium C source was defined by a slit in an evacuated tube and went on to strike a zinc sulfide screen, producing a narrow bright band. Scattering foils could be introduced between the slit and the screen. In their presence the band was broadened out.

To study this quantitatively, the slit was replaced by a circular aperture, and the rate n of scintillation per unit area at various distances r from the center of the image on the scintillator was observed. The most probable angle of scattering was defined as the angle at which the quantity $2\pi rn$ had its greatest value. Typical values for gold foils were around $7°$. The angle increased strongly with the atomic weight of the scattering material and, for thin foils, as the square root of the thickness, but this became linear with greater thicknesses. The square root dependence is what would be expected if the observed scattering angle were the result of many small-angle scatterings in different directions with no appreciable slowing down of the particles. The linear dependence was explained as being due to the slowing down of the particles with a consequent increase in scattering.

In the course of some preliminary work on α-particle scattering, Ernest Marsden, a Hatfield Scholar at Manchester, observed a few scintillations representing possible scatterings of α-particles through angles greater than $90°$. This was not to be expected. Such large angles could not be the result of many small-angle scatterings; the probability of enough of these being in the same direction to produce such a large resultant was totally negligible. If this large-angle scattering were real, it would have to involve single scatterings through large angles. Such scatterings could not be accounted for by any current model of atoms.

The only force that was known to act on α-particles was the electromagnetic force, primarily the coulomb force. In order to have single scatterings through large angles, two things would be required, a very strong electric field and a body much heavier than the α-particle with which this field was associated. The electrons in an atom could not deflect such a heavy projectile very far from its path. They would simply be knocked out of the road like table-tennis balls by a bowling ball. There might be many α-particle–electron collisions, but they would cause minute deflections of an α-particle in random directions that would tend to cancel out, leaving only small fluctuations about the original direction of motion. The chance of the effects of many collisions accumulating into a large-scale scattering was extraordinarily small. The large positive charge that neutralized the charges of all the electrons was, indeed, associated with the mass of the whole atom, but in the Thomson model, it was spread out over the entire volume of the atom and therefore could not produce a strong enough field at any point to cause this scattering. The reality of this large-angle scattering had to be established, and if it turned out to be real, its characteristics had to be studied in detail.

Geiger and Marsden did this and reported their results in 1909 in a paper entitled "On a Diffuse Reflection of the Alpha Particles." They established the effect as real.[5] Using very thin gold foils as the reflector, they found that the reflectivity increased with the thickness up to a point, which showed that the reflection was not a surface effect like the reflection of light from a sheet of glass. This confirmed the idea that the reflection was due to scattering, but the thickness of foil in which this scattering took place had to be very small:

> If the high velocity and mass of the α-particle be taken into account, it seems surprising that some of the α-particles, as the experiment shows, can be turned within a layer of 6×10^{-5} cm. of gold through an angle of 90°, and even more. To produce a similar effect by a magnetic field, the enormous field of 10^9 absolute units would be required.

Rutherford described the effect in this way:

> It was almost as incredible as if you fired a fifteen inch shell at a piece of tissue paper and it came back and hit you.[6]

This experiment did not lend itself to a study of the angular distribution of the scattered particles. Their source had to be close to the scatterer, and so there was no well-defined direction from which the α-particles struck it, and the angle of scattering of any particular α-particle could not be determined. Overcoming this would be a major project, not to be undertaken without a good reason.

The good reason soon emerged. In a paper published in 1911 Rutherford sought an explanation for the large-angle scattering seen by Geiger and Marsden.[7] He noted the success of the Thomson model in accounting for the scattering of β-particles by atoms. This was one of the three approaches that Thomson had used to estimate the number of electrons in an atom, and assuming that the distribution of the positive electricity in the atom was continuous, Crowther's experiments yielded a number about three times the atomic weight. The scattering that Crowther measured was interpreted as the result of many independent small-angle scatterings by many atoms. This did not fit the observations of large-angle scattering of α-particles. Rutherford proposed that the α-particle scattering must be the result of a single event involving a very intense electric field in the target atom. The way he got the intense field was to picture the atom as having a concentrated charge $\pm Ne$ at its center and an equal charge of opposite sign uniformly distributed over a sphere of the radius of the atom, R. The ambiguity of sign was to indicate that the result of his calculation did not depend on whether the force between the α-particle and the central charge was attractive or repulsive, and so if his model were correct, it would not distinguish between these two signs of the charge. This was an almost complete reversal of the Thomson model, in which the mass and positive charge were distributed over the entire atomic volume and the electrons were point particles embedded in this positive distribution. Rutherford concentrated the positive charge and most of the mass in a small central body and treated the electrons as a continuous negative charge distribution surrounding this central body or nucleus. This last was purely for simplicity; he knew that the electrons could not scatter an α-particle appreciably, so their exact distribution was unimportant.

Rutherford now had constructed a model from which he could make a definite prediction of the angular distribution of scattered α-particles. The mechanics of a particle moving under the influence of an inverse square law force had, after all, been treated by Newton, even if only for the attractive force case. First Rutherford calculated the distance b of closest approach that an α-particle from radium could make to a positive central charge, the distance at which all the kinetic energy of the particle would be converted into potential energy, using his knowledge of the mass and speed, 2.09×10^9 cm per second, of these particles and, presumably, Thomson's estimate of the number of electrons in an atom of gold to get the magnitude of the central charge. He obtained a minimum separation of 3.4×10^{-12} cm.

Since R is supposed to be of the order of the radius of the atom, viz. 10^{-8} cm., it is obvious that the α particle before being turned back penetrates so close to the central charge, that the field due to the uniform distribution of negative electricity may be neglected. In general, a simple calculation shows that for all deflexions greater than a degree, we may without sensible error suppose the deflexions due to the field of the central charge alone. Possible single deviations due to the negative electricity, if distributed in the form of corpuscles, are not taken into account at this stage of the theory. It will be shown later that its effect is in general small compared with that due to the central field.

Rutherford then went on to make the calculation. The angle through which a given α-particle was scattered depended on how closely it passed by the central charge. This could be specified by imagining the α-particle going straight past the central charge as if it were electrically neutral; the distance of closest approach of this straight-line path to the center is now called the *impact parameter* and was denoted by p by Rutherford. Since it was impossible to aim an α-particle to pass at a given distance from a target atom, Rutherford had to consider the probability of its being scattered through an angle ϕ from its original direction of motion so as to strike a particular element of area dS of a detector at a distance R away from the scatterer. Because the incident α-particle is equally likely to go through any element of area on its way to the target, the probability of its having the impact parameter between p and $p + dp$ is proportional to the area of a ring of radius p and width dp centered on the target atom. The scattered α-particle moved away from the scatterer along a radial line after the scattering process was completed, so the probability of its striking an element of area dS of a detector was proportional to the ratio of the area dS to the area of a sphere of radius R centered on the scatterer, and so it was proportional to the ratio dS/R^2. This ratio was the element of *solid angle* $d\Omega$ in which the scattered particle was detected. Rutherford's calculation determined the angle ϕ through which an α-particle approaching the scatterer at impact parameter p would be deflected according to classical mechanics. Doing this led to the famous Rutherford formula for the *cross section* σ for scattering through the angle ϕ with the beam direction.

$$\sigma = \frac{(2Ze^2)^2}{4m^2v_0^4 \sin^4\left(\dfrac{\phi}{2}\right)} = \frac{(2Ze^2)^2}{4m^2v_0^4} \csc^4 \frac{\phi}{2}.$$

Here m is the mass of the α-particle and v_0 is its incident velocity.

Having derived this expression, Rutherford pointed out the features that could be checked experimentally:

> We see from the equation that the number of α particles (scintillations) per unit area of zinc sulphide at a given distance ... from the point of incidence of the rays is proportional to
>
> (1) $\csc^4 \phi/2$ or $1/\phi^4$ if ϕ be small;
> (2) thickness of scattering material ... if this is small;
> (3) the magnitude of central charge Ne;
> (4) and is inversely proportional to $(mv^2)^2$, or to the fourth power of the velocity if m is constant.

A definitive report on the scattering of α-particles was published by Geiger and Marsden in 1913. In this monumental work over 100,000 scintillations were counted.[8] At the beginning of the paper they stated:

It may be mentioned in anticipation that all the results of our investigation are in good agreement with the theoretical deductions of Prof. Rutherford, and afford strong evidence of the correctness of the underlying assumptions that an atom contains a strong charge at the centre of dimensions, small compared with the diameter of the atom.

They studied the variation of scattering with the scattering angle over a range of angles such that the value of $1/\sin^4 \phi/2$ changed by a factor of more than 100,000. Their collected results are summarized in Table 7.1.

Table 7.1 Variation of Scattering with Angle (collected results)

I	II	III	IV	V	VI
		SILVER		GOLD	
Angle of deflexion ϕ	$\dfrac{1}{\sin^4 \phi}$	Number of scintillations, N.	$\dfrac{N}{1/(\sin^4 \phi/2)}$	Number of scintillations, N.	$\dfrac{N}{1/(\sin^4 \phi/2)}$
150°	1.15	22.2	19.3	33.1	28.8
135	1.38	27.4	19.8	43.0	31.2
120	1.79	33.0	18.4	51.9	29.0
105	2.53	47.3	18.7	69.5	27.5
75	7.25	136	18.8	211	29.1
60	16.0	320	20.0	477	29.8
45	46.6	989	21.2	1435	30.8
37.5	93.7	1760	18.8	3300	35.3
30	223	5260	23.6	7800	35.0
22.5	690	20300	29.4	27300	39.6
15	3445	105400	30.6	132000	38.4
30	223	5.3	0.024	3.1	0.014
22.5	690	16.6	0.024	8.4	0.012
15	3445	93.0	0.027	48.2	0.014
10	17330	508	0.029	200	0.0115
7.5	54650	1710	0.031	607	0.011
5	276300	. . .	. . .	3320	0.012

The last six rows represent observations at small angles where geometrical corrections were applied. The difference in the scale of the ratio in coluns IV and VI was due to a change in the size of the diaphragm defining the beam. They noted the following:

In Cols. IV. and VI. the ratios of the numbers of scintillations to $1/\sin^4 \phi/2$ are entered. It will be seen that in both sets the values are approximately constant. The deviations are somewhat systematic, the ratio increasing with decreasing angle. However, any slight asymmetry in the apparatus and other causes would affect the result in a systematic way so that, fitting on the two sets of observations and considering the enormous variation in the numbers of scattered particles, from 1 to 250,000, the deviations from constancy of the ratio are probably well within the experimental error. The experiments, therefore, prove that the number of α particles scattered in a definite direction varies as $\csc^4 \phi/2$.

They looked at the dependence on target thickness and found the ratio of counts to thickness sensibly constant when they used from one to nine gold foils.

To find the dependence of the scattering on the size of the scatterer's charge, they used foils of materials with different atomic number A, namely, tin, silver, copper, and aluminum, as well as gold. They found the ratio of the number of counts to $A^{3/2}$ to be nearly constant. The scattering by a single atom should be proportional to A^2. The determination of the number of atoms per unit area of the foils could not be found by weighing; they were too light and too nonuniform. Instead, their thickness was determined by finding their stopping power for an α-particle beam. W. H. Bragg and others had given numbers connecting the thickness of foils of various materials and their stopping power, showing that the numbers of atoms per unit area were inversely proportional to the square roots of the atomic weights. This led them to the relevant factor $A^{3/2}$.

Geiger and Marsden measured the velocity dependence of the scattering by inserting mica sheets in the incident beam to slow it down. The value of the product of the number of scintillations and the fourth power of the velocity was found to vary only between a low of 22 and a high of 28 when the zero to six sheets were inserted. This checked the predicted velocity dependence.

All these results gave enormously strong support to the Rutherford model, but in order to exploit the model fully, it was necessary to find the absolute number of particles scattered so that the numerator of the expression giving the scattering cross section could be evaluated and, with it, the charge of the scattering atoms. This was a very difficult measurement and could not be carried out with great accuracy. Geiger and Marsden used the α-particles scattered through 45° from radium C source whose strength was determined by measuring its γ activity. They found that the probability of these α-particles being scattered into unit solid angle at 45° was 3.7×10^{-7}. This led to the conclusion that the central charge was within 20 percent of being one half the atomic weight.

This nailed down the Rutherford model of the atom. It required the electrons to be outside the very compact positive charge and implied a planetary picture with all the difficulties of stability and radiation already encountered in the Nagaoka model. In fact they are worse here than in Nagaoka's case, because now the number of electrons must be just sufficient to make the atom neutral, namely, about $A/2$, a much smaller number than could be postulated in 1904, and hence radiating more strongly. This model was not, however, a speculative one but was founded on extraordinarily strong experimental evidence. It represented a real dilemma for classical physics.

Notes

1. E. Rutherford, *Nature* **68**, 366 (1903); *Collected Papers of Lord Lord Rutherford of Nelson*, ed. J. Chadwick (New York: Interscience, 1962), vol. 1, p. 149.

2. E. Rutherford, *Philosophical Magazine* **12**, 152 (1906); *Collected Papers*, vol. 1, p. 609.

3. E. Rutherford and T. Royds, *Philosophical Magazine* **17**, 281 (1909); *Collected Papers*, vol. 2, p. 163.

4. H. Geiger, *Proceedings of the Royal Society* **81**, 174 (1908); **83**, 492 (1909).

5. H. Geiger and E. Marsden, *Proceedings of the Royal Society* A **82**, 495 (1909).

6. *Rutherford at Manchester*, ed. J. B. Birks (New York: Benjamin, 1963), p. 68.

7. E. Rutherford, *Philosophical Magazine* **21**, 669 (1911); *Collected Papers*, vol. 2, p. 238.

8. H. Geiger and E. Marsden, *Philosophical Magazine* **25**, 604 (1913).

CHAPTER 8

Quantum Rules and Recipes

I n the early years of the twentieth century, the question of how the atom was built up out of electrons and the positive charge needed to make the structure neutral became acute. Thomson's model with the electrons distributed around inside an extended sphere of positive charge had had some successes but had been unable to account for the scattering of α-particles observed by Rutherford's group. Rutherford's nuclear model had accounted for this scattering very well but was perhaps even more violently in conflict with classical physics than Thomson's was. The Thomson model required some unknown mechanism to keep the sphere of positive charge from being blown up by the internal repulsion of its parts, while the Rutherford model required a suspension of the fundamentals of electromagnetic theory to keep it from collapsing from the loss of energy in the form of radiation by the orbiting electrons, aside from preventing the positive nucleus from exploding because of the coulomb forces between its parts, here much larger than in the Thomoson model because its parts were so much closer together.

Another difficulty with all models was the identity of all the atoms of a given element. This bothered Maxwell, and in his article "Atom" in the *Encyclopedia Britannica* he wrote:

> Whether or not the conception of a multitude of beings [atoms] existing from all eternity is in itself contradictory, the conception becomes palpably absurd when we attribute a relation of quantitative equality to all these beings. We are then forced to look beyond them to some common cause or common origin to explain why this singular relation of equality exists, rather than any one of the infinite number of possible relations of inequality.

In the Thomson model the size of the sphere of positive charge appeared

224

as an arbitrary parameter whose magnitude had no theoretical relation to any overarching principle relating the sizes of atoms of the same or of different kinds. In Rutherford's model the electron orbits could be of any size, just as in Newton's theory of planetary motion the radius of the earth's orbit around the sun is not specified by the law of gravitation but only by initial conditions. There was no way of accounting for the existence of some 10^{23} apparently identical helium atoms in a mole of helium. The only quantities governing the behavior of the electrons in a Rutherford atom were the central positive charge, the charge of the electrons, and the masses of the electrons and the central charged body. There was no way to combine electric charges and masses to form a quantity with the dimensions of a length, so there was no way to express the size of his atoms in terms of the quantities governing their structure. The only length that could be formed from fundamental physical quantities was the classical electron radius, e^2/mc^2, whose value of about 2×10^{-13} cm was five orders of magnitude too small to characterize an atom, and furthermore, there was no role for magnetic fields or effects due to the propagation of radiation in his model, so the speed of light could not influence the size of his atoms.

There were other difficulties. The number of optically active "dispersion electrons" per atom determined by X-ray scattering was much too small to account for the richness of the spectra, especially with the Thomson model. In the Rutherford model the emission and absorption of light of the well-defined colors of the spectrum could not be explained, especially since the model required that the radiation expected according to Maxwell–Hertz theory did not occur.

The idea of introducing Planck's constant h as a fundamental quantity relevant to atomic structure was not considered for some time, because h was associated only with thermal phenomena such as blackbody radiation and specific heats. In had as yet no status as a fundamental constant of nature but was a quantity to be explained rather than to provide explanations. It was, however, possible to construct a length from the quantities e, m, and h, namely, $h^2/me^2 = 2.09 \times 10^{-7}$ cm. This was of a nice atomic size. The trouble was that no one had shown how this combination of constants came to enter the theory of atomic structure.

Among the first to connect the quantum of action with atoms was Arthur Erich Haas, who considered the motion of an electron at the surface of an N-electron Thomson atom.[1] This electron was attracted to the center of the atom by the positive charge Ne and was repelled by the charges of all the other $N - 1$ electrons. If these were symmetrically distributed throughout the sphere, the electron at the surface would experience a coulomb force produced by the part of the sphere's positive charge not neutralized by the other electrons, namely, e. Haas equated this force to the centripetal force necessary to keep the electron moving around the atom and, in this way, derived a relation between the radius a of the orbit and the speed v of the electron. This was pure classical physics and did not specify either the orbit radius or the electron speed. To connect Planck's constant with this system, he then assumed that the potential energy of the electron in its orbit was equal to Planck's h times the frequency of revolution of the electron, giving him another relation between the radius and the speed. This second relation contained Planck's constant. What Haas did next showed the status of h at this time. He chose to eliminate the electron speed between his two relations and to solve the resulting equation for h! Because h was the mysterious thing that needed to be explained, he found an expression for it in terms of the electron charge and mass and the atomic radius a,

$$h = 2\pi\sqrt{e^2 ma}.$$

The size of an atom, though completely unexplained in 1908 when Haas did this, was at least a concept that fitted into the scheme of classical physics, while Planck's constant h certainly did not represent any such concept. So Haas expressed the mysterious in terms of the comprehensible, but this did not further atomic physics very much. Even so, this introduction of Planck's constant in connection with atomic structure was too advanced for his time, and Haas was criticized for mixing up totally unconnected ideas. Had he solved for a he would have obtained the Bohr radius of the hydrogen atom in terms of e, m, and h before Bohr did,

$$a = \frac{(h/2\pi)^2}{me^2},$$

though without any of the other features that made the Bohr atom so important.

John W. Nicholson was another to incorporate the quantum of action into atomic physics. He developed rather complicated models of atoms containing rings of electrons.[2] The angular momentum of a ring was constrained to change only by an integer multiple of Planck's constant. The emission and absorption of radiation were ascribed to oscillations of these rings at the frequency of the radiation. Nicholson's basic atoms were assumed to exist only under the extreme conditions present in stellar and nebular matter, not on earth. Terrestrial atoms were compounded from these primary ones. His primary atoms consisted of bands of electrons rotating around a small, heavy core. The angular momentum of these rings was set to an integer multiple of Planck's constant h. To prevent the circulating electrons from radiating, Nicholson required the vector sum of their accelerations to be zero. (Actually this would only make the radiation be of high-order multipole moment, not dipole, but it would reduce it drastically if the wavelength corresponding to the rotation frequency were much larger than the atom.) He called the atom with two electrons *coronium*, the atom with three electrons primary hydrogen, and so forth. His terrestrial hydrogen atom contained two primary hydrogen atoms. Although his atoms radiated definite spectra, the frequency emitted was the frequency of motion of the electrons in the atom as required by classical physics. Finally, Nicholson looked for and thought he found lines he predicted in stellar and nebular spectra.

These two important ideas of Haas and Nicholson incorporated Planck's constant into a scheme based on a classical description of the motion of the electrons ultimately responsible for the emission and absorption of radiation. They both sought to describe a state of motion of the atomic system that would reproduce some characteristic of the atom. Haas was concerned with static properties, in particular the size of the atom in its normal state, while Nicholson was concerned with spectral radiation. What Nicholson missed, as did everybody before Bohr, was the hint provided by the Ritz combination principle that *two* states were involved in radiative processes. This was indeed a veiled hint, since no mechanism was imagined connecting the emission of a train of waves with two states of motion rather than just one. As we shall see in this chapter, Bohr adapted Nicholson's idea of quantizing the angular momentum to a new situation and derived Haas's relation between the atomic radius *a* and the constants *e*, *m*, and *h* in a different context in which *h* was included among the fundamental constants and the atomic radius *a* was to be found.

Notes

1. W. Haas, *Physi,allische Zeitschrift* 11, 537 (1910).
2. J. W. Nicholson, *Philosophical Magazine* 22, 864 (1911).

The Hydrogen Atom According to Bohr

The development of Niels Bohr's ideas culminating in his model of the hydrogen atom that appeared in the July 1913 issue of the *Philosophical Magazine* has been the subject of a great deal of study and some speculation.[1] I do not want to recapitulate this development in detail, but I shall try to give an accurate account of what appeared in the published literature.

Like Haas and others, Bohr at first attempted to find a model that would account for the stability and identity of the atoms of a given element. The old difficulty that the classically relevant constants *e*, the electron charge, and *m*, the electron mass, could not define a length because no combination of charges and masses had the dimensions of a length plagued all atomic models, including the Rutherford model, which Bohr considered to be the most viable one. A considerable period of effort between 1911 and 1913 was spent on investigating the stability of the motion of rings of electrons around a nucleus, assuming that for some reason there was no radiation, in essence reexamining Nagaoka's Saturnian model. Some way of introducing Planck's constant *h* was needed, a need that both Nicholson and Bohr recognized. Among other ideas, Bohr was considering one in which Planck's quanta entered through the process of forming a multielectron atom. In a letter to Hevesy he wrote:

> If we thus assume that the systems considered are formed by a successive binding of the electrons by the nucleus until the whole system is neutral (comp. the formation of a helium-atom from an α-particle), and further assume that the energy emitted by this binding is equal to Planck's constant multiplied by the frequency of rotation of the electron considered in its final orbit, we get results which seem to be in conformity with experiments.[2]

According to his unpublished notebook calculations, Bohr used Planck's constant multiplied by a numerical factor close to unity, not the constant itself, for this purpose.

This binding radiation had nothing to do with spectral lines emitted by the atoms; rather, the purpose of introducing it was to arrive at definite sizes for the normal atoms. Spectra were not Bohr's concern. He had approached the Rutherford model from the point of view of the scattering of α-particles and β-particles by atoms, and especially the slowing down of these particles as they traversed thin foils of matter. This latter involved the mechanisms for transferring energy to the electrons and the nucleus, and these were model dependent. Bohr succeeded in relating these processes to the strength of the electrons' binding to their equilibrium states and, in this way, connected energy-loss processes to the optical properties of the foils' material through the theory of dispersion as formulated by Helmholtz, Reiff, and Drude.[3]

Spectroscopy was at this time a highly developed experimental science, and many facts and a few general rules, such as Rydberg's rules for multiplet series and Ritz's combination principle, were known. The spectroscopist's position was like that of a linguist who had discovered some rules of grammar in a language whose symbols he recognized but whose vocabulary was a total mystery. Such a linguist has trouble exciting a large community about his progress. Likewise, the physicists worrying about atomic structure were not following with rapt attention the spectroscopists' progress in identifying series and finding obscure similarities between them.

The fact that there did exist *simple* regularities in spectra—notably the Balmer formula for the lines in the hydrogen spectrum—apparently came to Bohr's attention early in 1913. By his own later accounts the Balmer formula was the clue he needed to arrive at a theory including both a specification of the possible states of motion within an atom and the frequencies of the spectral lines emitted or absorbed by that atom. By March of that year he had written a draft of the first part of his epic trilogy of papers, "The Constitution of Atoms and Molecules," that he sent to Rutherford for comment and eventual publication. Part I of this trilogy was entitled "Binding of Electrons by Positive Nuclei", Part II was entitled "Systems Containing Only a Single Nucleus", and Part III was entitled "Systems Containing Several Nuclei."[4] The radical assumptions leading to Bohr's theory of the hydrogen atom and other atoms and molecules were set out in Part I.

Bohr began the first section of this remarkable work by stating somey well-known properties of the Kepler motion that an electron would execute around a positive nucleus if radiation by the accelerated charge did not occur. Denoting the electron charge by $-e$, the nuclear charge by E, the electron mass by m, the orbital frequency (not the angular velocity) by ω, the major axis of the elliptical orbit by $2a$, and the energy required to remove the electron to an infinitely great distance from the nucleus by W, he wrote what he referred to as formula (1):

$$\omega = \frac{\sqrt{2}}{\pi} \frac{W^{3/2}}{eE\sqrt{m}}, \qquad 2a = \frac{eE}{W}.$$

He also noted that for the Kepler motion, W is the kinetic energy of the electron averaged over its orbit. He then made this crucial observation:

We see that if the value of W is not given, there will be no values of ω and a characteristic for the system in question.

The size of the atom and the frequency characterizing its internal motion could be specified only when the energy of this internal motion had a definite value or one of a set of values. The finding of this set of possible energy values was the next object of this paper.

After some discussion of the consequences of classical electrodynamics in requiring the emission of radiation in the course of this motion, Bohr reverted to his original system:

Returning to the simple case of an electron and a positive nucleus considered above, let us assume that the electron at the beginning of the interaction with the nucleus was at a great distance apart from the nucleus, and had no sensible velocity relative to the latter. Let us further assume that the electron after the interaction has taken place has settled down in a stationary orbit around the nucleus. We shall, for reasons referred to later, assume that the orbit in question is circular; this assumption will, however, make no alteration in the calculations for systems containing only a single electron.

Let us now assume that, during the binding of the electron, a homogeneous radiation is emitted of a frequency ν, equal to half the frequency of revolution of the electron in its final orbit; then, from Planck's theory, we might expect that the amount of energy emitted by the process considered is equal to $\tau h\nu$, where h is Planck's constant and τ an entire number. If we assume the radiation is homogeneous, the second assumption suggests itself, since the frequency of revolution of the electron at the beginning of the emission is 0. . . .

Putting

$$W = \tau h \frac{\omega}{2}$$

we get by help of the formula (1)

$$W = \frac{2\pi^2 m e^2 E^2}{\tau^2 h^2}, \qquad \omega = \frac{4\pi^2 m e^2 E^2}{\tau^3 h^3}, \qquad 2a = \frac{\tau^2 h^2}{2\pi^2 m e E}.$$

If in these expressions we give τ different values, we get a series of values for W, ω, and a corresponding to a series of configurations of the system. According to the above considerations, we are led to assume that these configurations will correspond to states of the system in which there is no radiation of energy; states which consequently will be stationary as long as the system is not disturbed from outside.

Here the idea of a series of stationary states other than the permanent or normal state that do not radiate appeared for the first time. They were seen as formed from the ionized state by the radiative capture of a distant electron by the nucleus in what we would now call a *multiphoton process*. The choice of the frequency to be emitted in the binding process was highly arbitrary, but the one made was the one that led to an expression for the Rydberg, giving close to the observed value. A peculiar feature of this picture of the way an electron was captured into an orbit was that according to it, the stationary state formed by the emission of a large number of quanta were less tightly bound than were those formed by the emission of fewer quanta. (Further on in the paper Bohr offered another way of achieving the desired result that was less arbitrary and did not have this pecularity. It was later regarded as much more satisfactory.) This picture was clearly an extension of Bohr's earlier one of the way that the permanent states of multiple electron atoms could be built up by successive electron captures, but here the state arrived at after the capture did not need to be the permanent or ground state of the atom. The state with $\tau = 1$, the normal or ground state, had an orbit diameter $2a = 1.1 \times 10^{-8}$ cm, an orbital frequency $\omega = 6.2 \times 10^{15}$ per sec, and an ionization potential $W = 13$ V, all of the right order of magnitude for atomic quantities. A key attraction of this picture was the energy's dependence on the integer τ as $1/\tau^2$, which led to the Balmer formula.

After a brief discussion of the relation of this development to that by Nicholson, Bohr stated his two principal assumptions:

(1) That the dynamical equilibrium of the systems in the stationary states can be discussed by the help of the ordinary mechanics, while the passing of the systems between different stationary states cannot be treated on that basis.

(2) That the latter process is followed by the emission of a *homogeneous* radiation, for which the relation between the frequency and the amount of energy emitted is the one given by Planck's theory.

He categorized the second of these as the more radical, but apparently necessary to account for experimental facts.

In the second section of the paper Bohr derived the Balmer formula using the preceding assumptions. He introduced the fundamental assumption connecting the energy difference between two of his stationary states and the frequency of the radiation emitted or absorbed when the atom made a transition between them, which he ascribed to Planck:

$$W_{\tau_2} - W_{\tau_1} = h\nu.$$

From this and his expression for the way that the energy W depended on the integer τ, he obtained

$$v = \frac{2\pi^2 m e^4}{h^3} \left(\frac{1}{\tau_2^2} - \frac{1}{\tau_1^2} \right),$$

which has the form of Balmer's formula. Using the then current values of the constants *e*, *m*, and *h*, Bohr found the value of the quantity in front of the parenthesis, now called the rydberg,* to be 3.1×10^{15}, while the value derived from measurement of the hydrogen spectrum was 3.29×10^{15}. This agreement was within the uncertainty of the constant's values. He went on to remark that his result for the major axis of the electron orbit, namely, that it is proportional to τ^2, could explain why only twelve lines in the Balmer spectrum had been seen in the laboratory, while thirty-three lines had been observed in some stars. The density of the gas in laboratory discharge tubes was so great that the major axis in states with large value of τ was comparable to the atomic mean free path, so that collisions could occur before emission. This was not necessarily true for stellar atmospheres where enormous volumes of very dilute gas could emit light of observable intensity.

Next Bohr discussed lines observed by E. C. Pickering in a stellar spectrum, as well as others observed by A. Fowler in powerful discharges through tubes containing a mixture of hydrogen and helium which were attributed to hydrogen. Bohr ascribed these lines to ionized helium which, having a single electron, should behave just like hydrogen except that the nuclear charge would be 2*e* rather than *e*. The series observed by Pickering he attributed to ionized helium with $\tau_2 = 4$, and that by Fowler to the helium series with $\tau_2 = 3$.

That Bohr's picture of just how spectra come about as described in this first paper was not quite what we have today is indicated in his discussion of the helium spectrum. In connection with the failure to see all the helium lines in a normal helium gas discharge tube, Bohr noted:

> The condition for the appearance of the spectrum is, according to the above theory, that helium atoms are present in a state in which they have lost *both* their electrons. Now we must assume that the amount of energy to be used in removing the second electron from a helium atom is much greater than that to be used in removing the first. (italics added)

This reflects his starting point in arriving at the stationary states of an atom; He started with the electrons and nucleus well separated. At this point he seems to have regarded this as a necessary part of the process of emitting spectral lines, but soon it was recognized that in order to see the lines of the spectrum of ionized helium, only one electron had to be removed from the helium atom, and the other one had to be excited from a lower state to a higher one.

Bohr continued this line of argument when he turned from helium to

---

* Usually the value of the rydberg is given by the formula with the wave number rather than the frequency on the left. The two differ by a factor of the velocity of light.

discuss the situation for heavier atoms with more electrons, and the reason for the constant K in the Rydberg–Ritz law's expression

$$\frac{K}{(\tau + a)^2}$$

for the spectral terms having the same value no matter what element was involved. In this discussion he wrote:

> Let us assume that the spectrum in question corresponds to the radiation emitted during the binding of an electron; and let us further assume that the system including the electron considered is neutral. The force on the electron, when at a great distance apart from the nucleus and the electrons previously bound, will be very nearly the same as in the above case of the binding of an electron by a hydrogen nucleus. The energy corresponding to one of the stationary states will therefore be very nearly the same as in the above case of the bonding of an electron by a hydrogen nucleus.

The conclusion reached is clear and correct. It is only the picture of how and when the spectral radiation occurs in the process that is obscure. At this very early stage in the theory's development, the distinction between the radiation associated with defining the stationary states and the radiation emitted in transitions between stationary states was blurred. In some places they seemed to be identified with each other, and in other places they seemed to be different.

After this elaboration of the particular cases of hydrogen and helium, Bohr continued his more general considerations in the third section, in which he presented a more physically realistic picture of the process by which a distant electron becomes bound to a nucleus. He did this by inventing what came to be known as the *correspondence principle*. The argument is both simple and profound.

In his first argument Bohr had assumed that the binding energy was emitted as *homogeneous* radiation whose frequency was the average of the initial (zero) frequency of the electron in its infinite orbit and the final orbit frequency. This did not involve quanta at all. He then further assumed that the energy of this radiation had to consist of τ quanta of that average frequency. This led him to an expression for the energy of the electron in its final orbital proportional to $1/\tau^2$ and to the Balmer formula. The numerical coefficient in the expression for the energy W of a stationary state depended on this particular choice of the frequency of the homogeneous radiation. Bohr had chosen the one that led to an expression whose value agreed with the observed one. In the alternative line of argument he weakened the specification of the frequency to be emitted during the binding, requiring only that it be of the form $f(\tau)h\omega$, with τ an integer. To obtain a Balmer-like formula, $f(\tau)$ had to be of the form $c\tau$, with c a constant to be determined. It was for the evaluation of c and therefore of the Rydberg constant that Bohr invented the correspondence principle.

Bohr argued that for very large N, $N \gg 1$, the radiation emitted in a transition from the stationary state with $\tau = N$ to one with $\tau = N - 1$ should be of the orbital frequency, because in this case the average of the initial and final orbital frequencies was nearly equal to the orbital frequency of either state. This meant that for motions in very large orbits, the classical Maxwell–Hertz theory should be valid. This led him to the value $c = 1/2$ and the same result he had obtained with his previous argument. On extending this new approach to a transition from $\tau = N \gg 1$ to $\tau = N - n$ with $n \ll N$, he arrived at $v = n\omega$. This was like emission at a harmonic of the frequency of the periodic Kepler motion, and it too agreed with classical expectations. In this way Bohr provided a correspondence between his quantum treatment of the loosely bound states of hydrogen atoms and the classical treatment of the radiation of accelerated charges when the charges moved in very large orbits.

At this point Bohr pointed out that his original assumption about the number of quanta emitted in the process of binding an electron initially very far away was not necessary.

> We are thus led to assume that the interpretation of the equation (?) is not that the different stationary states correspond to an emission of different numbers of energy-quanta, but that the frequency of the energy emitted during the passing of the system from a state in which no energy is yet radiated out to one of the different stationary states, is equal to different multiples of $\omega/2$, where ω is the frequency of revolution of the electron in the state considered. From this assumption we get exactly the same expressions as before for the stationary states, and from these by help of the principal assumptions . . . the same expression for the law of the hydrogen spectrum. Consequently we may regard our preliminary considerations . . . only as a simple form of representing the results of the theory.

After a little more discussion he concluded:

> Taking the starting point in the form of the law of the hydrogen spectrum and assuming that the different lines correspond to a homogeneous radiation emitted during the passing between different stationary states, we shall arrive at exactly the same expression for constant in question [Rydberg's constant] as that given [previously] if we only assume (1) that the radiation is sent out in quanta hv, and (2) that the frequency of the radiation emitted during the passing of the system between successive stationary states will coincide with the frequency of revolution of the electron in the region of slow vibrations.

Only the concept of spectral lines being emitted as quanta hv during transitions between stationary states and the correspondence principle for emission between adjacent loosely bound states was needed to arrive at the expression for the rydberg that agrees with observation.

For circular orbits Bohr next pointed out that the rule giving the energy of the stationary states could be rephrased as the following simple condition:

> The angular momentum of the electron round the nucleus in a stationary state of the system is equal to an entire multiple of a universal value [$h/2\pi$], independent of the charge on the nucleus. The possible importance of the

angular momentum in the discussion of atomic systems in relation to Planck's
theory is emphasized by Nicholson.

The third one, the angular momentun rule, and neither of those concerned
with the manner of formation of the excited state, is the one that is remembered
and taught as *the* basic element of Bohr's theory.

In the fourth section of this remarkable paper Bohr considered absorption
spectra. Kirchhoff had shown that systems that can emit radiation of some
frequency can necessarily absorb radiation of that frequency. If the emission
of a spectral line is associated with a transition from a stationary state A_1 to
another such that A_2 with $\tau_1 > \tau_2$, then for emission to take place there must
be atoms in the former state. For light of this frequency to be absorbed, there
must be atoms present in state A_2. In hydrogen gas there is no observed
absorption of the Balmer lines. Bohr accounted for this by the fact that in the
normal state of the atom $\tau = 1$ and that the final state into which the atoms
make transitions in emitting the Balmer lines go has $\tau = 2$. In a gas discharge
tube, the atoms in the final state with $\tau = 2$ that had emitted Balmer lines
quickly made a subsequent transition to the state with $\tau = 1$, emitting invisible
ultraviolet radiation in the process, and did not stay around long enough for
any appreciable absorption of Balmer radiation. In normal hydrogen gas the
atoms had combined to form molecules that changed the stationary states,
and so here, too, there was no absorption of Balmer radiation. The case of
sodium vapor was quite different. R. W. Wood had shown the presence of
many absorption lines in sodium vapor. Although the terms of the sodium
spectrum could not be calculated on the basis of his theory of the hydrogen
atom, Bohr interpreted the presence of this absorption as meaning that the
principle series of sodium, unlike that of hydrogen, had the permanent or
ground state as its final state so that there would be atoms in this state present
and able to absorb light of these frequencies.

He remarked at this point that

> How much the above considerations differ from an interpretation based on
> the ordinary electrodynamics is perhaps most clearly shown by the fact that
> we been forced to assume that a system of electrons will absorb a radiation
> of a frequency different from the frequency of vibration of the electrons
> calculated in the ordinary way.

This divorce between the frequency of radiation and the frequency of vibration
of the system emitting or absorbing that radiation was undoubtedly the most
radical and most important aspect of the Bohr theory. It was postponed for
so long—thirteen years after Planck's introduction of the quantum—because
of the concentration on harmonic oscillators. Even the present-day quantum
theory of the harmonic oscillator does not distinguish between the frequency
of the oscillator and the frequency of the radiation it emits and absorbs; the
level spacing is constant over the entire range of energy levels, so that only
the oscillator frequency and its harmonics appear.

Bohr also made a connection with Einstein's theory of the photoelectric effect. If light of a frequency v, such that hv was greater than the binding energy W of the electron in an atom, were absorbed by that atom, the electron would end up with a kinetic energy T given by Einstein's photoelectric equation

$$T = hv - W.$$

Millikan's accurate verification of this equation had not yet been made, but Bohr obviously thought this a valuable consequence of his theory.

In the fifth section of his paper Bohr returned to "the main object of this paper—the discussion of the 'permanent' state of a system consisting of nuclei and bound electrons," apparently dismissing the revolutionary and enduring discussion of spectra given in the previous sections to the role of preliminaries and asides. The problem he attacked was real and important, but the new tools just developed were not sufficient to permit decisive progress. The many-electron atom remained a mystery until the Pauli *exclusion principle* emerged many years later. The electron shells suggested by the periodic table of the elements were sought on the basis of stability conditions for the rings of electrons in Bohr's permanent states. I shall not trace the history of these efforts, as they were highly technical and in the end their goal was reached by very different means.

As with any genuinely new idea, Bohr's model of hydrogen and the mechanism by which spectral lines arose was not immediately accepted. There was skepticism concerning whether it really worked, and even if it did work, whether the assumptions on which it was based were necessary. One objection of the first kind concerned the helium spectrum, which was not recognized for what it was in all cases. Fowler observed series of lines, some of which came very close to Balmer lines, but there were members of this series between the Balmer lines that could be accounted for if half-integer numbers (halves of odd integers) were allowed alongside Planck's integers. Following Rydberg's idea, Fowler attributed these to series in hydrogen that he called the *first* and *second principal series* and the *sharp series*. Pickering had seen these lines in the spectrum of ζ-Puppis, and Fowler had seen them in discharges through a mixture of hydrogen and helium. We mentioned earlier the observation of these series, and Bohr's explanation of them as due to ionized helium. But Fowler had grave doubts that this explanation sufficed because the accuracy was not as good as it was for the Balmer lines. The value of the rydberg under discussion was known to six significant figures, and so small disagreements were to be taken seriously. Bohr's theoretical value could not, of course, be given with such precision, but whatever its value, it had to be the same for all lines of a series. Relative values had to be extremely accurate if the theory was to be satisfactory. Fowler objected that this was not so.

There were two responses to this criticism, one experimental and one theoretical. Rutherford persuaded E. J. Evans at Manchester to look for the helium spectrum in a discharge completely free of hydrogen. Evans saw the line at 4686 Å that Fowler had ascribed to another series of hydrogen lines without seeing any trace of the hydrogen Balmer lines that should have been present if this line really came from hydrogen. This did not completely persuade Fowler because of the slight numerical discrepancy in the relative frequencies of the hydrogen and helium lines. According to Bohr's original formulation, the only difference between the two series of spectra should have been a factor four coming from the charge on the helium nucleus being twice that on the hydrogen nucleus and the fact that this charge appeared squared in the formula for the rydberg.

In a letter to *Nature* Bohr pointed out that this formula also makes the rydberg proportional to the mass of the electron.[5] In a system containing two bodies, one massive and one light, a good approximation is to regard the massive body as immovable, as though its mass were infinite, and to use only the mass of the light body that does most of the moving, and Bohr had done this in his paper. The correct was to treat the problem is to regard both particles as moving around their center of mass. This two-body system is exactly equivalent to a system in which one mass is infinite and therefore remains fixed while the other particle has a mass called the *reduced mass* μ of the two-body system given by

$$\frac{1}{\mu} = \frac{1}{m} + \frac{1}{M} \quad \text{or} \quad \mu = \frac{mM}{m + M}.$$

For hydrogen $M = 1835m$, and for helium $M = 4 \times 1835m$, so the ratio of the two reduced masses is 1.00041. Replacing the electron mass with the reduced mass of the electron-nucleus system made the agreement between Fowler's lines and Bohr's formula just as good as the rest, with two exceptions that Fowler was ready to think might have experimental uncertainties. Fowler was persuaded that Bohr might be right but was not convinced that he was! He was bothered by the need to alter the value of the rydberg—the constant in front of the term containing the difference of squares in Rydberg's formula—when going from one series to another. The common value of this quantity for various series in various elements was one of the striking regularities observed in spectra, and changing it by a factor of 1.00041 seemed an undesirable abandonment of this principle, even though its basis was not understood. Fowler also pointed out that "it should be noted in conclusion that Dr. Bohr's theory has not yet been shown to be capable of explaining the ordinary series of helium lines."[6]

Multielectron atoms were a problem then just as they are today. Most of our quantitative knowledge of these systems is acquired through numerical calculation, although now we believe we understand the principles on which to base these calculations. But these principles were not yet known at the time I am describing. Recall Einstein's inattention to Kaufmann's electron mass measurements after he had formulated the special theory of relativity, because their results did not accord with the symmetry he had discovered. Likewise, Fowler was unwilling to accept Bohr's proposal that the rydberg was not a universal constant because it violated a regularity. The velocity of light did turn out to be a universal constant and the rydberg turned out not to be one, but in neither case was the choice obvious to those studying the issues. The rejection of classical electrodynamics in atomic emission processes was at least as hard to accept as was the statement that the speed of light is the same to observers moving relative to one another.

There is a famous letter from Georg Karl Hevesy to Bohr on this matter.† Hevesy had told Einstein at a meeting in Vienna of Bohr's successful attribution of Fowler's lines to helium. Einstein replied, according to Hevesy's memory, "Then the frequency of the light does not depend at all on the frequency of the electron! This is an enormous achievement! The theory of Bohr must then be right!" Not everybody shared this acceptance as early as September 1913, less than three months after Bohr's initial publication of his theory.

Another of the holdouts was Nicholson. In a series of papers and letters to the editor of *Nature* continuing through 1915, he insisted that the lines in the spectra of hydrogen and ionized helium could be fitted into Rydberg's scheme for hydrogen alone. Bohr responded to these in various publications, expressing his inability to agree with Nicholson's interpretations of the spectra observed by Evans and by Fowler.

The developments of the three years from 1911 to 1913 had produced the first clear outline of atomic physics. The Rutherford model of the atom, based as it was on strong experimental evidence, gave a picture of the distribution of electric charge and mass in an atom. Nearly all the mass was concentrated in the tiny nucleus, as was all the positive charge. There were electrons external to this nucleus whose configuration was a mystery because of the lack of any classical definition of the size of the atom and because of the classical inevitability of electromagnetic radiation from a system of orbiting charges. The Bohr picture of stationary states—inexplicable classically but strongly indicated by the regularities of atomic spectra—between which transitions could take place with the emission of radiation governed by Planck's idea of quanta provided a framework into which a great deal of experimental experience could be fitted. What it was that determined the stationary states of atoms more complex than those containing a single electron and what governed the making of transitions between these states remained hidden. Spectroscopy moved from its obscure place in the wings to take center stage. Now there were points of attack that had been lacking before. Now theory and experiment could deal with the same questions and could guide each other toward a real theory of atoms.

Bohr himself took one of the first steps toward applying his theory to more complicated phenomena connected with the hydrogen spectrum. Johannes Stark had detected the splitting of the spectral lines of hydrogen when an electric field was imposed, the *Stark effect*.[8] In attempting to explain some features of this splitting, after explaining the term schemes of Balmer, Rydberg, and Ritz, Bohr summarized his theory:

† It is reproduced in Niels Bohr's *Collected Works*, ed. U. Hoyer (Amsterdam: North–Holland, 1981), vol. 2, p. 532.

On the above view this can be interpreted by assuming:

(1) That every line in the spectrum corresponds to a radiation emitted by a certain elementary system during its passage between two states in which the energy, omitting an arbitrary constant, is given by $-hf_s(n_2)$ and $-hf_r(n_1)$ respectively;

(2) That the system can pass between any two such states during emission of a homogeneous radiation.[9]

Here there was one quantum number per state and no limitation on how much that number could change, but there was a series of function f_r that specified now the energy depended on the quantum number in each series. The label r for the series was not yet itself a quantum number.

Bohr then went on to consider the effect of applying an electric field to the atom:

A detailed investigation of the motion of the electron may be very complicated; but it can be simply shown that the problem only allows two stationary orbits of the electron. In these, the eccentricity is equal to 1 and the major axis parallel to the axis of the external field; the orbits simply consist of a straight line through the nucleus, one on each side of it.

He thought he had achieved some agreement with Stark's measurements of the two most widely split lines, but he did not account for the other lines in any clear manner. We shall see in the next section that the orbits Bohr used here were soon to be excluded as impossible because in these orbits the electron would collide with the nucleus.

Real progress in analyzing spectra other than that of the free hydrogen atom required further development of Bohr's ideas by Bohr and others.

Notes

1. Especially full and interesting accounts of this are to be found in U. Hoyer, ed., *Niels Bohr, Collected Works* (Amsterdam: North–Holland, 1981), vol. 2; and in J. Mehra and H. Rechenberg, *The Historical Development of Quantum Mechanics* (New York: Springer-Verlag, 1982), vol. 1. See also A. Pais, *Niels Bohr's Times* (New York: Oxford University Press, 1991).

2. Reprinted in vol. 2 of *Niels Bohr, Collected Works*, p. 531.

3. See P. Drude, *The Theory of Optics* (New York: Dover, 1959), chap. 5.

4. N. Bohr, *Philosophical Magazine* 26, 1, 476, 857 (1913).

5. N. Bohr, *Nature* 92, 231 (1913).

6. A. Fowler, *Nature* 92, 232 (1913).

7. N. Bohr, *Nature* 95, 6 (1915); *Philosophical Magazine* 30, 394 (1915).

8. J. Stark, *Berlin Academy* (1913), p. 932.

9. N. Bohr, *Philosophical Magazine* 27, 506 (1914).

The Hydrogen Atom, Continued

There was only one kind of hydrogen atom, and so a radical theory that accounted for its spectrum and no other while discarding some of the most important aspects of classical electromagnetic theory had rather limited support. Interpreting the spectra observed by Pickering and Fowler as belonging to ionized helium helped make Bohr's theory convincing, but as Fowler had pointed out, the spectrum of even the next element in the periodic table, neutral helium, was beyond the new theory's ability to describe quantitatively. Strong evidence that came from elements higher in the periodic table soon gave Bohr's theory a credibility that was hard to question. The major contribution was furnished by the work of Henry G. J. Moseley on the characteristic X-ray spectra of the elements and by the interpretation of this research by W. Kossel. Another important contribution was the experiments by J. Franck and Gustav Hertz on the energy loss of electrons going through mercury vapor.

In Manchester the Rutherford model of the atom stimulated much experimental work. The study of α particle scattering that had engendered the Rutherford model in the first place had shown that the charge on the nucleus was about half the atomic weight of the element doing the scattering. Barkla's study of X-ray scattering had shown that the number of electrons in an atom was about half the atomic weight of the atom, and according to the Rutherford model this was the same as the charge on the nucleus. The question of exactly what this charge was moved to the center of attention.

It was known that X-rays produced by the bombardment of a material by energetic cathode rays were of two kinds. One was a broad-band or heterogeneous radiation not strongly dependent on the material making up the bombarded target or anticathode of the X-ray tube. The other was rather homogeneous radiation characteristic of the target material. At this time all wavelength determinations were quite crude, being based mostly on the ability of the radiation to penetrate aluminum foils. The prediction by Max von Laue that crystals could be used as diffraction gratings for X-rays was verified by W. Friedrich and P. Knipping in 1912,[1] and in 1913 W. H. Bragg and his son W. L. Bragg in Leeds turned this idea into a practical X-ray spectrometer.[2] This gave Moseley in Manchester a means for accurately measuring the wavelengths of the X-rays characteristic of a variety of elements. His first results appeared in the volume of the *Philosophical Magazine* following the one containing Bohr's first paper on the constitution of atoms and molecules, a paper he obviously must have seen and studied before it was published.[3] More of his work appeared the next year.[4]

What Moseley did was measure the wavelengths of the most intense component of the highest-frequency X-ray lines emitted by twelve elements, ten of them forming a continuous sequence of entries in the periodic table extending from calcium to zinc, except for a gap where scandium was missing. He measured them using both second-order and third-order reflections from a

Table 8.1   Values of Q for a Series of Elements

Element	Q	Atomic number	Atomic weight
Calcium	19.00	20	40.99
Scandium	. . .	. . .	44.1
Titanium	20.99	22	48.1
Vanadium	21.96	23	51.06
Chromium	22.98	24	52.0
Manganese	23.99	25	54.93
Iron	24.99	26	55.85
Cobalt	26.00	27	58.97
Nickel	27.04	28	58.68
Copper	28.01	29	63.57
Zinc	29.01	30	65.37

"fine specimen" of platinum ferrocyanide. The two orders gave consistent results. In order to give these wavelengths a physical meaning. Moseley introduced a quantity Q that he defined as

$$Q = \sqrt{\frac{v}{\frac{3}{4}v_0}},$$

where v is the observed frequency of the X-ray line and v_0 is the "fundamental frequency of ordinary line spectra" obtained from Rydberg's constant; Q^2 was just the ratio of the observed X-ray frequency to the frequency of the first Balmer line in the hydrogen spectrum according to Bohr. The results of this measurements included the entries given in Table 8.1. The nickel–cobalt pair was especially noted because it was one of the cases in which the arrangement of elements in the periodic table on the basis of chemical properties conflicted with that based on atomic weights. Moseley described this result as follows:

> It is at once evident that Q increases by a constant amount as we pass from one element to the next, using the chemical order of the elements in the periodic system. Except in the case of nickel and cobalt, this is also the order of the atomic weights. While, however, Q increases uniformly the atomic weights vary in an apparently arbitrary manner, so that an exception in their order does not come as a surprise. We have here a proof that there is in the atom a fundamental quantity, which increases by regular steps as we pass from one element to the next. This quantity can only be the charge on the central positive nucleus, of the existence of which we already have definite proof. Rutherford has shown from the magnitude of the scattering of α particles by matter, that this nucleus carries a + charge approximately equal to that of $A/2$ electrons, where A is the atomic weight. Barkla, from the scattering of X-rays by matter, has shown that the number of electrons in an atom is roughly $A/2$, which for an electrically neutral atom comes to the same thing. Now atomic weights increase on the average by about 2 units at a time, and this strongly suggests the view that N increases from atom to

atom always by a single electronic unit. We are therefore led by experiment
to the view that N is the same as the number of the place occupied by the
element in the periodic system. . . . This theory was originated by [A. J. van
den] Broek and since used by Bohr.

Having shown Q to be an interesting quantity that increases by unity at
each step up the periodic table, Moseley went on to relate it to Bohr's theory
of the hydrogen atom. He cited an argument by J. J. Thomson that X-rays
come from the innermost ring of electrons. If this ring is held in mechanical
equilibrium, the centripetal force required to keep them in their orbits is
provided by the central nuclear charge Ne diminished by a small correction
due to the repulsion of the electrons among themselves. By a short but
somewhat obscure argument Moseley arrived at the conclusion that his
experiments combined with Bohr's theory require that the angular momentum
of an electron in these innermost rings be the same for all atoms (provided
that the number of electrons per ring is the same). During this argument he
used the equality of the radiated frequency with the rotational frequency of
the ring, an equality that Bohr used only for very large rings and that did not
hold in his theory for any but the very large rings or orbits, certainly not for
the innermost ones. Pursing this argument, Moseley decided that the innermost
ring for these elements contained four electrons. On assuming that the
radiation whose wavelength he measured came from a transition from level 2
to level 1, its frequency according to Bohr's version of the Balmer formula
would be

$$v = \left(\frac{1}{1^2} - \frac{1}{2^2}\right) \frac{2\pi^2 e^4 m}{h^3} (N - \sigma_n)^2,$$

where σ_n is the small correction mentioned earlier coming from the mutual
repulsion of the electrons in the ring. The first parenthesis is the source of the
factor $\frac{3}{4}$ in Moseley's formula for Q.

This gave enormous support to Bohr's theory because it showed that his
expression for the Rydberg gave the unit increase in the nuclear charge,
measured in terms of the electron charge, over a sequence of many steps in
the periodic table.

Moseley's work was analyzed by Kossel in Munich. He interpreted the
X-ray lines as being produced by the transitions of single electrons rather than
by whole rings of them making transitions in the interior of the atom.[5] This
made the analogy with hydrogen much closer. The most penetrating of the
characteristic lines, the K series, Kossel ascribed to transitions ending on the
stationary state with $n = 1$, the various lines in the series coming from electrons
falling from the various outer rings into the innermost one. The next, L, series
he attributed to transitions from rings outside the one with $n = 2$ to that ring,
and he even suggested the presence of an M series ending on states with $n = 3$.
For these X-rays to be emitted, there had to be a vacancy in the final ring,
$n = 1$, $n = 2$, or $n = 3$, into which an electron from one of the higher rings
or from outside the atom could fall. The role of the cathode rays that

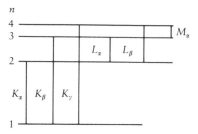

Fig. 8.1 Kossel's scheme of states for a single tightly bound inner electron in a Bohr atom. The transitions giving rise to X-ray lines observed by Moseley are shown.

produced the characteristic X-rays was to remove an electron from one of these inner rings so that an electron from a higher ring could fall into it and fill it.

This picture of Kossel's made the Ritz combination principle apply to X-ray spectra just as it did to optical spectra. In particular it led to relations among the frequencies of various characteristic X-ray lines from a given element. Labeling the successive lines of a series by α, β, and so forth, the combination principle led immediately to relations such as

$$\nu_{K_\beta} - \nu_{K_\alpha} = \nu_{L_\alpha}$$

$$\nu_{K_\gamma} - \nu_{K_\beta} = \nu_{L_\beta} - \nu_{L_\alpha} = \nu_{M_\alpha},$$

as can be seen by looking at the energy-level diagram shown in Fig. 8.1. Kossel showed that Moseley's data together with an extrapolation of his empirical formula for L_α satisfies the first of these relations quite accurately for the elements from calcium to zinc. In a paper written in 1915 Bohr cited this work and also that by Malmer, who extended measurements over a wider range of elements, as supporting his theory of spectra.[6] Moseley found some lines he attributed to the L series that did not fit Kossel's scheme.

Pursuing this subject, Bohr also cited W. H. Bragg's work:

> In a recent paper W. H. Bragg has shown that, in order to excite any line of the K radiation of an element, the frequency of the exciting radiation must be greater than the frequency of all the lines in the K radiation. This result, which is in striking contrast to the ordinary phenomena of selective absorption, can be simply explained on Kossel's view. The simple reverse of the process corresponding to the emission of, for instance, K_α, would necessitate the direct transfer of an electron from ring 1 to ring 2, but this will obviously not be possible unless at the beginning of the process there was a vacant place in the latter ring. For the excitation of any line in the K radiation, it is therefore necessary that the electron should be completely removed from the atom.

In stating the "obvious" fact that an extra electron cannot be accommodated in a ring unless there is a "vacancy" in that ring, Bohr introduced the idea of filled rings or closed shells of electrons and gave what was effectively a primitive statement of the Pauli exclusion principle, that no completely defined state can be occupied by more than one electron. What he lacked, and what remained lacking for almost a decade, was a way of specifying the state of an electron in an atom completely; he thought that the closed shells were required in order for the system of electron motions to be stable.

Another important set of experiments was proceeding in this period, those by Franck and Hertz. They were looking at the ionization of gas atoms by cathode rays using mercury vapor, which is monatomic, as their gas.[7] They found that electrons did not lose energy when passing through mercury vapor as long as their energy was less than 4.9 electron volts but that electrons of just above this energy could lose nearly all their energy on colliding with an atom. When this happened, light in the ultraviolet at 2536 Å, characteristic of mercury, was detected. They pointed out that the energy quantum associated with this wavelength was 4.9 eV. Franck and Hertz interpreted this energy as being the ionization energy of mercury. They also pointed out that not all these energy-losing collisions resulted in ionization but that many of them resulted in the emission of the characteristic radiation. They did not connect their results to the possible existence of stationary states until Bohr suggested it in his 1915 paper.[8] There Bohr questioned their interpretation, pointing out that the ionization potential should be obtainable from the limit of a mercury series measured by Friedrich Paschen. This particular series was a good one to use because its high-frequency line at 1850 Å could be seen in absorption. This meant that the lower state involved must be the normal or ground state of the atom and that the ionization energy must be greater than the energy of the highest state occurring in the series. (Mercury not being hydrogen, there was no way of theoretically identifying the ring occupied by the outermost electrons in the normal or ground state of mercury. In order to be able to estimate the ionization potential from a series limit, one had to establish that the lower level involved in the transition was the ground state. The series limit of the Balmer series in hydrogen, for example, is not the ionization energy of hydrogen. It falls short by the energy associated with the $n = 2$ to $n = 1$ transition.) The limit of this series gave a value of 10.5 eV, considerably higher than 4.9 eV. Bohr conjectured that the electron must be transferring enough energy to an atomic electron to raise it to an unoccupied excited state that was still bound, not ionized. The radiation resulted from the subsequent transition of the electron back to the ground state.

With these confirming experiments appearing, the acceptance of the Bohr atom was rapid in comparison with the acceptance of Planck's original theory of quantized oscillators and comparable to that of Einstein's special relativity. It was not instantaneous, however. F. Hund quoted Otto Stern as saying in 1914 that he and Max von Laue would give up physics if there was "anything in this nonsense of Bohr's."[9] Some twelve years later Bohr's fundamental idea was still producing strong reactions. Werner Heisenberg reported Erwin Schrödinger as saying in a 1926 conversation with Bohr, "If we are still going to have to put up with these damn quantum jumps, I am sorry that I ever had anything to do with quantum theory." It was shortly after the developments that I have just described that a major part of the physics community was at work on Bohr's ideas.

Notes

1. W. Friedrich, P. Knipping, and M. Laue, *München Berichte*, June 8, 1912, p. 303, and July 6, 1912, p. 363; *Annalen der Physik* **41**, 971 (1913).

2. W. H. Bragg and W. L. Bragg, *Proceedings of the Royal Society* A **85**, 285 (1911); **87**, 277 (1912).

3. H. G. J. Moseley, *Philosophical Magazine* **26**, 1024 (1913).

4. H. G. J. Moseley, *Philosophical Magazine* **27**, 703 (1914).

5. W. Kossel, *Verhandlungen der Deutsche Physikalische Gesellschaft* **16**, 953 (1914).

6. N. Bohr, *Philosophical Magazine* **30**, 394 (1915).

7. J. Franck and G. Hertz, *Verhandlungen der Deutsche Physikalische Gesellschaft* **15**, 34 (1913).

8. N. Bohr, *Philosophical Magazine* **30**, 394 (1915).

9. F. Hund, *The History of Quantum Theory*, trans. G. Reece (London: Harrap, 1974), p. 74.

What to Quantize?

The idea of imagining the variation of some continuous quantity as taking place in discrete steps and then making the size of the steps approach zero goes back at least to the Greek geometers. In this way they could consider a circle to be the limit of a regular polygon as the number of sides approached infinity and the length of each side approached zero. Boltzmann had done this with the energy of a gas molecule when he required the kinetic energy of a molecule to be an integer multiple of ε so that the possible energy states of the gas could be counted. Making a continuous quantity into a discrete one was always considered a formal device, and in the end the limit as the size of the discrete elements approached zero was always taken. Boltzmann followed this procedure, and his introduction of quanta did not disturb the basis of classical physics.

In his study of blackbody radiation Planck had followed Boltzmann's footsteps in introducing his energy quanta, but he had found that he could not take the final step and make the size of the quanta approach zero. Because of the Wien displacement law, this size had to be proportional to the frequency of the oscillator whose energy was quantized, and the constant of proportionality was one of the two parameters appearing in his blackbody radiation formula needed to fit the experimental data.

Some six years after his original derivation, Planck had pointed out that if the motion of one of his oscillators was represented on a phase diagram im which the oscillator momentum p was plotted as a function of the oscillator coordinate q, the resulting plots were ellipses and that the effect of his quantization was to allow only those ellipses whose areas were integral multiples of his constant h to represent possible motions. Still later he allowed all of these ellipses to represent actual motions, but radiation from the oscillator could take place only when the motion was along one of the previously allowed ellipses. There was considerable uncertainty in how to apply the idea of quantization, and a desire to minimize its effects.

Einstein had not introduced quanta into his theory of blackbody radiation as had Boltzmann and Planck. He had found them already there when he analyzed the consequence of taking literally, at least for high frequencies, the Wien law, which accurately described the high-frequency end of the observed blackbody radiation spectrum. Einstein had then used this presence of energy quanta in his discussion of the "creation and conversion" of light, as I described in Chapter 5. Later he extended the idea of the existence of quanta of wave energy to the thermal vibration of solids in his treatment of the specific heats of solids.

Haas had quantized the potential energy of an electron skimming the surface of a Thomson atom, requiring it to be just, hv, with v the orbital frequency of the electron, and so finding a connection between Planck's constant and the size of atoms.

Nicholson had imposed the condition that rings of electrons in his model of the atom considered as a whole must have a quantized angular momentum. Most of his predecessors had quantized energy associated with a definite frequency; Planck had quantized action, the area of the ellipse traced out in phase space during the motion of the phase point describing the motion of an oscillator. Nicholson had not applied a quantization condition to individual electrons and had not connected the quantization directly with the process of emission of spectral lines by the atom. Angular momentum has the same dimensions as energy divided by frequency or energy multiplied by time, and so this quantization also was in units of Planck's constant h or, more accurately, $h/2\pi$.

In his momentous 1913 paper establishing his theory of the hydrogen atom, Bohr had quantized the energy emitted in radiative processes much as Einstein had done in 1905 when he applied his discovery of energy quanta in the radiation field to Stokes' rule and suggested that radiation is "created and converted" in units of size $(R/N)\beta v$ or hv. Bohr required the energy radiated by an electron on being captured by a hydrogen nucleus to be an integral multiple of $h\omega/2$, with $\omega/2$ the average of the initial zero frequency and ω the frequency of the final orbital motion of the captured electron. He had then shown the connection of this quantization with classical theory by looking at the radiation emitted during a transition between two states in both of which the electron was at a great distance from the nucleus. Finally, he had demonstrated that this radiative energy quantization led to results similar to

Nicholson's, only here the angular momentum quantized was that of a single electron and not a ring of them, provided that the orbit was circular. Having arrived by a variety of means at quantized energy levels, Bohr then made the crucial assumption that the spectral lines were emitted by electrons making transitions between these quantized stationary states and emitting the corresponding energy in the form of a single Planck quantum.

When the problem of specifying the states of atoms more complicated than hydrogen and the kind of transitions that were responsible for these atoms' emitting spectral lines was addressed, it was not at all clear which of these quantization principles to apply or to which physical quantities they should be applied. In the second paper of his 1913 trilogy on the constitution of atoms and molecules, Bohr described the normal state of multielectron atoms as containing rings of electrons, each electron having angular momentum $h/2\pi$. It was this requirement that he credited with removing the instability against disturbances in the plane of the orbits that classical mechanics predicted. The excited states of these atoms was not treated in this paper because they appeared to be too complicated. Spectroscopy did not figure here.

The possibility of elliptical orbits caused trouble. Bohr's original proposal for quantization worked for elliptical orbits. Even though there are both angular and radial motions present in elliptical motion, in elliptical orbits they have the same frequency, and so Bohr could say that the frequency of the radiation emitted in the capture of a free electron had to be an integer multiple of half of this frequency and would obtain the same stationary-state energies for elliptical orbits as for circular ones. He could not, however, expect angular momentum quantization to give the energy of motion in an elliptical orbit, because an electron moving in an elliptical orbit had energy associated with its radial motion as well as with its angular motion. Angular momentum quantization alone sufficed only for circular orbits.

The situation got much worse when atoms with two or more electrons were considered. For example, when Moseley interpreted his characteristic X-ray data that led him to the simple ordering of elements in the periodic table by nuclear charge, he assigned the K series of lines to transitions involving a ring of four electrons in the innermost shell. The quantum conditions he considered seemed to apply to transitions by the ring as a whole. In a multielectron atom an electron moves not in a pure coulomb field but in a complicated one arising from the nucleus and the other electron(s). The simplest description of the resulting orbits is as ellipses whose major axes precessed so that the electrons trace out rosettes rather than closed ellipses. This is like the orbits of the planets whose motion about the sun is described in celestial mechanics as being in Keplerian ellipses whose major axes precess at various rates because of the presence of the other planets. When an electron moves in a rosette, the frequency of the radial motion differs from that of the angular motion. How is such a motion to be quantized? Which frequency should the radiation emitted in electron capture have? The quantization of nonperiodic orbits posed problems for Bohr's picture that he recognized but could not solve.

Another puzzle that was emerging about this time was the suspected doublet structure of the Balmer lines, a suspicion that became better and better established as the precision of spectroscopy improved. The splitting was very small, its present value being 0.365 wave numbets out of 15,237 for the first line of the Balmer series. There was no room for a doubling of this spectrum on Bohr's model of a point electron moving around a point nucleus. Bohr suspected that if the doubling actually were there, it would have to be ascribed to the nucleus's not acting quite like a point particle. It was natural to blame the nucleus because it was heavy and essentially nothing was known about it except its mass and its charge, while the electron was light and was thought to have a mass that was totally electromagnetic in nature, so that it was considered a relatively well understood particle.

The question of what kinds of classically defined quantities could be quantized without invalidating all the classically derived relations of those quantities to the rest of physics had arisen even before Bohr's stunning application of the process to atoms. As early as 1911 Paul Ehrenfest had been puzzled by Planck's success in imposing the classically derived Wien displacement law on his quantized oscillators to achieve a totally nonclassical result that the size of the energy quanta had to be proportional to the frequency.[1] He sought the feature of the displacement law that made it retain its validity even in the quantum regime. His method of search was similar to the one used earlier by Einstein when he understood Planck's blackbody work in his own way, as described in Chapter 5. Einstein had separated the process of calculating an average into two parts, a dynamical one depending on the quantity to be averaged and a statistical one specifying the weight to be given to each value of the dynamical quantity, by explicitly introducing a function representing this weight into the expression for the average of that physical quantity. Ehrenfest introduced a weight function just where Einstein had, defining the weight to be given to the states of a system—in his case the modes of the electromagnetic field in a cavity—in calculating mean values. Unlike Einstein, Ehrenfest used this weight function directly in making the combinatorial calculation of how many equally probable ways there were for the system to be in a specified macroscopic state. The equilibrium state was, according to Boltzmann, the one that could be achieved in the greatest number of ways.

The question that Ehrenfest posed to himself was what restrictions there would be on his weight function if he required the entropy of the equilibrium state to remain unchanged in value if the volume of a cavity containing radiation were changed adiabatically—"infinitely" slowly with no heat entering it or leaving it in the process—as Wien had done in his derivation of the displacement law. Rayleigh had studied the effect on the normal modes of transverse vibraton of a string fixed at both ends as the length of the string was adiabatically shortened and had shown that the energy of a mode and its frequency increased in the same proportion. The energy gain was supplied by the work necessary to shorten the string in the presence of the "radiation pressure" exerted on the support. Rayleigh had imagined that the end of the string was kept from moving transversely by a small ring through which it went. The string on the active side of the ring was generally not parallel to the equilibrium direction of the string, while that on the other side was. The tension in the string therefore produced an unbalanced force on the ring, tending to move it so as to lengthen the active part of the string. This was the radiation pressure. The frequency gain was caused by the need to keep the same number of half-wavelengths of the wave in the reduced length available. Rayleigh extended this argument to sound waves in a cavity that was slowly compressed, with the rest that again the energy of any mode and the frequency of that mode increased in proportion, their ratio remaining unchanged. Applying this result to the electromagnetic field modes in a cavity, Ehrenfest saw that the ratio of the mode energy to the mode frequency was an *adiabatic invariant* and that this being so, his weight function had to be a function of the invariant ratio E/v, and not of the two variables separately. The Wien displacement law imposes a similar condition on the function specifying the blackbody spectrum, and so Ehrenfest had provided a new derivation of this law that avoided having to treat complicated details such as the Doppler shift in the frequency of a wave reflected obliquely from a moving mirror that entered Wien's derivation. He pursued this further, demonstrating that the observed falling off of the spectrum at both high and low frequencies places further limits on the weight function, but this is outside our present concern.

This new understanding of the displacement law led Ehrenfest to investigate adiabatic invariants in more general cases. Eventually he was led to what Einstein called the *adiabatic principle*. According to this principle, only adiabatically invariant quantities can be subjected to quantization. Ehrenfest delayed publishing a full account of it until 1916, by which time other methods of establishing quantum rules had been developed, especially those by Sommerfeld and his colleagues in Munich, so that the impact of this work was greatly reduced. Throughout this period Ehrenfest shared the opinions of Stern and von Laue about the Bohr model, considering it "monstrous."

In 1915 a massive series of contributions to the solution of the problem of what to quantize began to come from Munich. In a letter to Rutherford dated November 29, 1916, Bohr referred to "an uninterrupted stream of German papers on this subject." Sommerfeld had made a proposal at the first Solvay Conference in 1911 regarding the quantization of nonperiodic motions using the action integral that Hamilton had introduced in connection with his form of the classical equations of motion. This integral was one with time as the variable of integration, and Sommerfeld had suggested that its value over the interval during which the physical event of concern took place was the quantity that should be subject to quantization in units of Planck's constant *h*. This proposal had not met a favorable response, and no important results were obtained from it, but its fundamental idea may have remained in Sommerfeld's mind. In 1915 a paper by Debye and Sommerfeld on the dispersion of light by atoms using a Bohr-like model for the atoms but treating the effect of the electromagnetic field on the atom classically had appeared, but it too did not lead far. The stream referred to by Bohr started with Sommerfeld's reading a paper to the Bavarian Academy in December 1915, entitled "On the Theory of the Balmer Series."

In this paper Sommerfeld made a major advance by giving a method for quantizing noncircular Kepler orbits.[2] These orbits needed two parameters to specify them, for example, their angular momentum and their eccentricity. Bohr's angular momentum rule provided only one quantum condition, so with the eccentricity not quantized and not restricted to the value zero it had for a circular orbit, the energy of Bohr's stationary states could depend in a continuous way on it, and the sharp spectral lines might not be there. Sommerfeld suggested a way to quantize the eccentricity that was to prove extraordinarily useful and could be extended to other situations.

A particle moving in a circular orbit of radius a required only one coordinate to specify its position. This was usually chosen to be the angle ϕ between the radial line connecting the particle to the center and some arbitrary fixed radial line. The kinetic energy T of the particle was then

$$T = \tfrac{1}{2}ma^2\left(\frac{d\phi}{dt}\right)^2 = \tfrac{1}{2}ma^2\omega^2.$$

The momentum p_ϕ conjugate to ϕ is the angular momentum of the particle around the center of the circle and had the value

$$p_\phi = ma^2\omega,$$

which was a constant for this motion. The Bohr–Nicholson condition that the angular momentum have one of the values $nh/2\pi$ with integer n was equivalent to requiring that the area of the phase plot of p_ϕ as a function of ϕ over one period of the motion, over an interval of length 2π, be an integer multiple of Planck's constant h. In mathematical notation,

$$\int_0^{2\pi} p_\phi\, d\phi = 2\pi p_\phi = nh.$$

This was reminiscent of Sommerfeld's earlier concern with quantizing action, where he would have written the integral as one over one period T of the time instead of over the angle, as

$$\int_0^T p_\phi\left(\frac{d\phi}{dt}\right) dt,$$

which is exactly the same thing. Whatever the motivation, for the case of elliptical motion in which the radial coordinate r is not fixed, Sommerfeld imposed a similar condition on the radial motion, requiring that

$$\int_0^T p_r\left(\frac{dr}{dt}\right) dt = n'h,$$

where again the integral is taken over one complete cycle of the electron motion. Because the motion is completely periodic, the value of T appearing as the limit of the time integration is the same in both the angular and radial quantum conditions. On expressing the energy of the electron in the orbit subject to these two quantum restrictions in terms of the two quantum numbers n and n', Sommerfeld was surprised at the answer he got. The Balmer formula appeared as before, except that the original angular momentum quantum number n was replaced by the sum $n + n'$ of it and the new radial quantum number:

> It seems to me excluded that such a precise and fruitful result could be ascribed to an algebraic accident; rather I see in it a persuasive justification for the extension of the quantization assumption to the radial coordinate, for the separate application of this assumption to both the degrees of freedom of our problem.

Nonetheless it was almost an accident. If the hydrogen atom had been as complicated as other atoms, the discovery of the quantum principles that govern atomic physics would have been much more difficult. The Kepler motion and the harmonic oscillator are systems whose unusual simplicity provided guidance on how to apply the new quantum principles. It was the failure of these quantum numbers always to combine in this simple way that caused the complications in the spectra of atoms with more than one electron, as Sommerfeld himself pointed out later in his paper.

There was now more than one stationary state of the hydrogen atom with the same energy, namely, all those with the same value of $n + n'$ but different values of n and n' separately. Those with $n' = 0$ where the radial momentum was zero were those with the circular orbits treated by Bohr. Those with $n = 0$ where the angular momentum was zero straight-line orbits extending out from the nucleus. Following one of them, the electron would collide with the nucleus. Sommerfeld excluded them from consideration, referring to Rutherford's experiments on the scattering of α-particles without explaining exactly how they pertained to this issue. The lower state reached on emitting a Balmer line, a state with $n' + n = 2$, could now be either one of two, the state with $n' = 0$, $n = 2$ having a circular orbit or that with $n' = 1$, $n = 1$ having an elliptical orbit with the same major axis.

A month later Sommerfeld presented another major report and announced another remarkable result.[3] In the earlier paper he had introduced the phase integral over the radial motion and evaluated it in the usual nonrelativistic way. In this second paper he evaluated it relativistically, taking into account the dependence of the electron mass on the velocity. This altered the motion, and for elliptical orbits the radial and angular periods were no longer equal.

This brought up a question about the range of the coordinates to be used in defining the quantum conditions. Originally Sommerfeld had, as mentioned earlier, considered the action integral over some appropriate time interval. Since the two periods, angular and radial, were now unequal, what was the appropriate time interval? This question had not come up in the nonrelativistic case. Sommerfeld chose to integrate both over the period of the radial motion so that the angular integration extended not over the range from 0 to 2π but over the longer range from 0 to $2\pi/\gamma$, where γ was a number less than unity. This caused some complications that he managed to ignore as physically unimportant.

The result of this calculation was that the lines of the Balmer series should be doublets because the two possible orbits with $n + n' = 2$, namely, those with $(n, n') = (2, 0)$ and $(1, 1)$ had different energies. In the first orbit (the circular one) the electron velocity was constant, and in the second one the velocity and therefore the mass varied during the motion. The orbit, $(0, 2)$, had been excluded. The higher states were also affected but not nearly as much, so that the observable splitting was produced by the doublet character of the final state. The doublet character of these lines was just being established observationally, and it had posed a problem for the Bohr model. Sommerfeld turned it from a problem into a pillar of the model.

In a note added in proof, Sommerfeld made an important change. The observations of the doublet spectrum of hydrogen had become more precise in the meantime, and they led him to alter the quantum condition giving the quantum number n. He gave up integrating the two quantum conditions over a common time interval and instead integrated each over a complete cycle of its values, such as from the minimum value of r to its maximum and then back to its minimum, as shown in Fig. 8.2. The quantum condition became one on the area of the resulting figure that had to be an integer multiple of h. This changed rule became the basis for all quantizations from then to the time ten years later when a consistent quantum theory began to be developed.

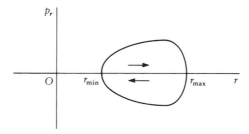

Fig. 8.2 The Sommerfeld scheme for the quantizing the radial motion of an electron in an atom. The radial momentum p_r was plotted as a function of r for a fixed energy E. The area enclosed in the plot, which depended on the energy, was required to be an integer multiple of Planck's constant h.

I have mentioned only the highlights of these two papers. The published version of the first one ran thirty-four pages, and that of the second took the next forty-two pages of the academy's proceedings. They were full of detailed treatments of many aspects of their subject, in particular the theory of the Stark effect, the splitting of spectral lines by applying an electric field to the emitting atoms. Their contents were published in 1916.[4]

Other scientists also made similar proposals for quantization rules. W. Wilson in England and J. Ishiwara in Japan had suggested these phase integral quantum rules, but neither of them pursued their implications for spectra as far or analyzed them as deeply as did Sommerfeld.

Sommerfeld had proposed his quantum conditions with certain reservations, even though they were so successful. What was not clear to him in his presentation was how to be sure to select the right coordinate system in which to do the calculations. For hydrogen, the use of polar coordinates was intuitively obvious, but in more complicated cases the choice might not be so easy. Two astrophysicists working independently cleared this up.

K. Schwarzschild and P. Epstein showed that the theories developed to calculate planetary motions could be applied to the hydrogen system and to any other conditionally periodic or multiply periodic system.[5] A system is conditionally or multiply periodic when each of its coordinates repeats the same sequence of values in the same times over and over, even though different coordinates take different times to complete one of their cycles. The solar system is multiply periodic if interplanetary perturbations are neglected; each planet executes its own periodic motion around the sun without the period of one's motion being related to the period on another's. As early as 1870, sophisticated mathematical methods had been developed to permit the calculation of the effects of one planet's motion on that of another. Among them was the introduction of a set of variables to describe the motion known as *action and angle variables*. These variables could not be defined for all mechanical systems, but for most multiply periodic systems they could be, and they made the formulas describing the motion of the system look like the formulas describing uniform motion in a circle. The angle variable w increased uniformly with time and the action J, the momentum conjugate to w, was constant. The way to find these coordinates was complicated, but the criteria they must meet were well defined. Schwarzschild and Epstein pointed out that the phase integrals that Sommerfeld had quantized were exactly the action variables for the systems he was studying. They also showed how to treat the Stark effect in an unambiguous way by introducing a parabolic coordinate system, a method that had been used in celestial mechanics to study satellite problems. This use of methods developed for celestial mechanics in atomic physics gave the subject a great impetus. The atom became a miniature solar system, but a quantized one.

Sommerfeld's papers produced a great impression on Bohr. He withdrew a major paper that he had submitted to the *Philosophical Magazine* for publication, as needing drastic revision in the light of Sommerfeld's contributions. Two year later he published a major memoir entitled "On the Quantum Theory of Line Spectra" in which all the latest developments of Sommerfeld, Einstein, Schwarzschild and Epstein, and Ehrenfest played a large part.

From this time on, the general scheme of deriving quantum conditions from classical mechanics was established. There were many questions that arose, and much detailed calculation was done in particular cases. Great difficulties were encountered in applying the methods derived from celestial mechanics to atomic systems. The greatest of these were in what were called *degenerate* systems, systems in which different motions with the same frequency were present. Ehrenfest's adiabatic principle was an important tool in resolving these problems. Only those atomic systems that resembled hydrogen could be treated with any success. These included ionized helium, as we have seen, and the alkali elements in which the spectrum could be accounted for by assuming the presence of a single *optically* active electron outside a core composed of the atomic nucleus and the other electrons. The core did not participate actively in the emission and absorption of light but supplied the electric field that determined the motion of the optical electron that had to be quantized. Another kind of problem was to determine the physical meaning of some of the quantum numbers that were needed to account for the observed spectral series and the behavior of spectral lines when external electric and magnetic fields were applied.

Notes

1. P. Ehrenfest, *Annalen der Physik* 36, 91 (1911).
2. A. Sommerfeld, *Bavarian Academy Sitzungs Berichte*, December 6, 1915, pp. 425, 459.
3. Ibid., p. 459.
4. A. Sommerfeld, *Annalen der Physik* 51, 1 (1916).
5. K. Schwarzschild, *Berlin Berichte* (1916), p. 548; P. Epstein, *Annalen der Physik* 50, 489 (1916); *Annalen der Physik* 51, 168 (1916).

Chance Enters: Einstein's Transition Probabilities

In 1916, after having just published his paper "The Foundation of the General Theory of Relativity" that established the present-day basis for our understanding of the cosmos, Einstein gave a report to the Zurich Physical Society that set the quantum theory of radiation on a new course. The substance of this report was published in 1917.[1] B. L. van der Waerden stated that "all subsequent research on absorption emission and dispersion of radiation was based upon Einstein's paper."[2] It was here that Einstein completed his work of 1905 by considering the aspects of emission and absorption involving momentum as well as energy. In so doing he introduced an entity possessing both energy and momentum, replacing the somewhat vaguely defined "energy quantum" with what became the photon. But above all, he introduced an element of probability into the discussion of radiative processes.

The only part of quantum theory that Einstein used was the idea introduced by Bohr that a molecule had

> a discrete set of states Z_1, Z_2, ... Z_n, ... with (internal) energies ε_1, ε_2, ... ε_n, ... apart from its orientation and translatory motion. If such a molecule belongs to a gas at temperature T, the relative frequency W_n of such states Z_n is given by the formula
>
> $$W_n = p_n \, exp(-\varepsilon_n/kT)$$
>
> which corresponds to the canonical distribution of states in statistical mechanics. For the present basic investigation, a detailed determination of the quantum states is not required.

Here p_n was the number of states with the same energy ε_n, called the statistical weight of states with that energy, and the exponential was the ubiquitous Boltzmann factor relating to the probability of finding a system in a state with a given energy at that temperature. With this straightforward combination of Bohr's stationary states and standard statistical mechanics, Einstein proceeded to give a new derivation of Planck's radiation law and Bohr's connection between energy differences and frequency.

Following Bohr, Einstein considered transitions between pairs of these stationary states to be accompanied by the emission or absorption of radiation of a definite frequency v whose energy was equal to the energy difference of the states. He made no assumption about how the frequency of this radiation was related to the energy difference. Bohr had given no indication of how he envisioned radiative transitions between stationary states taking place except to say that classical electrodynamics was incapable of describing them. Einstein made the remarkable assumption that the emission of light by a molecule was similar to the emission of a γ-ray by a radioactive atom. A molecule in a state Z_m was assigned in a fixed probability per unit time of going to a state Z_n of lower energy, emitting light in the process. He denoted this probability by an *A* coefficient:

> The statistical law which we assumed, corresponds to that of a radioactive reaction, and the above elementary process corresponds to a reaction in which only γ-rays are emitted. It need not be assumed here that the time taken for this process is zero, but only that the time should be negligible compared with the times which the molecule spends in states Z_1, etc.

This process, now called *spontaneous radiation*, he took as the quantum equivalent of the emission of radiation by an oscillator according to Maxwell–Hertz electrodynamics.

Next Einstein considered absorption. This required an energy source that he took to be the equilibrium radiation present at temperature T. He assumed that the molecule in a lower state Z_n had a probability per unit time of making an upward transition to a higher state Z_m stimulated by this radiation and absorbing energy from it. This probability he took as proportional to the density of the radiation present at the frequency corresponding to the energy difference of the levels, and he wrote it as the product of this energy density and a *B* coefficient. To this upward transition there corresponded a downward one stimulated by the presence of this same radiation. This was in addition to the probability for spontaneous downward transitions. The probability per unit time of this additional downward transition he took to be proportional to the relevant energy density multiplied by a different *B* coefficient.

With this foundation, Einstein could make a short and elegant derivation of the Planck law and Bohr's frequency condition. In equilibrium the rate of upward transitions from states of lower-energy ε_n to states of higher-energy ε_m of the molecules of a gas had to be the same as the rate of downward transitions between these states. The rate of upward transitions was the product of the number of the molecules in states with the lower-energy ε_n and the probability per unit time of any one of them making the transition. Since upward transitions required a source of energy, this rate was proportional to the energy density ρ. The rate of downward transitions was the product of the number of molecules in states with the higher-energy ε_m and the probability per unit time of any one of them making the transition. Here there were two ways of making it, spontaneously or by stimulation. The spontaneous term did not contain the energy density, while the stimulated term did. The ratio of the numbers of molecules in the upper- and lower-energy states was given by the Boltzmann factor. It was this that introduced the temperature into Einstein's considerations. Equating these two rates produced an equation connecting the energy density ρ at this frequency, the *A* and *B* coefficients, and the temperature *T*.

In order to analyze the consequences of the equation obtained in this way, Einstein assumed that at sufficiently high temperatures the energy density became so high that the stimulated processes overwhelmed the spontaneous ones. (Recall that this would not happen if the Wien radiation law were correct. According to the Wien law, the intensity at any fixed frequency approached a finite limit as the temperature increased indefinitely. Einstein had used the Wien radiation law when it was appropriate to the situation being studied, as it was in his introduction of the quantum, but here it was not.) Then the upward and downward stimulated rates by themselves had to be equal, a requirement that reduced the number of independent *B* coefficients.

Einstein next solved this simplified equation for the energy density at this frequency. Like Planck in 1900, he then put this equation in the form required by the Wien displacement law, namely,

$$\rho = \alpha v^3 \phi(v/T).$$

As we saw in Chapter 4, Planck's equation contained the temperature only in a term $\exp(\varepsilon/kT) - 1$, with ε the size of his energy quantum, and in order to make this conform to the displacement law he had to introduce his quantum by making $\varepsilon = h\nu$. Einstein's equation contained the temperature only in a term of the form $\exp[(\varepsilon_m - \varepsilon_n)/kT)] - 1$ coming from the Boltzmann factors. To make this conform to the displacement law he had to make

$$\varepsilon_m - \varepsilon_n = h\nu,$$

which was the relation between the energy difference of the two atomic levels and the frequency of the light emitted in a transition between them that Bohr had postulated. It is now often referred to as the *Bohr–Einstein relation*.
Einstein pointed out:

> To compute the numerical value of the constant α, one would have to have an exact theory of electrodynamic and mechanical processes; for the present, one has to confine oneself to a treatment of the limiting case of Rayleigh's law for high temperatures, for which the classical theory is valid in the limit.

Eight years later S. N. Bose showed how to derive the Planck law in an even more direct way without appealing to either the Wien displacement law or the detailed electrodynamic or mechanical processes. I shall describe this in the next chapter.

By this wonderful exploitation of the properties of thermal equilibrium, Einstein had simplified and deepened the understanding of the problems of spectroscopy. This was the fourth time that he had used this approach; his previous uses had revealed the energy quantum and explained the Brownian motion in 1905 and had explained the behavior of atomic heats of crystals in 1907. He used it once more in 1924 to give the quantum theory of a gas of identical particles.

The greater part of Einstein's paper was devoted to the momentum problem. Using methods taken from his treatment of Brownian motion, Einstein examined the motion of a molecule exchanging energy with thermal radiation by emitting and absorbing quata. Confining his attention to motion along a line and to the effect of just two internal states of the molecule, he argued as follows:

> The momentum $M\nu$ of a molecule undergoes two different types of change during the short time interval τ. Although the radiation is equally constituted in all directions, the molecule will nevertheless be subjected to a force originating from the radiation, which opposes the motion. Let this equal $R\nu$, where R is a constant to be determined later. This force would bring the molecule to rest, if it were not for the irregularity of the radiative interactions which transmit a momentum Δ of changing sign and magnitude to the molecule during time τ; such an unsystematic effect, as opposed to that previously mentioned, will sustain some movement of the molecule.

The calculation of R was analogous to that of the viscosity in the case of Brownian motion, only somewhat more complicated, and that of the mean square $\overline{\Delta^2}$ was analogous to that of $\overline{x^2}$. Einstein required the average kinetic energy of the molecule to remain that corresponding to the temperature and demonstrated that this would not happen if the emission of radiation by the molecule did not produce a recoil momentum. The balance would be maintained if this recoil momentum were that corresponding to the concentration of all the energy emitted in a transition into a single direction and so producing the maximum recoil. He came to the following conclusion:

> If the molecule undergoes a loss in energy of magnitude $h\nu$ without external excitation, by emitting this energy in the form of radiation (outgoing radiation), then this process, too, is directional. *Outgoing radiation in the form of spherical waves does not exist.* During the elementary process of radiative loss, the molecule suffers a recoil of magnitude $h\nu/c$ in a direction which is only determined by "chance," according to the present state of the theory. (italics added)
>
> These properties of the elementary processes, imposed by Eq. (12) [connecting $\overline{\Delta^2}$ with R], make the formulation of a proper quantum theory of radiation appear almost unavoidable. The weakness of the theory lies on the one hand in the fact that it does not get us any closer to making the connection with wave theory; on the other hand it leaves the duration and direction of the elementary processes to "chance." Nevertheless I am fully confident that the approach chosen here is a reliable one. . . . A theory can only be regarded as justified when it is able to show that the impulses transmitted by the radiation field to matter lead to motions that are in accordance with the theory of heat.

What impressed Einstein was the unidirectional nature of the emission process that maintaining the Maxwell velocity distribution for the molecules in the presence of the radiation required. The nonexistence of "outgoing radiation in the form of spherical waves" was truly revolutionary. But what impressed the rest of the world, which did not believe in light quanta anyway, was the introduction of the transition probabilities. The calculations or estimation of these, and their connection with the intensities of spectral lines, became a principal subject of interest among the atomic physicists, but the existence of atomic or molecular recoil was either denied or thoroughly ignored until the observation of the Compton effect in 1922 forced it on the physics community.

Notes

1. A. Einstein, *Physikalische Zeitschrift* **18**, 121 (1917). An English translation is included in B. L. van der Waerden, *Sources of Quantum Mechanics* (Amsterdam: North–Holland, 1967), p. 63.

2. Van der Waerden, *Sources of Quantum Mechanics*, p. 4.

Amplification

Einstein wrote the probability that a molecule in state Z_m would make a spontaneous transition to a state of lower-energy Z_n in the time interval dt as

$$dW = A_m^n \, dt$$

where the A coefficient was independent of the temperature. Similarly, he wrote the probability that a molecule in state Z_m would make an induced transition to state Z_n, which might be of either lower or higher energy, as

$$dW = B_m^n \rho \, dt$$

where ρ was the energy density of the radiation of the frequency corresponding to the energy difference of the two levels. This frequency was not assumed to be given by Bohr's rule but was left to be determined later.

The rate of upward transitions from state Z_n to state Z_m, all of which were induced, was then proportional to

$$p_n \, exp(-\varepsilon_n/kT)B_n^m \rho,$$

while the rate of downward transitions from Z_m to Z_n with the emission of radiation, either spontaneous or induced, was proportional to

$$p_m \, exp(-\varepsilon_m/kT)(A_m^n + B_m^n \rho)$$

with the same proportionality constant. Equating these two expressions gave

$$p_n \exp[[(\varepsilon_m - \varepsilon_n)/kT]B_n^m \rho = p_m(B_m^n \rho + A_m^n).$$

Now assuming that increasing T meant increasing ρ without limit (which was not true for the Wien law of blackbody radiation), at high temperatures the A coefficient on the right became negligible in comparison with the term containing the B coefficient, and the exponential became unity, so that Einstein was left with

$$p_n B_n^m = p_m B_m^n$$

which reduced the number of independent B coefficients. Using this result and solving for the energy density ρ gave

$$\rho = \frac{A_m^n/B_m^n}{exp[(\varepsilon_m - \varepsilon_n)/kT] - 1}.$$

The Wien displacement law required this to be of the form

$$\rho = \alpha v^3 \phi(v/T).$$

In this way Einstein found that

$$\frac{A_m^n}{B_m^n} = \alpha v^3$$

and

$$(\varepsilon_m - \varepsilon_n) = h v.$$

More Quantum Numbers and Rules

In the early sections of this chapter I described the successful introduction of quantum ideas into theories of the structure of the hydrogen atom. Now I shall turn to the troubles and travails encountered in trying to understand things more complicated than the Balmer series.

Sommerfeld introduced a second quantum number into the description of each of the Balmer levels, as we discussed earlier, in the second of his two memoirs on the Balmer series. In the version of these reports published in the 1916 Sommerfeld added a third quantum number.[1] The necessity for this appeared when he treated the motion of the electron in the hydrogen atom as one taking place in three dimensions, that is, when he did not tailor his coordinate system to the orientation of the electron orbit in space so that the plane of the orbit was a coordinate plane. Now he required three coordinates to locate the electron relative to the nucleus. He used spherical coordinates, the radial distance r, the colatitude θ (measured from the pole, not the latitude that geographers use measured from the equator), and the longitude or azimuth ψ. Corresponding to these three coordinates were the three momenta p_r, p_θ, and p_ψ. On each q-p pair Sommerfeld imposed his quantum condition that the area of the q-p plot traced out during one complete cycle of that particular coordinate should be an integer multiple of h. If the motion happened to be in the equatorial plane where $\theta = \pi/2$ and $P_\theta = 0$, nothing would be different from the case treated earlier. In general, the motion would not be in this plane, and Sommerfeld had to proceed carefully.

In order to impose his quantum condition, he had to be able to express the momentum being considered as a function of its conjugate coordinate alone, with no other coordinate or momentum appearing in the expression for it. This was the case for p_ψ which, according to the equations of motion, was a constant. The range of variation of ψ was 2π, and so the quantum condition was simply

$$p_\psi = k_1 h/2\pi.$$

This would now be described as a quantization of the component of the angular momentum along the polar or z-axis and could be of either sign. But this was not how Sommerfeld thought of it. His attention was on the area of the ψ-p_ψ phase plot, and here he did not distinguish between the two possible directions of the electron's motion around the polar axis. All the quantum numbers were taken to be positive.

The expression for the momentum p_θ depended on θ but also on p_ψ. In the particular case of a central force, however, this latter is a constant. The plot of p_θ as a function of θ contained the value of p_ψ as a parameter, but it was still meaningful to require the area under one complete cycle of the p_θ-θ plot to be an integer multiple of h. In this way Sommerfeld arrived at a second condition,

$$\oint p_\theta \, d\theta = k_2 h / 2\pi.$$

Because the angle θ was not measured in a fixed plane, the physical meaning of the quantum number k_2 was not easily seen. What was physically clear was the meaning of the combination $k = k_1 + k_2$. Here k was the value that k_1 would have if the equatorial plane of the coordinate system had been chosen as the orbit plane, and so was the magnitude of the angular momentum vector. If α was the angle between the orbit plane and the equatorial plane, then

$$k_1 = k \cos \alpha.$$

The quantization of the radial motion went through as before and again yielded the Balmer states, in which the integer labeling the states was the sum $n = (n' + k_1 + k_2)$ of the three quantum numbers. This sum was called the *principal* quantum number. Now every combination of the three quantum numbers adding up to the same principal quantum number led to a state of the same energy. As before, states with $k = 0$ were excluded because, having no angular momentum, they would entail collisions of the electron with the nucleus.

Because both the magnitude of the angular momentum and the z-component of the angular momentum had to be integer multiples of $h/2\pi$, the angle α between the angular momentum vector and the z-axis was restricted to those angles for which

$$\cos \alpha = \frac{k_1}{k}$$

is the ratio of two integers. The angle α was quantized! This *spatial* quantization was important to analyzing the effect of external fields on the states of the atom.

Five years later this received a direct experimental confirmation in the experiments by Otto Stern and W. Gerlach.[2] These two passed a beam of silver atoms through an inhomogeneous magnetic field that exerted a force on any magnetic moment the atoms had, tending to pull it into or repel it away from the stronger field, depending on whether the moment had a component in the direction of increasing or decreasing field (see Fig. 8.3). Classicaly, one would anticipate that the presence of this inhomogeneous field would simply spread out the beam along the direction in which the field

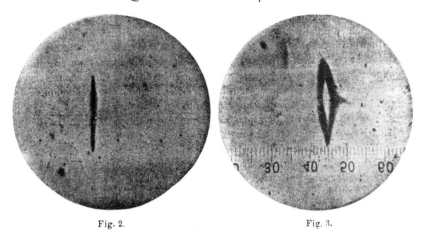

Fig. 2.   Fig. 3.

Fig. 8.3 The pattern of a narrow beam of silver atoms deposited on a glass plate. In Fig. 8.2 there was no magnetic field present. In Fig. 8.3 there was a strongly inhomogeneous magnetic field present, with the horizontal field strength varying rapidly in the horizontal direction. The splitting of the beam into two well-defined components showed the presence of only two orientations of the magnetic moment of the atoms relative to the direction of the magnetic field. This was what spatial quantization predicted.

varied, as classically the magnetic moment could be in any direction and therefore could have a component of either sign and any size between zero and the maximum allowed by the size of the moment. Instead of this, however, they observed a clean division of the beam into two parts. The atoms were either attracted to or repelled from the region of stronger field with the maximum possible force; there was no middle ground. They had shown that the magnetic moment was spatially quantized and that only the values $\cos \alpha = \pm 1$ occurred. Careful measurements of the magnetic field and of the trajectories of the silver atoms established that the magnitude of the magnetic moment was what would be expected for a single electron in a circular orbit with $k = 1$, namely, $eh/4\pi mc$, called a *Bohr magneton*. The reason that the value $\cos \alpha = 0$ was missing was not clear at the time of the experiments. By what mechanism moments aligned themselves with the field also was not clear.

Sommerfeld and Bohr excluded the orbits with $k_1 = 0$ even when k was greater than zero, on the ground that in these orbits the electron would eventually approach the nucleus in the presence of an electric field along the polar axis, and as this was the case of arbitrarily weak fields, it should also hold for a zero field.[3] This is another illustration of the troubles encountered in trying to visualize the meaning of spatial quantization. In the absence of any electric field, which orbits were excluded by this condition depended entirely on the choice of direction of the coordinate system's polar axis.

The unsatisfactory nature of this exclusion was recognized, but without it there was no way to account for the number of states that were seen to be present in spectra.

Sommerfeld applied these ideas to an analysis of the Zeeman effect in hydrogen before there were good experimental data with which to compare the theory.[4] He used a long-known result of Joseph Larmor that the principal effect of a magnetic field on the motion of an electron in a coulomb field is to produce a motion that is almost exactly like that in the absence of the magnetic field if only it is described in a coordinate system rotating around the direction of the field at the rate

$$o = \frac{eB}{2mc}$$

If the motion of the electron in a hydrogen atom were described in a coordinate system rotating around the polar axis at this rate, it would be just the normal Kepler motion, and so Sommerfeld could easily find it in the laboratory coordinate system not rotating around the field and could apply the quantum rules to this perturbed motion. He found that the energy W of this perturbed motion differed from the energy W_0 of the same motion in the absence of the magnetic field,

$$W = W_0 + \frac{o}{4\pi} k_1 h.$$

This change of energy would change the frequencies of the radiation emitted in a transition to this state. (Bohr had speculated that although this would happen in the presence of an electric field, as in the Stark effect, the presence of a magnetic field might alter his fundamental relation between emitted frequency and energy difference. Sommerfeld did not agree with this, and it was soon seen to be an unnecessary complication.)

Earlier, Sommerfeld had proposed that all, or nearly all, transitions that were observed had changes in quantum numbers from initial values to final values that were not larger than the initial ones. In applying this "quantum inequality" to the Zeeman effect, he had to recognize the possibility of two directions of motion of the electron in its orbit; one would be with an increased energy and the other with a decreased energy. In effect this introduced negative values of k_1 corresponding to motions in which the component of angular momentum about the polar axis was negative and made the splitting of spectral lines symmetric about the original line. He then found the number of components into which the various lines of the Balmer series would be split. Labeling the states with the quantum numbers n', k_1, and k_2, Sommerfeld found that the Balmer α-line emitted in transitions between states with principal quantum number 3 and those with principal quantum number 2 obeying his quantum inequality would indeed split into just three lines with the separation of the lines of a Lorentz triplet. Only transitions in which k_1

changed by zero or unity occurred. For the Balmer β-line, from 4 to 2, however, there would be five components, and transitions in which k_1 changed by two also seemed possible. In general the greatest change in k_1 permitted by his inequality was the difference of the initial and final principal quantum numbers. Those transitions in which k_1 changed by more than unity Sommerfeld called *überzählige* (excess) components. He considered their reality very dubious, especially as it was known that for sufficiently strong fields the spectrum did consist of triplets, with the Lorentz separation where the change in k_1 was unity. The existence of physically based *selection rules* restricting the change of various quantum numbers was not yet contemplated. He then proceeded to a long calculation of the effect of relativity on the Zeeman effect but, to his disappointment, found no significant change.

In 1918 the implications of conservation laws for radiative transitions began to be explored. Einstein had already done this for blackbody radiation in 1916. In his analysis of thermal equilibrium between molecules with discrete energy states and the radiation field, he had shown that if the molecules were to maintain the Maxwell–Boltzmann speed distribution while emitting and absorbing radiation, each emission process had to be accompanied by a recoil coming from the conservation of momentum corresponding to the emission of the entire quantum of energy in a single direction. His remarkable conclusion that "outgoing radiation in the form of spherical waves does not exist" (see the previous section) was a consequence of the energy quanta, the present day *photon*, that he alone recognized as present in the electromagnetic field and that was completely ignored in all the studies of spectra by Bohr, Sommerfeld, and others. Emission in spherical waves was always assumed, and the possibility of recoil was denied. The conservation law that became important in connection with spectra was that of *angular* momentum.

Two accounts of the implications of angular momentum conservation for atomic transitions were published in 1918, one by Bohr and the other by A. Rubinowicz. At the end of Part 1 of his paper entitled "The Quantum Theory of Line Spectra," Bohr considered the angular momentum contained in the field radiated by a charge rotating about the polar axis according to classical electrodynamics and showed that

> the amount of energy and of angular momentum round an axis through the
> centre of the field perpendicular to the plane of the orbit, lost by the electron
> in unit of time as a consequence of the radiation, would be equal to $2\pi v a F$
> and aF respectively. Due to the principles of conservation of energy and of
> angular momentum holding in ordinary electrodynamics, we should therefore
> expect that the ratio between the energy and the angular momentum of the
> emitted radiation would be $2\pi v$, but this is seen to be equal to the ratio
> between the energy hv and the angular momentum $h/2\pi$ lost by the system
> considered above during a transition for which we have assumed the radiation
> is circularly polarized.

The transition considered was one in which his quantum number n_3 (Sommerfeld's k_1) changed by either plus or minus unity, while the other two quantum numbers could change by arbitrary amounts and in which light that was circularly polarized as seen along the axis of symmetry was emitted. If the light were linearly polarized instead, this quantum number would not change. So here was a physically based *selection rule* governing the amount by which the the quantum number associated with motion around this axis could change.

Rubinowicz made a calculation similar to Bohr's, whose priority be acknowledged in a footnote added just after completing his paper. His approach led to a selection rule for the quantum number, k in Sommerfeld's notation, specifying the magnitude of the electron's angular momentum as well as for k_1. Without considering the polarization of the emitted light, Rubinowicz showed that angular momentum conservation required that the change in the azimuthal quantum number k in a transition from an initial term i to a final term f was restricted by the inequality

$$|k_i - k_f| \le 1.$$

This was correct only for dipole radiation, but this was the only kind that played any role in the theory. Then taking polarization into account, Rubinowicz essentially rederived Bohr's result for the k_1 selection rule. He proposed that these rules replace Sommerfeld's quantum inequalities in limiting the kinds of transitions between stationary states that should be seen. They eliminated the "excess" lines in the Zeeman spectrum, and they appeared to eliminate some faint lines seen in the helium spectrum. The presence of these lines could, however, be explained as the effect of an electric field from the discharge-producing spectrum on the helium atom and so presented no difficulty for the rules.

As I recounted earlier, Rydberg had extended Balmer's idea to the spectrum of the alkalies in the first column of the periodic table of the elements, writing the Balmer formula with the addition of a "quantum defect" to the integer in Balmer's formula for the frequency. Ritz had carried this further, stating that the frequency of the lines in a spectral series could always be expressed as the difference between two terms, each of which was labeled by an integer. According to Sommerfeld ,

> The object of spectroscopy is to determine the atomic states and their energy values. . . . Only when the spectral lines have been developed in series, and have been resolved into terms, may the object of spectroscopy be said to have been attained.[5]

The helium spectrum presented formidable complications, about which I quote Bohr:

> In a detailed discussion of these spectra it seems necessary to take the mutual perturbing effect of the orbits of the inner electrons and of the outer electrons into account. In general this constitutes a very intricate problem due to the fact that already when the outer electron is absent, the system of inner electrons will in general be unstable for small displacements, if the effect of

such displacements is calculated by means of ordinary mechanics. In case of helium, however, where there are only two electrons in the neutral atom, this is different since the motion of the inner electron will be mechanically stable for any shape or position of its orbit, if the outer electron is removed to infinite distance from the nucleus. Just in this property of the helium atom an explanation may be sought for the fact that helium, besides its simple spark spectrum, . . . , possesses two complete series spectra of the first order, the so called orthohelium and parhelium spectrum, for which no mutual combination lines are observed. This, which is seen to be in striking contrast to what should be expected for the spectrum of a simple central system, must be ascribed to the existence of two different types of motion of the inner electron.[6]

The nature of these two different types of motion remained unknown.

The spectra of the elements in the first three columns of the periodic table of the elements bore enough similarity to the hydrogen spectrum to permit the identification of several series of lines. They differed from the hydrogen lines by being multiplets; the two yellow *D*-lines in the spectrum of sodium, for example, corresponded to a single line of hydrogen. The spectra of the alkali earths in the second column of the periodic table showed triplets. An enormous amount of work was done by Ritz, Paschen, and others in organizing the observed lines into series and identifying the terms whose differences gave their frequencies, all without the benefit of any theoretical guidance from a good physical picture of what was going on. There was an ever-increasing mass of data that the Bohr–Sommerfeld theory had to confront and fit into a scheme of atomic energy states labeled by quantum numbers that had a physical meaning in the theory.

According to the theory developed up to 1918, the ordinary or *arc* spectra of elements were emitted in transitions between stationary states of a single electron moving in the electric field of the atomic nucleus and all the other electrons. The field of these other electrons was assumed to be a central field to a very good approximation so that the angular momentum of the active or optical electron would be conserved. Under these circumstances the energy of an atom's stationary states should be determined by two quantum numbers, the principal quantum number n and the azimuthal quantum number k in Sommerfeld's notation, if no external field were present. The "spatial" quantum number k_1 specified only the orientation of the orbit in space and could not affect the energy unless there were an external field to define a preferred direction. The study of multiplet spectra soon showed the necessity for an additional quantum number.

Several series of terms or states had been identified and given the designations *sharp* or *s*, *principal* or *p*, *diffuse* or *d*, and *fundamental* or *f* (originally *Bergmann*). Sommerfeld had successfully correlated these labels with values of the azimuthal quantum number k, *s* with $k = 1$, *p* with $k = 2$, *d* with $k = 3$, and *f* with $k = 4$. These terms were identified by analyzing series of spectral lines, principal series from *p*-states to an *s*-state, first subordinate series from d-states to a p-state, second subordinate series from s-states to a p-state, and fundamental series from f-states to a d-state. These assignments were consistent with the selection rule for k that it should change by one unit in either direction. More important, terms became more and more like Balmer terms with increasing k. This was persuasive because large k-values corresponded to wide orbits, with the electron staying away from the nucleus and the inner electrons and, therefore, behaving much as though it were in a hydrogen atom. Sommerfeld arranged these series of terms in an array

1s	2s	3s	4s	5s	6s	. . .
	2p	3p	4p	5p	6p	. . .
		3d	4d	5d	6d	. . .
			4f	5f	6f	. . .
				5g	6g	. . .
					6h	. . .

that showed the principal quantum number or running index of a series, starting with 1 for the s-series, 2 for the p-series, and so on. Beyond f the series were labeled alphabetically. The principal series had $1s$ as its constant term; both subordinate series had $2p$ as their constant term; and the fundamental series had $3d$ as its constant term.

This scheme explained the gross feature of the observed spectra. The lines observed led, however, to doublet or triplet terms for elements in the first and second columns of the periodic table. There were term series consisting entirely of doublets and others consisting entirely of triplets, but not all transitions between these multiplet terms were seen. To account for this, Sommerfeld introduced an additional index j, which he described in the following way:

> Clearly it cannot be the azimuthal quantum number which is effective here. The azimuthal quantum number is associated with the angular momentum of the entire atom, so-to-say its outer rotation; it has, according to our interpretation of the series, the value 3 for all the d-terms and the value 2 for all the p-terms. Every transition from a d-term to a p-term satisfies the selection rule. The distinguishing index of the different d- and p-terms must be an inner quantum number, perhaps corresponding to a hidden rotation. We know as little about its geometrical meaning as we do about the difference between the orbits that underlay the multiplicities of the series terms.[7]

This "hidden rotation" was conceived of as associated with the atomic core, as distinct from the active electron. The s-term did not need this new quantum number, but the others did. The doublet p-terms were now labeled mp_j, with the principal quantum number m and p_j, with $j = 1, 2$ and the triplet p-terms with $j = 1, 2, 3$ (which later became 0, 1, 2). The higher series of triplets received similar labels differing only in the three values of j used. Many features of the series were explained by this, and so it was not just an arbitrary labeling even if its physical basis was unclear:

> The separation of the doublets in the two subordinates series ending on the 2p-term, one coming from the s-terms and the other from the d-terms, was the same for all lines in the series. The s-terms were all simple and the d-terms were expected to have very small splittings because of the large width of their orbits, so the splitting could be attributed entirely to that of the fixed 2p-term.

> The separation of the doublets in the principal series decreased with increasing frequency of the lines, it all coming from the p-terms. The 1s-term on which the transitions ended was simple and the splittings of the p-terms decreased with increasing principal quantum number where the orbits became more hydrogen-like.

There were many other less simple features of the spectra that also fitted this pattern.

The selection rule for *j* that emerged from the first subordinate series of the elements in the second column of the periodic table, the alkaline earths, was the value of *j* could not change by more than unity. Both the p-series and the d-series of these elements were triplets, so if there were no restrictions on which components of one series could combine with those of the other, there should be 3 × 3 = 9 lines. Only six were seen, and so three combinations were excluded for some reason. The selection rule suggested did exclude three combinations. Things were getting complicated, but a complicated set of data was being organized if not explained.

The Zeeman effect continued to defy all efforts to understand it. The only way that a magnetic field could influence the terms was by the atom's acting like a small magnet. This it was expected to do because the circulating electrons constituted an electric current, and a current loop produces a magnetic moment. If the current were composed only of electrons, the magnetic moment μ would be proportional to the angular momentum of all the electrons with a calculable constant of proportionality:

$$\mu = \frac{e}{2mc} P$$

where *P* is the total angular momentum of all the electrons in the atom. This led inevitably to Lorentz triplets and gave no hint about how the anomalous Zeeman splittings arose.

The innermost electron shell, the K-shell containing just two electrons, investigated by Moseley in his X-ray studies, presented a puzzle and caused a disagreement between Bohr and Sommerfeld. Sommerfeld presented the situation in this way:

> While for the L-shell with its eight electrons symmetrical arrangements offer themselves in which the angular momenta cancel each other out, for the K-shell and its symmetry complete uncertainty remains. The author is—in

contrast to Bohr—of the persuasion that also in the K-shell a compensation of the angular momentum must occur and that therefore the resultant angular momentum of the K-shell, even as that of the L-, M-, . . . -shells, must vanish. This means that both electrons of the K-shell must be arranged in the same orbit plane with opposite directions of circulation. Two congruent . . . circular orbits of this kind are impossible, as in them the electrons moving in opposite directions would collide. . . . In contrast to this Bohr sees it as settled that the two electrons in the K-shell describe 1_1 [circular] orbits which must be crossed in space and whose angular momenta cannot cancel out. It appears to me unthinkable, however, that in the middle of every atom there sits a top that does not make itself noticeable from the outside. Especially everything we know about atomic magnetism appears to require that the K-shell can have no resultant angular momentum.[8]

There were real difficulties associated with atomic models in which the electrons were treated at all times as classical particles with well-defined positions and velocities. The way out to be provided by quantum mechanics— the denial of these classical attributes to electrons in atoms—was beyond the wildest speculations of the physicists struggling with these problems at this time. Not only the avoidance of collisions between the electrons in a given shell caused difficulty; the whole question of why there were shells remained unanswered. The only reason advanced for the existence of closed electron shells was that they provided stable configurations when the classical motion was restricted by the quantum conditions, but no calculation could establish that this was the case.

Finally in 1921 the fatal simplicity of Larmor's theorem on the effect of a magnetic field on the motion of the electrons in an atom was overcome by a suggestion by Alfred Landé, and some order was brought into the anomalous Zeeman effect.[9] Landé proposed that the proportion between angular momentum and magnetic moment was not always that required by the Larmor theorem but could depend on the quantum state of the system. In particular, the magnetic moment associated with the core angular momentum could be twice that required by Larmor. The most significant consequence of this was that the total angular momentum and the total magnetic moment of the atom were no longer necessarily in the same direction, a fact that opened up new possibilities that could encompass the anomalous Zeeman splittings. Landé gave no derivation of the formulas that he proposed for the way to combine the quantum numbers to achieve agreement with the observed splittings. He also found himself forced to introduce some half-integer quantum numbers in order to obtain the correct number of Zeeman components. The picture he ended up with could be described in this way: The vector total angular momentum of all the electrons in the atom, which would be fixed in the absence of a magnetic field, precessed about the direction of the field at the rate given by the Larmor theorem, but in addition, there was a precession at a different rate of the atom about the precessing angular momentum. The physical cause of this additional precession remained unknown. The implication was that the total angular momentum of, say, a sodium atom consisted of a contribution from the outer, optically active electron *and* a contribution from the "atomic core." The core was the site of the "hidden rotation" that Sommerfeld described when he introduced the inner quantum number.

A momentous advance was made in 1925 when Wolfang Pauli stated his *exclusion principle*. The mystery surrounding the existence of closed shells of electrons in atoms was mentioned earlier. These shells were necessary to account for the arrangement of the elements in the periodic table, especially to account for the properties of the noble gases. The chemical and spectroscopic properties of an element were attributed to the outer electrons that moved under the influence of the nucleus and the closed shells of inner electrons, in contrast with the X-ray spectra that depended on the structure of the inner, closed shells. The number of electrons in each shell was known, but the reason for these numbers was not. Pauli introduced a great deal of order into this in a paper with the formidable title "On the Connection of the Closing of the Electron Shells in Atoms with the Complex Structure of Spectra."[10] His argument hinged on proposals by Bohr and Edmond C. Stoner to systematize the structure of the atoms as they appeared in the periodic table of elements.

Bohr had long been concerned with the structure of elements other than hydrogen and helium. As early as 1921 he had proposed the first scheme of building up the periodic table by considering the successive attachment of electrons to a nucleus carrying a charge equal to the atomic number of the element being considered.[11] Combining knowledge of the chemical behavior of the element and the spectrum of that element, he arrived at a possible picture of the normal state of the noble gases and provided a physical framework in which to probe the structure of the heavier atoms. In this picture the noble gases played a special role; they were chemically inert and were located between elements in the first column of the periodic table, the alkalies, which easily parted with one electron, and the elements in the seventh column of the periodic table, the halogens, which easily acquired an extra electron. This suggested that in the noble gases the electrons formed closed shells, in the sense that the electrons in them were so arranged that they were particularly stable and could not easily either accept another electron or give one up. Bohr labeled these closed shells by the principal quantum number n and the azimuthal quantum number k of the electrons in them. In Bohr's scheme, the number of electrons in a closed shell of specified principal quantum number and azimuthal quantum number was not uniquely determined by those quantum numbers. For example, the shell in argon with $n = 3$, $k = 1$ acted closed when it contained four electrons, but the shell with these same quantum numbers in krypton was called closed only when it contained six electrons.

Bohr's pioneering effort was modified three years later by Stoner.[12] What Stoner added was a consideration of subshells labeled by two quantum numbers, equivalent to k_1 and k_2, instead of larger shells labeled only by k, the sum of k_1 and k_2. What he gained by this was a scheme describing the noble gas shells in such a way that a shell once filled remained filled as atoms

of higher atomic number were considered, so that the closed shells acquired
a more universal significance than they had had in Bohr's model. A feature of
both models was that the number of electrons in a closed shell was always
twice the number arrived at by just looking at the number of terms associated
with the quantum numbers associated with the shell.

When analyzing Stoner's version of Bohr's structure of the noble gases,
Pauli saw that this structure could easily be described by a simply stated but
totally inexplicable principle. First he needed to make the fourth quantum
number labeling a term characterize the electron itself and not the atomic core,
as had been Sommerfeld's idea. In this way the inner quantum number became
a fourth quantum number for each electron, rather than one characterizing
the atom as a whole. Pauli identified this quantum number with a magnetic
quantum number m_1 that could take on only two values that had been found
necessary to label the states of an atom in a strong magnetic field where the
classical picture of Lorentz triplets worked. He then formulated his exclusion
principle in the following way:

> There can never be two or more equivalent electrons in an atom for which
> in strong magnetic fields the values of all the quantum numbers n, k_1, k_2,
> m_1 are all the same. If an electron is present in an atom for which these
> quantum numbers (in an external field) have specified values, then this state
> is "occupied."

Pauli went on to point out that even though the principle was formulated in
connection with the quantum numbers appropriate to the presence of a strong
external magnetic field, on thermodynamic grounds it could be extended to
the case in which such fields were absent, by using the adiabatic principle and
imaginging the magnetic field to be slowly turned off. It was the restriction of
m_1 to a range of just two values that doubled the number of electrons in each
state labeled by definite values of the three quantum numbers n, k_1, k_2. In the
abstract of his paper concerning the alkali doublets, Pauli wrote:

> The interpretation will be suggested that these doublets and their anomalous
> Zeeman effect are the result of a double-valuedness of the quantum theoretic
> properties of the optical electron that cannot be described classically, without
> involving the atomic core with its noble gas configuration either through a
> core momentum or as the location of the magneto-mechanical anomaly of
> the atom.

This last referred to the failure of Larmor's theorme to apply to the Zeeman
effect. Here Pauli explicitly put all responsibility for the magnetic behavior of
the atom on the outer electrons. But he did not go further and describe how
this double-valuedness arose.

Pauli examined the application of his principle to a variety of situations.
One of special interest for the future was his observation that an atom with
one electron outside a noble gas core had states that were in one-to-one
correspondence with those of an atom lacking only one electron from having
the next closed shell. That is, a single electron outside a noble gas core acted

similarly to a vacancy in the almost-complete shell around that same noble gas core. Paul Dirac extended this argument to relativistic quantum mechanics to predict the positron as a "hole" in a filled sea of electrons. It also is used in condensed-matter physics, especially in connection with semiconductors, to explain the existence of two types of conduction, that by electrons and that by "holes," the two corresponding to the *n*-type and *p*-type semiconductors forming the basis of solid-state electrons.

With the statement of Pauli's exclusion principle, the periodic table finally made sense. Not that it was possible to calculate the properties of the elements, or even to analyze their spectra completely except in relatively simple cases, but a framework of rules and principles now existed within which the observed phenomena could be expected to fit. The source or foundation of the rules was as mysterious as ever; the relationship between the new quantum rules and the classical mechanics and electromagnetic theory that they modified in some places while requiring them in others was totally hidden. The patchwork had enormous successes but remained a patchwork.

The final act in this long process of finding rules that accounted for the regularities observed in spectroscopy and a few other branches of physics was the discovery of the electron spin by G. E. Uhlenbeck and S. A. Goudsmit in 1925. This was announced in a brief note published in late 1925.[13] In it they proposed to reinterpret the four quantum numbers (aside from the principal quantum number) used to label an atomic term along the lines suggested by Pauli, who had just shown that in the alkalies they all must be associated with the electron and not with the atomic core. According to Landé's vector model, these quantum numbers, using one of the many notations that were then in use, had been taken to be (1) R, the angular momentum of the atomic core; (2) K, the angular momentum of the optical electron; (3) J, the vector sum of these two angular momenta; and (4) m_1, the component of J along the direction of an external magnetic field. In this assignment it was necessary to assume that the magnetic moment of the atomic core was twice as large as it should have been according to the Larmor theorem; otherwise all the Zeeman splittings would have been into triplets. Uhlenbeck and Goudsmit pointed out that if as Pauli had shown, four quantum numbers were required to specify the state of an electron in a magnetic field, the electron should have four degrees of freedom. The center-of-mass motion accounted for three; the *z*-component of angular momentum of the electron, fixed in magnitude but but variable in direction, could provide the fourth. According to them, R represented an intrinsic electron angular momentum or *spin*. The other quantum numbers retained their original meanings. The idea of spatial quantization applied to the spin allowed only a discrete set of orientations of the spin, and to achieve the double-valuedness required by Pauli there had to be only two orientations, one parallel and the other antiparallel to the field direction. This required the magnitude of the spin angular momentum to be one half of the quantum unit $h/2\pi$. Uhlenbeck and Goudsmit required the magnetic moment of the spinning electron to have twice the Larmor value, just as the atomic core had had in the earlier model, in order to give the correct splittings in a magnetic field.

The electron spin model embodied all the features that Pauli had demanded in such a way that the Landé vector model could be used, not only for the alkalies, but also for all atoms. It was a model to which Bohr's correspondence principle could not be applied. Because the spin angular momentum was fixed in magnitude, with the value $(1/2)h/2\pi$, there was no way of taking the limit as the quantum number approached infinity and so no way of approaching classical behavior. Interpreting the model classically, its proposers showed that an electron of the classical electron radius e^2/mc^2 would have to have a surface velocity exceeding that of light in order to produce the required magnetic moment. The spin was indeed inexplicable classically.

Notes

1. A. Sommerfeld, *Annalen der Physik* **51**, 1 (1916).

2. O. Stern, *Zeitschrift für Physik* **7**, 249 (1921); W. Gerlach and O. Stern, *Zeitschrift für Physik* **8**, 10 (1922); **9**, 349, 352 (1922).

3. A. Sommerfeld, *Atombau und Spektrallinien*, 4th ed. (Braunschweig: F. Viweg, 1924), p. 143–44. See also N. Bohr, *Collected Works* (New York: North–Holland, 1976), vol. 3, p. 75.

4. A. Sommerfeld, *Physikalische Zeitscrift* **17**, 491 (1916).

5. Ibid.

6. N. Bohr, *D. K. D. Vikensk. Selsk. Skr., naturvidensk og mathem. Afd.*, 8. Raekke. IV, 1; also Bohr, *Collected Works*, vol. 3, p. 171.

7. A. Sommerfeld, *Annalen der Physik* **63**, 221 (1920).

8. A. Sommerfeld, *Atombau und Spektrallinien*, p. 197.

9. A. Landé, *Zeitschrift für Physik* **5**, 231 (1921), and **7**, 398 (1921).

10. W. Pauli, *Zeitschrift für Physik* **31**, 765 (1925).

11. N. Bohr, *Zeitschrift für Physik* **9**, 1 (1922).

12. E. C. Stoner, *Philosophical Magazine* **48**, 719 (1924).

13. G. E. Uhlenbeck and S. A. Goudsmit, *Naturwissenschaften* **13**, 953 (1925).

CHAPTER 9

Particles and Waves:
A Dissolving Distinction

There were many difficulties with the semiclassical model of the atom consisting of a number of particles, electrons, that moved rapidly around another particle, the nucleus. An obvious one was the apparent ability of these electrons to avoid collisions with one another even when their orbits crossed. Collisions would disrupt the steady state that was at the heart of the Bohr picture. Another, seldom-mentioned problem was the flatness of the atoms. If there was an outermost electron in hydrogen and in all the alkali atoms whose orbit had to avoid the nucleus, there had to be an orbital plane in in which this electron moved. Yet there was no direct evidence for there being such a plane. Kinetic theory calculations achieved good results assuming spherical atoms. Even though it was clear that classical mechanics and classical electrodynamics had to break down somewhere, classical mechanics provided the only way to carry out the calculations required to specify the stationary states. If the electrons followed classically determined orbits, how could the possibility of collisions between them be excluded? Such collisions occur in the solar system, as the craters on the moon testify. Why did the flatness apparently not affect the behavior of atoms, especially as they formed molecules or crystals?

Just as bad for electrodynamics was the existence of energy quanta. Maxwell–Hertz theory predicted a continuous radiation process, but spectroscopy found a discrete one. This discreteness was confirmed by experiments such as those by Frank and Hertz on the energy loss of electrons going through mercury vapor, as described in Chapter 8 in the section "The Hydrogen Atom, Continued," and even more strongly by Millikan's on the photoelectric effect, in which Einstein's photoelectric equation was finally shown to be correct to within the narrow range of experimental errors finally achieved.[1] Einstein's finding that there must be a recoil when an atom emitted a quantum of spectral light made things even worse for him, though not for the vast majority of physicists, who simply ignored that effect. Einstein wrote in his paper reporting this work that

> The weakness of the theory lies on the one hand in the fact that it does not get us any closer to making the connection with wave theory.[2]

This was a confirmation that radiation had a dual nature, part wavelike and part particlelike. As early as 1909 Einstein had found that the energy fluctuations in blackbody radiation consisted of a part like that to be expected from the random interferences of waves and a part like that to be expected from fluctuations in the number of randomly moving particles in a given volume. So the dual nature of radiation was clearly on the record, if not appreciated, by the beginning of the 1920s.

Until the work of Louis de Broglie in the middle 1920s, however, the duality was one sided. No one had detected it on the particle side. There was no hint that the electrons in an atom also had a dual nature and that this duality might account for the nonappearance of the difficulties just mentioned that followed from a purely particlelike picture of electrons. In this chapter I shall discuss the attempts to reconcile the discontinuous nature of atomic transitions with the continuous nature of electrodynamic processes perceived by everyone except Einstein. The most important path of attack was through the study of dispersion relations, the relations describing the scattering of light by atoms and relating the optical properties of gases (and other transparent media) to the properties of the atoms composing the gas. Another major factor was the discovery of the Compton effect and its consequences for the dispersion theories.

De Broglie's work was independent of dispersion theory. It was based on the idea that the energy quanta radiated in an atomic transition had to be described in a way consistent with the theory of special relativity. Lorentz transformations mixed time and space coordinates with each other, so that a timelike quantity like $h\nu$ had to have spacelike parts associated with it. The result was de Broglie waves.

This chapter on the dissolving of the sharp distinctions between waves and particles brings us to the end of our story. The next step was the formulation of a real theory of quantum mechanics. One form, due to Werner Heisenberg, was a direct outcome of the study of dispersion relations; another form due to Erwin Schrödinger was developed from the de Broglie wave picture. The two forms looked very different, but their equivalence was soon established.

The invention, or discovery, of quantum mechanics represented the triumph of the concepts of the electron charge e, the quantum of action h, Boltzmann's constant k, and the speed of light c. The electron charge e measured the strength of the coupling between atoms and radiation, the coupling that led to the existence of both blackbody radiation and atomic spectra. The quantum of action h measured the size of the radiation quanta

needed in the description both of these phenomena. Boltzmann's constant k came in less directly, but the formula in which Planck's constant first appeared, his law of blackbody radiation, was also the first to contain k in a way that led to its experimental determination. All of Einstein's enormous contributions to quantum theory came through his study of thermodynamic equilibrium, in which k played an essential role (even though at first Einstein always wrote R/N rather than k). Finally, the speed of light c had first appeared in atomic physics in Sommerfeld's calculation of the relativistic fine structure of the Balmer states in hydrogen. It played more dramatic roles later when de Broglie used special relativity to associate wave properties with particles, and again when relativistic mechanics was necessary to describe the Compton effect, the effect that persuaded the world that Einstein had been right all along and that there were light quanta.

Notes

1. R. A. Millikan, *Physical Review* 7, 315 (1916).
2. A. Einstein, *Physikalische Zeitschrift* 18, 121 (1917); B. L. van der Waerden, *Sources of Quantum Mechanics* (New York: Dover, 1967), p. 76.

Radiation and Matter Revisited: The Compton Effect

X-ray scattering had played an important part in the development of atomic models. The scattering of hard X-rays from the lighter elements had given Barkla the means to estimate the number of electrons per atom and had led to the then surprising result that this number was only about half the atomic weight of the scattering element. The demonstration of X-ray diffraction by crystals by von Laue and the Braggs had demonstrated the atomic character of crystals beyond any shadow of a doubt. Moseley's studies of the X-ray spectra of the lighter elements had provided important confirmation of Bohr's model of the atom.

The simplest X-ray process conceivable was the scattering of an X-ray beam by free electrons. The analysis of this scattering was based on J. J. Thomson's 1906 theory which applied classical electrodynamics and classical mechanics in a straightforward way. The motion of the electron in the electric field of the X ray beam was found, and then the radiation produced by this accelerated motion was evaluated according to Maxwell–Hertz theory. The electron oscillated at the frequency of the incident X-ray beam and radiated the scattered X-rays of this same frequency. The only possibility for a difference in frequency between the scattered and the incident X-rays admitted by this theory would be a Doppler shift if the scattering electron had some linear motion in addition to its oscillation. A free electron target was, of course, almost impossible to obtain. If, however, the frequency of the incident X-rays

was much higher than the natural frequency of vibration of electrons in an atom, the electrons would respond to the incident rays as though they were almost free. The combination of hard X-rays and scatterers of low atomic weight offered an opportunity to use the theory in analyzing scattering data. The scattering of X-rays by atoms as in Bragg reflection from crystals also produced scattered radiation of the frequency of the incident beam.

Doubts about the exact equality of the frequency of the incident and scattered rays surfaced early in the 1920s. By 1921 A. H. Compton had begun to look at scattered X-rays using a Bragg spectrometer and had found that some of the scattered radiation had a slightly longer wavelength than did the incident radiation. He reported that

> in addition to scattered radiation there appeared in the secondary rays a type of fluorescent radiation, whose wavelength was nearly independent of the substance used as a radiator, depending only upon the wavelength of the incident rays and the angle at which the secondary rays are examined.[1]

The "scattered radiation" was the radiation of a wavelength indistinguishable from that of the incident ray; the "fluorescent radiation" was of a distinguishably longer wavelength, the terminology coming from the emission of light of longer wavelength by atoms illuminated by visible or ultraviolet light. In atomic fluorescence the wavelength of the fluorescent light depended on the kind of atom being illuminated.

Compton continued to work on his problem. Using the K X-rays from molybdenum as his source and graphite as the scattering material because it was light and therefore contained electrons that were not too tightly bound, he measured the wavelengths of scattered X-rays at $45°$, $90°$, and $135°$ (see Fig. 9.1). A large part of them showed no change in wavelength, but there was a component that showed a definite increase. By 1923 Compton had concluded that the explanation for his observations lay in assuming that the scattering at increased wavelength was the result of an energy- and momentum-conserving collision between a quantum and an almost free electron that was ejected from the atom in the scattering process. The scattering at unchanged wavelength ejected no electron; the atom as a whole recoiled and its great mass made negligible any change in wavelength due to this recoil. On this basis he derived his famous equation[2] for the increase $\Delta\lambda$ in wavelength of X-rays scattered by a free electron through an angle θ,

$$\Delta\lambda = \left(\frac{h}{mc}\right)(1 - \cos\theta).$$

The scale of this phenomenon was set by the *Compton wavelength*, h/mc, whose present value is 2.426×10^{-3} nanometers. The same formula was independently derived by Peter Debye at about the same time, perhaps stimulated by Compton's earliest report on the softening of X-rays on scattering from light materials.[3] Compton's announcement of this result met with considerable skepticism among X-ray workers. Other experimenters at

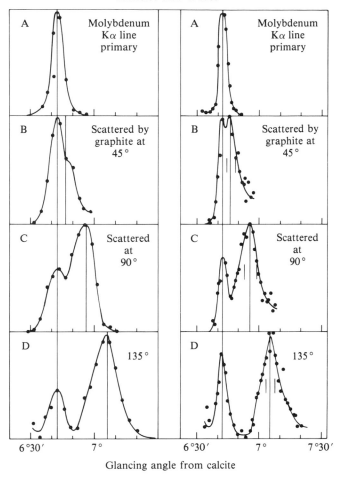

Fig. 9.1 Plots of the intensity of X-rays scattered from graphite at angles of 45°, 90°, and 135° as a function of their wavelength. The wavelength was determined by the angle of Bragg reflection from a calcite crystal used as a spectrometer. The two sets of plots correspond to two different widths of the incident X-ray beam.

first obtained different results when measuring the same effect. However, by the end of 1924 when the precision techniques needed for this work were more widely available, the agreement was all but universal that this equation accurately described the wavelength of scattered radiation. A new puzzle was presented to a physics community that had not believed in radiation quanta.

The reluctance to accept the idea of light quanta carrying energy and momentum was illustrated in a paper by Niels Bohr, H. A. Kramers, and John C. Slater, which was very important for a point of view it proposed but not for the conclusions drawn from it and which had this to say about the Compton effect:

As mentioned, Compton has reached a formal interpretation of this effect on the theory of light-quanta by assuming that the electron may take up a quantum of the incident light and simultaneously re-emit a light quantum in some other direction. By this process the electron acquires a velocity in a certain direction, which is determined, just as the frequency of the re-emitted light, by the laws of conservation of energy and momentum, an energy of hv and a momentum hv/c being ascribed to each light quantum. In contrast to this picture the scattering of the radiation by the electrons is, on our view, considered as a continuous phenomenon to which each of the illuminated electrons contributes through the emission of coherent secondary wavelets. Thereby the incident virtual radiation gives rise to a reaction from each electron, similar to that to be expected on the classical theory from an electron moving with a velocity coinciding with that of the above-mentioned source and performing forced oscillations under the influence of the radiation field. That in this case the virtual oscillator moves with a velocity different from that of the illuminated electrons themselves is certainly a feature strikingly unfamiliar to the classical conceptions. In view of the fundamental departures from the classical space-time description, involved in the very idea of virtual oscillators, it seems at the present state of science hardly justifiable to reject a formal interpretation as that under consideration as inadequate. On the contrary, such an interpretation seems unavoidable in order to account for the effects observed, the description of which involves the wave-concept of radiation in an essential way.[4]

As we shall describe more fully in the next section, this paper included a demotion of the laws of conservation of energy and momentum from dynamical to statistical status because this seemed necessary to reconcile the discrete jumps of the quantum atom with the continuous wave nature of the light emitted as a consequence of the jump. Even though this demotion did not stick, the idea of virtual oscillators referred to in the quotation led to important developments in the theory. These struggles to progress from a set of rules to a consistent theory were being undertaken in earnest. During them it was not at all clear what of the old would remain and what would be replaced. The wave nature of light as embodied in Maxwell's equations was the last thing that most scientists were willing to relinquish. For example, it was suggested that because the general theory of relativity predicted frequency shifts in light traveling through gravitational fields, the presence of strong gravitational fields in atoms might explain the Compton effect. The abstract of a paper to this effect stated that "the possibilities of the wave hypothesis are not yet exhausted."[5]

The Compton–Debye picture of the collisions between X-ray quanta and electrons was put to experimental test. It would be confirmed only if both particles, the scattered quantum and the recoil electron, resulting from a collision could be detected. Two methods were used, one by Compton and A. W. Simon and another by Walter Bothe and Geiger. Both showed a correlation between the scattered quantum and a recoil electron that could be explained only by the conservation of energy and momentum in each collision.

Compton and Simon used a cloud chamber. X-rays do not leave cloud chamber tracks as charged particles do, but their presence can be detected by electrons that they produce by either the photoelectric effect or Compton scattering. Using stereoscopic photography of electron tracks in a cloud chamber into which a well-defined X-ray beam was admitted, they observed the tracks of "R particles" (electrons) and concluded:

> It follows that the momentum acquired by an R particle is not merely that of the incident quantum, but is the vector difference between the momentum of the incident and that of the scattered quanta.[6]

The magnitude of the momentum of an "R particle" was determined by the curvature of the track it left in the cloud chamber, caused by an applied magnetic field. This magnitude was found to depend on the direction of the recoil in just the way required by the statement quoted. In later experiments they succeeded in observing the direction of some of the scattered X-ray quanta by finding the tracks of electrons ejected from gas molecules by these quanta.[7] They verified that this direction correlated with the direction of recoil of the target electron in the way required by the conservation of energy and momentum in the collision between the X-ray quantum and the electron. With this additional check, the conservation of energy and momentum in individual scattering events was established. Compton and Simon concluded the following:

1. Scattered X-rays proceed in directed quanta of radiant energy.
2. Energy and momentum are conserved during the interaction between radiation and individual electrons.

During this same time Bothe and Geiger performed what was probably the first coincidence-counting experiment to demonstrate the simultaneous production of a recoil electron and a scattered quantum. A diagram of their scattering chamber is shown in Fig. 9.2. A beam of X-rays came in at the top. A magnetic field swept out any electrons produced by this beam before it entered the scattering volume that was filled with hydrogen at atmospheric pressure. On the right was an ionization chamber to count electrons, and on the left, one to count X-ray quanta. The angular resolution of the counters was not very good. Counts from the two chambers were registered on a moving photographic film. In their last two runs of 260 minutes in duration Bothe and Geiger obtained coincidences of counts at a rate of one every 11 minutes. A picture of one of their coincidences is shown in Fig. 9.3. The vertical background stripes represent intervals of 1/1000 of a second, and the white lines represent the responses of the two chambers. On the basis of this work Bothe and Geiger decided:

> The experiments described are not consistent with Bohr's interpretation of the Compton effect. Indeed they do not prove strictly that energy and momentum conservation are valid in the elementary process, because for this it would be necessary to select out part of the scattered radiation and part

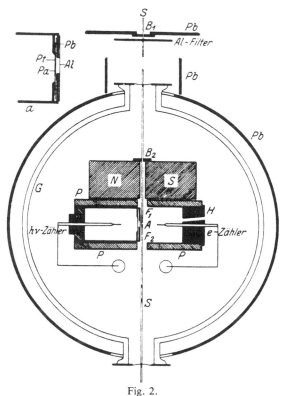

Fig. 2.
Versuchsanordnung im Horizontalschnitt. a: Stirneite des *h v*–Zahlers.

Fig. 9.2 The coincidence-counting apparatus of Bothe and Geiger. On the right is an ionization chamber to detect electrons, and on the left is one to detect electrons ejected by the scattered X-rays from a thin foil of heavy metal at the entrance to the chamber.

of the recoil rays corresponding according to the Compton theory by direction. One can, however, hardly doubt that even under these restricted condition, whose realization would be extremely difficult, that the coincidences would appear. It seems that until something further comes up one should retain the original picture of Compton and Debye. But not only with regard to the Compton effect but also in general it seems that the results obtained here raise very great difficulties, because Bohr, Kramers and Slater's interpretation of the Compton effect is very closely related to the statistical interpretation of energy and momentum conservation. One must therefore assume that the concept of the light quantum has a greater probability of reality than is assumed in this theory.[8]

Between them the experiments of Compton and Simon and those of Bothe and Geiger established the conservation laws for energy and momentum on an event-by-event basis. Compton and Simon had good angular resolution for individual events but poor time measurements; Bothe and Geiger had good

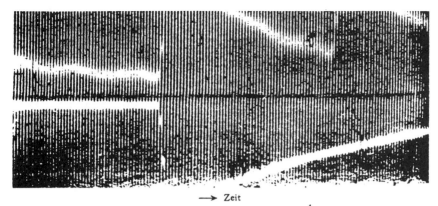

⟶ Zeit

Fig. 7. Beispiel einer Koinzidenz. Streifenabstand $^1/_{1000}$ Sekunde.
Oben *e*-Ausschlage, unten *h v*-Ausschlag.

Fig. 9.3 The photographic film running past the light spots of two electrometers registering the detection of counts in the two chambers of their apparatus shows the occurrence of a coincidence. The bars on the film are separated by about one millisecond.

simultaneity determinations but poor angular resolution. The statistical view of these laws could not be maintained after these results were established.

Even though the Compton effect resurrected the strict interpretation of the conservation laws, the idea of virtual oscillators suggested by John C. Slater and developed in the paper by Bohr, Kramers, and Slater survived, as will become apparent in the next sections.

Notes

1. A. H. Compton, *Bulletin of the National Research Council*, no. 20 (1922): 18: Washington, DC, meeting of the American Physical Society, April 1923, paper 49.

2. A. H. Compton, *Physical Review* **21**, 715 (1923).

3. P. Debye, *Physikalische Zeitschrift* **24**, 161 (1923).

4. N. Bohr, H. A. Kramers, and J. Slater, *Philosophical Magazine* **47**, 785 (1924), reprinted in B. L. van der Waerden, *Sources of Quantum Mechanics* (New York: Dover, 1967), p. 159.

5. C. Eckhart, *Physical Review* **24**, 591 (1924).

6. A. H. Compton and A. W. Simon, *Physical Review* **25**, 306 (1925).

7. A. H. Compton and A. W. Simon, *Physical Review* **26**, 289 (1925).

8. W. Bothe and H. Geiger, *Zeitschrift für Physik* **32**, 639 (1925).

Radiation and Matter Revisited: The Dispersion of Light

The dispersion of light on passing through transparent materials has been an important subject of research ever since Newton had reported:

> In a very dark chamber, at a round hole, about one third of an inch broad, made in the shut of a window, I placed a glass prism, whereby the beam of the sun's light, which came in at that hole, might be refracted upwards toward the opposite wall of the chamber, and there form a colour's image of the sun.[1]

The variation of the index of refraction with the frequency of the refracted light was responsible for the formation of the "colour'd image" reported by Newton. According to Paul Drude, writing in 1900,

> A theory which accounts well for the observed phenomena of dispersion may be obtained from the assumption that the smallest particles of a body (atoms or molecules) possess natural periods of vibration. These particles are set into more or less violent vibration according as their natural periods agree more or less closely with the periods of the light vibrations which fall upon the body.[2]

In a footnote to this work Drude pointed out that Rayleigh had discovered that Maxwell had used this basis for dispersion theory in setting a question on the Cambridge Mathematical Tripos Examinations for 1869, apparently not bothering to publish it in the usual way. The basis was used independently later by W. Sellmeier, H. von Helmholtz, and E. Ketteler. So the existence of a connection between dispersion and spectra had been established early, using classical theories of the systems involved. Things then changed for the worse. Bohr had stated in his first paper on the hydrogen atom that

> the passing of the systems between different stationary states cannot be treated on that basis [the ordinary mechanics].[3]

The quantum model of the atom put into question the previous theories of dispersion. All the investigation of the spectra of the elements, the fine structure, the Stark effect, and the Zeeman effect done after Bohr had concerned only the stationary states or spectral terms of the atoms involved; the actual process by which homogeneous electromagnetic radiation was emitted when the atom went from one stationary state to another was beyond anybody's comprehension. It was Einstein's introduction of transition probabilities, both spontaneous and induced, for the emission and absorption of light by atoms that first suggested how to treat transition processes and that provided a way for describing other processes in which atoms were coupled to radiation. Among these was the scattering of light by atoms.

The scattering of light by the atoms of a transparent material caused a slowing down of the propagation of a light wave through it, and the variation of this slowing down with frequency was the cause of the dispersion of light by a prism. In solids such as glass, the scattering had to be from bound atoms, and in gases, from the individual free atoms. The classical picture of this process described by Drude was simple: Each atom exposed to the incident light wave was made to oscillate at the frequency of that wave by the electric field of that wave and consequently emitted spherical outgoing waves of that same frequency. The spherical waves from all the uniformly distributed atoms of the material interfered with one another and the original wave in such a way that the original wave kept on going in its original direction with its original frequency but at a reduced speed c/n, with n the index of refraction of the material. The way that the index of refraction depended on the frequency of the incident light wave could be described by treating the atoms as containing electrons bound quasi-elastically. Then the amplitude of the wave scattered from an atom would depend on how close the incident frequency was to the natural frequency of the electrons. The greater the amplitudes of the scattered waves were, the greater the resulting slowing down of the wave would be. If the frequency of the incident wave was extremely close to a natural frequency of the atom, the incident light would be strongly absorbed, leading to *anomalous dispersion*. If the material was a gas, the natural frequencies could be identified with the spectral frequencies of the individual gas atoms or molecules.

This picture had no clear counterpart in Bohr atoms. Here there were no quasi-elastically bound electrons; indeed, here nothing moved with the frequency of the radiation emitted or absorbed by the atom. If Einstein's statement that "outgoing radiation in the form of spherical waves does not exist" had been taken seriously, the whole classical picture would have had to have been discarded. The people who tackled this problem did not pay any attention to this conclusion of Einstein's and assumed as a matter of course the existence of outgoing spherical waves. They did, however, adopt Einstein's suggestion of transition probabilities, made in the same paper, as the basis for studying the problem of how to fit the frequency dependence of the index of refraction into the Bohr model of atoms.

One of the first to publish on these problems was Rudolph Ladenburg. At the start of his paper entitled "The Quantum-Theoretical Interpretation of the Number of Dispersion Electrons" he wrote:

> According to classical electron theory, the absorption of isolated spectral lines is characterized above all by the number $\mathcal{N}$ of dispersion electrons per unit volume. . . . In quantum theory, on the other hand, the absorption is produced by a transition of the molecules from a state i to a state k, and the strength of the absorption is determined by the probability of such transitions $i \to k$. This result follows from Einstein's well-known considerations. . . . Einstein's above-mentioned theory now leads to an important relation between this probability factor and the probability of the spontaneous (reverse) transition from state k back to state i. It will be shown that it is this latter probability, multiplied by the ratio of the statistical weights of the two quantum states occurring in Einstein's relation, which takes the place of the dispersion constant $\mathcal{N}$, so that it is directly obtainable from absorption measurements.[4]

In other words, Ladenburg replaced the picture of atomic electrons being put into vibration by the incident wave with a picture in which the atom made an induced transition from its original state to a higher state and then made a spontaneous transition back down to the original state. An effect that classically had been proportional to the number of electrons put into vibration became proportional to the probability of a spontaneous transition back to the original atomic state. Ladenburg examined several sets of absorption and other data to verify his conclusion, including data from hydrogen and the alkalies. In the latter he found that the quantum value for $\mathcal{N}$ approximated the number of atoms per unit volume, so that the quantum decay time was approximately the same as that of a set of $\mathcal{N}$ oscillators radiating according to classical electrodynamics. He took this as a check on his theoretical work. In this paper Ladenburg did not study the frequency dependence of the index of refraction, having confined himself to absorption. But in 1923 he and F. Reiche included scattering and looked at the index of refraction also.[5]

A year later Kramers took up the subject and carried it considerably further. His paper was based on a remarkable one by Bohr, Kramers, and Slater containing just one equation in seventeen pages of text. What else made it remarkable was their proposal to give up the laws of conservation of energy and momentum for individual events and to reduce them to the status of statistical laws. This was as drastic as the similar treatment of the second law of thermodynamics at the hands of Boltzmann, but they deemed it necessary in order to match a discontinuous atomic process to a continuous electro-magnetic one:

> At the same time this author [Einstein] put forward his well-known theory of "light quanta," according to which radiation should not be propagated through space as continuous trains of waves in the classical theory of light, but as entities, each of which contains the energy hv, concentrated in a minute volume, where h is Planck's constant and v the quantity which in the classical picture is described as the number of waves passing in unit time. Although the great heuristic value of this hypothesis is shown by the confirmation of Einstein's predictions concerning the photoelectric phenomenon, still the theory of light quanta can obviously not be considered as a satisfactory solution of the problem of light propagation. This is clear even from the fact that the radiation "frequency" v appearing in the theory is defined by experiments on interference phenomena which apparently demand for their interpretation a wave constitution of light.[6]

Earlier, Slater had associated a set of "virtual" oscillators with an atom, these oscillators having the frequencies of the possible transitions the atom could make, and had thought of them as producing a "virtual" radiation field that kept atoms of the same kind in communication with one another.[7] The three authors used this idea:

> We will assume that the occurrence of transitions processes for the given atom itself, as well as for the other atoms with which it is in mutual communication, is connected with the mechanism by probability laws which are analogous to those which in Einstein's theory hold for the induced transitions between stationary states when illuminated by radiation. On the one hand, the transitions which in this theory are designated as spontaneous are, on our view, considered as induced by the virtual field of radiation

which is connected with the virtual harmonic oscillators conjugated with the atom itself. On the other hand, the induced transitions of Einstein's theory occur in consequence of the virtual radiation in the surrounding space due to other atoms. . . .

. . . As regards the occurrence of transitions, which is the essential feature of the quantum theory, we abandon . . . any attempt at a causal connexion between the transitions in distant atoms, and especially a direct application of the principles of conservation of energy and momentum, so characteristic for the classical theories.

This abandonment was made despite an explicit recognition, mentioned in the preceding section, that these conservation laws had been used by others to explain the Compton effect. The profound nature of the difficulties presented by the need to apply quantum ideas and classical theories to the same systems as the same time is demonstrated by these efforts.

Kramers pursued the ideas developed in this paper. He considered an atom in a stationary state not necessarily the normal or ground one so that transitions either upward or downward could be induced by an incident light wave. He then found an expression for the magnitude of the electric polarization of the atom produced by an incident light wave containing contributions from processes in which the atom is raised by the incident light to a state of higher energy from which it then decays spontaneously to the initial state, and from processes in which the atom decays spontaneously to states of lower energy and is returned to the initial state by a transition induced by the incident light. The contributions of the first kind could be correlated with what was expected classically, while those of the second kind were new. There was a characteristic difference in sign between the two kinds of contributions needed to make the theory consistent with the correspondence principle. Kramers summarized his result:

The reaction of the atom against the incident radiation can thus formally be compared with the action of a set of virtual harmonic oscillators inside the atom, conjugated with the different possible transitions to the stationary states.[8]

The old picture of an atom as containing a set of quasi-elastically bound electrons was replaced by one in which it contained a set of "virtual oscillators" whose frequencies were those of the spectral lines. There were many more virtual oscillators than there were electrons, and they were not necessarily all alike, so the richness of the spectrum and the small number of electrons were no longer contradictory. This gain in flexibility to fit the observed spectra was countered by the loss of a physical picture of what was responsible for the processes being studied.

The next year Kramers and Heisenberg extended the work just described. Kramers had considered the contributions to the scattered radiation emitted by an atom in a light beam coming from "virtual" processes in which the atom made transitions out of its initial state to states of either higher or lower energy and then returned to the initial state, emitting a quantum of the frequency of the incident light beam in the process. Kramers and Heisenberg found that the correspondence principle was better fulfilled if additional processes were included:

> Under the influence of irradiation with monochromatic light, an atom not only emits coherent spherical waves of the same frequency as that of the incident light: it also emits systems of incoherent spherical waves, whose frequencies can be represented as combinations of the incident frequency with other frequencies that correspond to possible transitions to other stationary states.[9]

This meant that the electrical moment of the atom possessed many frequencies of oscillation, not just the frequency of the incident light. Only the emitted light of the same frequency as the incident light could interfere with it and contribute to the dispersion or slowing down of the total wave; the incoherent radiation would be diffusely scattered light of different frequencies.

The electrical moment of the system, if the system were treated classically, could be expressed as a multiple Fourier series, the frequencies being the harmonics of the various frequencies of motion of the system. Kramers and Heisenberg showed how in the presence of an incident light beam this series could be transformed into one that contained the frequency of the incident light and the sums and differences of the frequencies of the multiply periodic system, together with the harmonics of these frequencies. Treating the same system quantum mechanically, they found they could write the quantum expression for the moment in a way similar to the classical one. It was a sum of terms, all proportional to the strength of the incident wave, with the sum extending over all frequencies that the atom could absorb and over all frequencies that the atom could emit spontaneously, that is, over frequencies that were the differences of atomic terms:

> In particular, we shall obtain, quite naturally, formulae which contain only the frequencies and amplitudes which are characteristic for the transitions, while all those symbols which refer to the mathematical theory of periodic systems will have disappeared.

The significance of this for Heisenberg's future introduction of his form of quantum mechanics cannot be exaggerated. That form would be based on the idea that only observable quantities should be represented in the mathematics describing physical process. A little further on Kramers and Heisenberg remarked:

> There is no point here in considering two stationary states separately. In the first place, one would not obtain in this way a quantity which could be related naturally to the reaction of the atom in a given stationary state. Secondly, one cannot ascribe a special meaning to the values of the amplitudes $\mathscr{C}$ themselves, in the stationary states. Instead it is a symbolic average of $\mathscr{C}$, taken over the region between two stationary states.

States were to be considered only in pairs between which transitions were, in principle at least, possible.

In order to discuss dispersion as distinct from scattering, Kramers and Heisenberg restricted their attention to those terms in the moment for which the emitted light had the same frequency as the incident light and the atom returned to its original state. Here in an especially simple case they obtained Kramers's previous result. They then proceeded to discuss scattering.

This was perhaps the last major paper on the subject of the interaction of light with atoms that preceded the invention or discovery of a consistent theory of quantum mechanics by Heisenberg later in 1925. This invention or discovery will be described briefly in the last chapter, but only to indicate how it grew out of the long development that I have been outlining up to now.

Almost simultaneously with Kramers and Heisenberg's paper two short notes were published, one by W. Kuhn and the other by W. Thomas.[10] They both derived what is now known as the Thomas–Kuhn sum rule. Kuhn's note was a little more explicit, and so I shall follow its presentation. He wrote an equation for the electric moment $\mathscr{P}$ of an atom in its ground state produced by an oscillating electric field $\mathscr{E}$ in the form of the classical one,

$$\frac{\mathscr{P}}{\mathscr{E}} = \frac{1}{4\pi^2}\frac{e^2}{m}\sum_{i=1}^{r}\frac{p_i}{v_i^2 - v^2}.$$

Here the coefficients p_i were called the "the number of dispersion electrons correlated with the transition from the ground state 0 to the state i." Kuhn then considered the limit as the incident frequency v became very large so that the dependence of the denominator on i could be ignored, yielding

$$\frac{\mathscr{P}}{\mathscr{E}} = -\frac{e^2}{4\pi^2 m}\frac{1}{v^2}\sum_{i=1}^{r}p_i.$$

On calculating the radiation from this moment for a single active electron atom he found agreement with the J. J. Thomson result if he put the sum equal to unity. But since the p_i had nothing to do with the frequency,

$$\sum_{i=1}^{r}p_i = 1$$

had to be true in general. The sum went over all the states to which a transition can be made, here always to higher ones.

If the atom was not initially in the ground state, Kuhn showed that a similar result was obtained, except that the p_i corresponding to downward transitions had to be negative. He mentioned that his p_i were related to the $\mathscr{N}$ of Ladenburg and hence to the Einstein transition probabilities. If the atom contained more than one active electron, then the sum should be the number of such electrons. Kuhn referred to work on the refraction of X-rays by B. R. V. Nardorf and by M. Siegbahn in which they had been able to measure the number of electrons in inner shells by this means and hoped that this would furnish experimental confirmation of his sum rule.

All this work on dispersion relations was based on the correspondence principle. There was no assurance that the results obtained were the only ones that would conform to that principle; they were not derived from a basic theory but, rather, were guesses that turned out to have been inspired. The work did lead to the proposal of a theory from which these dispersion relations could be derived.

Notes

1. I. Newton, *Opticks*, bk. I, pt. I, experiment 3 (New York: Dover, 1952).

2. P. Drude, *The Theory of Optics*, trans. C. R. Mann and R. A. Millikan (New York: Dover, 1959), p. 382.

3. N. Bohr, *Philosophical Magazine* **26**, 1 (1913).

4. R. Ladenburg, *Zeitschrift für Physik* **4**, 451 (1921). An English translation is included in B. L. van der Waerden, *Sources of Quantum Mechanics* (New York: Dover, 1967), p. 139.

5. R. Ladenburg and F. Reiche, *Naturwissenschaften* **11**, 584 (1923).

6. N. Bohr, H. A. Kramers, and J. Slater, *Philosophical Magazine* **47**, 785 (1924); in German in *Zeitschrift für Physik* **24**, 69 (1924); reprinted in van der Waerden, *Sources of Quantum Mechanics*, p. 159.

7. J. C. Slater, *Nature* **113**, 307 (1924).

8. H. A. Kramers, *Nature* **113**, 673, and **114**, 310 (1924).

9. H. A. Kramers and W. Heisenberg, *Zeitschrift für Physik* **31**, 681 (1925); an English translation is included in B. L. van der Waerden, *Sources of Quantum Mechanics*, p. 223.

10. W. Kuhn, *Zeitschrift für Physik* **33**, 408 (1925); W. Thomas, *Naturwissenschaften* **13**, 637 (1925).

Radiation and Matter According to Bose and Einstein

In 1924 S. N. Bose proposed a new derivation of the Planck radiation law that avoided any appeals to theorems depending on classical calculations such as the Wien displacement law or on the pre-Bohr correspondence principle in its most primitive form, requiring the blackbody spectrum to approach the Rayleigh form at low frequencies. From India Bose sent the manuscript of his paper to Einstein with the request that it be published if Einstein deemed it worthy.* Einstein translated it into German and it was published,[1] with an added note at the end saying that it presented an important forward step that could be extended to give the quantum theory of a monatomic gas, which Einstein proceeded to give in separate papers.[2]

* An interesting account of this incident is given in A. Pais, *Subtle Is the Lord* (New York: Oxford University Press, 1982), chap. 23.

As published, Bose's paper was only four pages long. He started by noting his dissatisfaction with the need to use classical concepts at any point in the derivation of such a purely quantum result. He then gave a brief but explicit account of how to specify the phase space of the radiation system, how to divide it into cells of volume h^3, how to count the distributions of phase points over these cells, and how to maximize the corresponding entropy, and he finally arrived at the Planck law complete with all the constants without recourse to classically derived results like the displacement law. Bose's paper reads like a recipe with no discussion of where the ingredients came from or why the suggested procedure worked.

Bose took Einstein's 1916 assignment of momentum as well as energy to a radiation quantum to its logical conclusion and treated quanta as relativistic point particles of zero rest mass. The momentum **p** of a quantum was then related to its energy E by

$$p_x^2 + p_y^2 + p_z^2 = \frac{E^2}{c^2} = \frac{h^2 v^2}{c^2}.$$

Because blackbody radiation fills the available space uniformly, the coordinates of the quantum could have with equal probability any values corresponding to points within the occupied volume V. An element of volume of the six-dimensional phase space of a quantum containing points with quantum energy between hv and $hv + h\, dv$ he then wrote as the product of the spatial volume V and an element of momentum space volume:

$$V \times 4\pi\left(\frac{hv}{c}\right)^2 \frac{h\, dv}{c} = 4\pi \frac{h^3 v^2}{c^3}\, V\, dv.$$

To apply the principles of quantum statistical mechanics as Planck formulated them on the basis of Boltzmann's ideas, Bose divided this phase space volume element into cells of phase volume h^3, so that the number of such cells in the element of phase volume just given was the preceding expression divided by h^3. To allow for the two polarization states of a quantum, he then doubled this number to get

$$8\pi V \frac{v^2\, dv}{c^3}$$

cells in the phase space for quanta with frequency in the range dv. This is exactly Rayleigh's expression for the number of modes of a transverse wave in this frequency range in a volume V, but it was obtained from the relativistic energy of a massless particle, not from considering standing waves in a cavity. This could have been regarded as associating wave properties with particles, but Bose did not suggest this.

Having defined his cells, Bose next had to choose the distributions of a set of quanta over these cells that were to be counted. Here, too, he broke new ground, not following the classical method of counting and yet not commenting on the novelty of his new method:

> Now it is simple to calculate the thermodynamic probability of a (macroscopically defined) state. Let N^s be the number of quanta belonging to the frequency range dv^s. In how many ways can these be distributed over the cells belonging to this frequency range? Let p_0^s be the number of vacant cells, p_1^s the number of those that contain one quantum, p_2^s the number of cell containing two quanta, etc. The number of possible distributions is then
>
> $$\frac{A^s!}{p_0^s! p_1^s! \cdots}, \qquad \text{where} \qquad A^s = \frac{8\pi v^2}{c^3} dv^s$$
>
> and where
>
> $$N_s = 0 \cdot p_0^s + 1 \cdot p_1^s + 2 \cdot p_3^s \cdots$$
>
> is the number of quanta belonging to dv^s.
>
> The probability of the state determined by the collection of p_r^s is obviously
>
> $$\prod_s \frac{A^s!}{p_0^s! p_1^s! \cdots}.$$

The number of possible distributions that Bose offered was the number of ways choosing p_0^s cells from the total number A^s to contain no quanta, p_1^s other cells to contain one quantum, and so forth. This clearly required no identification of the quanta that were contained in a cell. Having made this specification, his expression for the probability was "obvious," but the specification had not been so. Bose then followed the standard procedure to evaluate the entropy and to maximize it so as to determine the equilibrium macroscopic state. It turned out to be one with the Planck distribution. The novel feature of Bose's counting was the total denial of individual identities to the quanta. All that mattered was how many quanta there were in each cell. The interchange of quanta between cells was not considered to lead to any change in the macroscopic state as long as the number of quanta per cell was not affected. In classical statistical mechanics, interchanges of molecules of the same kind between cells had been regarded as leading to new states because molecules were taken to be identifiable particles; it was considered permissible to ask *which* of a set of molecules of a given kind were in particular cells. This had led to difficulties when mixtures of gases were considered, as J. W. Gibbs had noted. Interchanges of unlike molecules were clearly different from interchanges of similar ones. If one imagined the molecules becoming more and more alike, at what point did the difference between these two kinds of interchanges disappear? Einstein pointed out that Bose's method of counting avoided this difficulty completely.

If the counting method that Bose had used to arrive at Planck's law was new, how had Planck arrived there twenty-four years earlier? As I described in Chapter 4 in "Blackbody Radiation According to Planck; The Appearance of *h*," Planck had considered the energy distributed over N oscillators of a given frequency to be divided into P equal parts each of size ε purely for counting purposes, and he had not attributed any individuality to these separate parts. They were not considered to be "particles" in any sense, and they were not associated with the radiation field in any way, and so his counting was correct. Conversely, Bose treated Einstein's radiation quanta as particles and so should have given them identities according to classical ideas, but he did not. He made them anonymous. To an extent he was doing what Gibbs had done earlier when he introduced "generic phases" that were "not altered by the exchange of places between similar particles" in order to avoid the mixing paradox just mentioned and to make the entropy of a gas proportional to the number of molecules present.[3] Gibbs had not considered quantum effects at all.

Einstein appreciated what Bose had done for quanta and immediately proceeded to derelativize Bose's work and to apply his idea to a nonrelativistic gas of massive molecules. Even using Bose's notation but starting with the nonrelativistic expression for the energy of a monatomic molecule,

$$E = \frac{1}{2m} (p_x^2 + p_y^2 + p_z^2),$$

he followed Bose's steps and arrived at the expression for the entropy of the gas

$$S = -k \log \sum_{r,s} (p_r^s \log p_r^s).$$

Next Einstein found the equilibrium state of the gas by varying the p_r^s while keeping fixed the values of the total number of gas molecules and the total energy of the gas. The details of the result need not concern us. In his first paper he found deviations from ideal gas behavior, always in the direction of a reduction in the average energy of the molecules and the pressure at a given temperature. He remarked that here the entropy of the gas did approach zero as the temperature dropped toward zero, as required by Nernst's third law of thermodynamics, unlike the entropy calculated in the usual classical way.

In the last paragraph of this first paper Einstein described a "paradox" that he had not been able to overcome. His calculation could be carried through for a mixture of gases just as easily as for a pure gas, and the two gases acted entirely independently of each other, so that the entropy of the mixture was the sum of the entropies of the separate gases, which was fine. If the molecules were considered to become more and more alike, the pressure and the distribution of the molecules over the states of the mixture differed from those of a pure gas with the same number of molecules. This new version of Gibbs's paradox Einstein found "as good as impossible."

In the second of his three papers on this subject Einstein carried the discussion further. Ehrenfest and others had faulted the argument of the first paper, complaining that Einstein had not treated Bose's quanta and Einstein's molecules as statistically independent entities and also had not indicated that he had done this. Einstein admitted they were right and pointed out that if they were treated as statistically independent, the quanta would obey the Wien law rather than the Planck law and the molecules would behave like a classical ideal gas. His whole point was to apply to the gas molecules the statistics that worked for the quanta. He emphasized the difference between specifying a microscopic state *completely* (a) by giving the *number* of quanta or molecules in each phase space cell and (b) by identifying *which* quanta or molecules were in each cell. Only method (b) offered statistical independence, but method (b) did not yield the Planck law, which method (a) did, and so (b) was not physically correct.

The most remarkable result of this second paper was the prediction of what is now known as the Bose–Einstein condensation. Einstein stated in it in the form of the following theorem:

> According to the equation of state of an ideal gas developed here at every temperature there is a maximum density of the molecules that find themselves in motion. On exceeding this density the excess molecules drop out as stationary molecules ("condensed, without attractive forces"). Its noteworthiness lies in the fact that the "saturated ideal gas" represents not only the state of maximum density of moving molecules but also the density at which the gas is in thermodynamic equilibrium with the "condensed matter." A analog of "supersaturated vapor" does not exist for ideal gases.

In other words, the quantum gas at any temperature would, if the density were sufficiently large, form two phases. One phase would consist of a number of moving molecules, their motion providing the pressure of the gas, and the other would consist of the remaining number of molecules at rest in a condensed phase, a condensate. There would be no way to distinguish which molecules were in the gas and which were in the condensate, the molecules being indistinguishable from one another.

Toward the end of the second paper Einstein speculated that his theory might apply to the electrons in metals. Free electrons were supposed to account for both the electrical and thermal conductivity of metals, but they did not contribute to the specific heat. The specific heat was fully accounted for by the vibration of the atoms composing the lattice, as shown by the Dulong—Petit law. Einstein proposed here that the electrons were effectively a saturated ideal quantum gas and that most of the electrons were in the condensate. Being stationary, they would not contribute to the specific heat. The small number of uncondensed electrons would be enough to provide the electrical and thermal conductivity. This proposal was made at just about the time that the Pauli exclusion principle was being formulated. If one interpreted the state of an electron specified by its four quantum numbers as constituting a cell, then Pauli limited the population of this cell to either zero or unity. Bose and Einstein allowed up to the entire number of quanta or particles to be in any cell. It remained for Enrico Fermi and Paul Dirac to develop the statistics, now called *Fermi–Dirac statistics*, appropriate to particles obeying the exclusion principle, such as the electrons in metals. Not all particles do obey Pauli's principle, and so there are known examples of Bose–Einstein gases, helium being one.

Bose's argument depended on Einstein's attribution of particle properties to light waves, but it was purely a particle-based argument. When Einstein adopted Bose's statistics for a material gas, he also made a purely particle-based argument. He did not reverse the argument, as he had done so remarkably in his light quantum papers, and attribute wave properties to the molecules of his quantum gas. This was left for de Broglie to do.

Notes

1. S. N. Bose, *Zeitschrift für Physik* **26**, 178 (1924).

2. A. Einstein, *Preussische Akademie der Wissenschaften, Berlin Beriche*, pp. 261–67, 3–14, 18–25 (1925).

3. J. W. Gibbs, *Elementary Principles of Statistical Mechanics* (New York: Scribner, 1902).

Matter as Radiation According to de Broglie

The particle aspect of waves, both light waves and acoustical waves in solids, had been known, though not fully appreciated, for more than fiften years at the time I am now discussing. In 1922 the wave aspect of particles began to be discovered in stages as Louis de Broglie pondered the problem of Einstein's light quantum. It was outlined in three short notes published by the French Academy and appeared fully developed in de Broglie's doctoral dissertation submitted to the University of Paris in the fall of 1924. The work was published in a relatively short paper in English and then in extended form in French.[1] As is usually the case with radical ideas, de Broglie's thesis was at first not taken to apply to the real world but was accepted as evidence of scientific ability worthy of a degree, which was awarded. It was the communication of the thesis to Einstein by Paul Langevin, a member of the doctoral committee, and Einstein's positive reaction to it that led to both it results and its methods being regarded seriously.

De Broglie started out by thinking of blackbody radiation as a gas of quanta having finite but very small masses so that at laboratory temperatures they would move with highly relativistic speeds. He showed that such a gas would have a pressure given very nearly by the radiation pressure that follows from Maxwell theory. The small deviation was due to the small rest mass of the quanta. Applying the usual quantum statistics to these quanta, that is, by dividing the phase space of a quantum into cells of phase volume h^3, de Broglie derived the Wien spectrum for blackbody radiation. To get the Planck law, "I was obliged to suppose some kind of quanta aggregation." This conclusion was hardly new; Ehrenfest had arrived at in 1914.[2] Before going on to analyze this further de Broglie offered an argument based on relativity, which led to his famous association of a wave with a particle:

> Let us consider a moving body whose 'mass at rest" is m_0; it moves with regard to a given observer with velocity $v = \beta c$ ($\beta < 1$). In consequence of the principle of energy inertia, it must contain an internal energy equal to $m_0 c^2$. Moreover, the quantum relation suggests the ascription of this internal energy to a periodical phenomenon whose frequency is $v_0 = m_0 c^2/h$. For the fixed observer, the whole energy is $\dfrac{m_0 c^2}{\sqrt{1 - \beta^2}}$ and the corresponding frequency is $v = \dfrac{1}{h} \dfrac{m_0 c^2}{\sqrt{1 - \beta^2}}$.
>
> But if the fixed observer is looking at the internal periodical phenomenon, he will see its frequency lowered and equal to $v_1 = v_0\sqrt{1 - \beta^2}$, that is to say this phenomenon seems for him to vary as $\sin 2\pi v_1 t$. The frequency v_1 is widely different from the frequency v; but they are related by an important theorem which gives us the physical interpretation of v.
>
> Let us suppose that, at time 0, the moving body coincides in space with a wave whose frequency v has the value given above and which spreads with velocity $\dfrac{c}{\beta} = \dfrac{c^2}{v}$. This wave, however, cannot carry energy according to Einstein's ideas [because it travels with speed $> c$].
>
> Our theorem is the following:—"*If, at the beginning, the internal phenomenon of the moving body is in phase with the wave, this harmony of phase will always persist.*" In fact, at time t, the moving body is at a distance from the origin $x = vt$ and its internal phenomenon is proportional to $\sin 2\pi v_1 \dfrac{x}{v}$; at the same place the wave is given by
>
> $$\sin 2\pi v\left(t - \frac{\beta x}{c}\right) = \sin 2\pi v x\left(\frac{1}{v} - \frac{\beta}{c}\right).$$

The two sines will be equal; the harmony of phase will again occur if the following condition is realized:

$$v_1 = v(1 - \beta^2),$$

a condition clearly satisfied by the definitions of v and v_1.

This important result is implicitly contained in Lorentz's time transformation. If τ is the local time of an observer carried along with the moving body, he will define the internal phenomenon by the function $\sin 2\pi v_0 \tau$. According to Lorentz's transformation, the fixed observer must describe the same phenomenon by the function

$$\sin 2\pi v_0 \frac{1}{\sqrt{1 - \beta^2}}\left(t - \frac{\beta x}{c}\right)$$

which can be interpreted as the representation of a wave frequency v spreading along the x axis with velocity c/β.

> We are then inclined to admit that any moving body may be accompanied
> by a wave and that it is impossible to disjoint motion of body and propagation
> of wave.

This was as remarkable a joining of quantum and relativistic ideas as
Sommerfeld's when he calculated the hydrogen fine structure using the
relativistic mass in his quantum condition.

The wave velocity obtained in this way was the velocity with which a
point where the wave had a given phase traveled through space. For a pure
sinusoidal wave such as the preceding, this velocity has little physical meaning.
It does not correspond to the rate at which any physical attribute of the
wave travels. For example, a perfectly sinusoidal radio wave transmits no
information; it must have been traveling since time began, and it must
continue to travel until time ends. If it undergoes any change whatever, it will
deviate from being exactly sinusoidal, and this deviation will travel at a speed
less than that of the phase, a speed called the *group velocity*. De Broglie showed
that the particle with which his wave was associated traveled at the group
velocity of that wave an that this was just the right speed for the particle to
have the kinetic energy that was assumed in the beginning. There was a large
measure of self-consistency in his picture. What the wave "really" represented
was still mysterious, and even today its interpretation is a subject of debate
among those concerned with the deep meaning of quantum mechanics.

De Broglie then went on to compare the principle of least action, which
governs the motion of particles, with Fermat's principle, which governs the
propagation of light rays. This was not quite new; Hamilton had shown the
similarity in the forms of the equations for the motion of light rays and
particles in the first half of the previous century. His canonical equations of
motion were first derived in connection with geometrical optics and then found
to be applicable to particles. De Broglie, however, was the first to use this
commonality of descriptions to support an argument that particles and waves
were associated with each other in an indissoluble union.

Among other things de Broglie proposed that the electron in a stationary
state of a hydrogen atom had to have a wavelength such that an integral
number of them would fit around the Bohr orbit in order for the state to be
stationary. For circular orbits, this led directly to the angular momentum
quantization rule. De Broglie also considered the propagation of these waves
when the electron did not move with constant speed:

> Since an intimate connexion seems to exist between motion of bodies and
> propagation of waves, and since the rays of the phase wave may now be
> considered as the paths (the possible paths) of the energy quanta, we are
> inclined to give up the inertia principle and to say: "A moving body must
> always follow the same ray of its phase wave." In the continuous spreading
> of the wave, the form of the surfaces of equal phase will change continuously
> and the body will always follow the common perpendicular to two infinitely
> near surfaces.

When Fermat's principle is no longer valid for computing the ray path, the principle of least action is no longer valid for computing the body path. I think these ideas may be considered as a kind of synthesis of optics and dynamics.

He was right! When this synthesis was given mathematical form by Erwin Schrödinger two years later, the wave-mechanical form of quantum mechanics was born.

Notes

1. L. de Broglie, *Philosophical Magazine* **47**, 446 (1924); L. de Broglie, *Annales de Physique* **3**, 22 (1925).

2. P. Ehrenfest and H. Kammerlingh Onnes, *Proceedings of the Amsterdam Academy* **17**, 870 (1914).

CHAPTER 10

A New Order Emerges: Quantum Mechanics

All the groping for ways to fit the mass of knowledge that had accumulated concerning atoms, radiation, and their mutual interactions into a consistent and understandable framework began to bear fruit in 1925. With Werner Heisenberg's paper of that year, one new approach was proposed that promised to make the definition of the stationary states of atoms a matter of straightforward calculation following new but well-defined rules. These new rules of calculation were almost immediately shown by Max Born and Pascual Jordan to involve the mathematics of matrices. Then Born, Heisenberg, and Jordan joined forces in the "Dreimännerarbeit" to give a systematic account of the new mechanics. Its language was unfamiliar and the physical interpretation of the quantities appearing in it was not clear, but a coherent whole was emerging from the welter of rules that preceded it. The principal mathematical tool used in the theory was the algebra of matrices, a tool that physicists knew existed but that had not played a central role in classical theories.

Meanwhile, in another part of the field, Louis de Broglie's idea of associating waves with particles led to startling results. In 1926 Erwin Schrödinger found an equation for de Broglie's waves much as Maxwell had found the equations for Faraday's electromagnetic fields almost three quarters of a century earlier. The Schrödinger equation was a wave equation that differed from Maxwell's in important ways: It dealt unavoidably with complex functions, and it was of first order in the time while Maxwell's wave equations were of second order in both time and space variables. The solution of such equations was a standard task for theoretical physicists, and so Schrödinger's presented much less of a shock to the physics community, at least at first. The puzzle here was the meaning of the waves that satisfied the equation, not the mathematics used in finding the solution.

The contributions by Heisenberg and Schrödinger dealt with the quantum mechanics of an electron in an atom; they did not treat the radiation field. It was a third major contributor, Paul Dirac, who showed how to include radiation under the quantum umbrella.

The elaboration of quantum theory was a long and complicated process, and it continues even now. It has a vast literature of its own, and my only goal in this chapter is to show how the ideas whose development has been the subject of this book actually culminated in the emergence of a radically new and coherent theory of physics.

Heisenberg's Matrix Mechanics

The Rutherford–Bohr–Sommerfeld model of the atom was composed of point electrons moving around a compact nucleus in orbits that were subject to certain quantum rules. The coupling of these electrons to radiation was suspended most of the time, but there was always a probability that an atom would emit or absorb radiation while making a transition between two of its energy states. The similarity of the electron orbits to planetary orbits was taken for granted, and all these orbits involved some angular momentum so that the electrons would not collide with the nucleus. The experimental investigation of these orbits posed difficulties. In 1923 D. R. Hartree introduced his paper on X-ray reflection:

> The object of the present paper is . . . to enquire whether and to what extent it may be possible to obtain from X-ray reflexion evidence on the orientation of the orbits in the atom and on the relative phases of the electrons in the different orbits.[1]

Such evidence never was found in the sense implied here, and it gradually was recognized that such information simply could not be obtained experimentally by any known means. What did this unprecedented situation imply for the theory of atoms?

In Chapter 9, in "Radiation and Matter Revisited: Dispersion Relations," I discussed the paper by Kramers and Heisenberg, emphasizing the importance for the future of their statement:

> In particular, we shall obtain, quite naturally; formulae which contain only the frequencies and amplitudes which are characteristic for the transitions, while all those symbols which refer to the mathematical theory of periodic systems will have disappeared.[2]

At the end of their paper, no reference to electron orbits survived. On January 5, 1925, the *Zeitschrift für Physik* received this paper, and on July 29, 1925, the same journal received a paper by Heisenberg entitled "Quantum-Theoretical Re-interpretation of Kinematic and Mechanical Relations" whose abstract read in its entirety:

The present paper seeks to establish a basis for theoretical quantum mechanics founded exclusively upon relationships between quantities which in principle are observable.

The introduction to this paper expressed Heisenberg's new view that classical mechanics was in a sense irrelevant to atomic physics:

> It has become the practice to characterize this failure [to describe multielectron atoms] of the quantum-theoretical rules as a deviation from classical mechanics. This characterization has, however, little meaning when one realizes that the *Einstein–Bohr* frequency condition (which is valid in all cases) already represents such a complete departure from classical mechanics, or rather (using the viewpoint of wave theory) from the kinematics underlying this mechanics, that even for the simplest quantum-theoretical problems the validity of classical mechanics simply cannot be maintained. In this situation it seems sensible to discard all hope of observing hitherto unobservable quantities, such as the position and period of the electron, and to concede that the partial agreement of the quantum rules with experience is more or less fortuitous. Instead it seems more reasonable to try to establish a theoretical quantum mechanics, analogous to classical mechanics, but in which only relations between observable quantities occur. One can regard the frequency condition and the dispersion theory of *Kramers* together with with its extensions in recent papers as the most important first steps toward such a quantum-theoretical mechanics.[3]

Here Heisenberg concluded that it was impossible to associate an electron with a point in space within an atom or to determine the period of motion of such an electron. It was, however, possible to associate an electron with the radiation emitted or absorbed by an atom, and the emission of radiation depended on the velocity and acceleration of the emitting electron, so that these variables had to occur in the theory and have some sort of representation there. He set about finding a representation that depended only on observable characteristics of the system.

Heisenberg's guide in the search was provided by classical mechanics, but what he was looking for was something new, not derivable from classical physics. The correspondence principle imposed severe restrictions on what would constitute an acceptable representation, for classical physics was not to be discarded altogether, but just as it applied to atomic systems. His approach was very abstract, asking what kind of representation would be consonant with the Ritz combination principle and the Bohr–Einstein frequency condition. Heisenberg's starting point was the formula for the emission of radiation by a moving charge. Terms in this formula contained the product of the emitting particle's velocity v and acceleration dv/dt. Classically this product was evaluated by assigning numerical values to the velocity and the acceleration and then multiplying these numbers. If the position, and therefore also the velocity and acceleration, of an electron in an atom could not be determined experimentally, then the coordinates of the electron should not be represented by numerical functions. What else was there? Heisenberg stated the question succinctly:

> We may pose the question in its simplest form thus: If instead of a classical quantiy $x(t)$ we have a quantum-theoretical quantity, what quantum-theoretical quantity will appear in place of $x(t)^2$?[4]

What would replace the classical process of multiplying numbers together?

To begin to answer this question Heisenberg had to formulate how frequencies entered classical mechanics, because the nonclassical combination principal contributed to the difficulties in quantum theory. If $x(t)$ were a classical quantiy representing the coordinate of a particle executing some periodic motion with frequency $v(n) = \omega(n)/2\pi$ characterized by a label n, which might be the energy and in general might vary continuously, it could be expressed as a Fourier series, a sum of various terms depending on the time through functions such as

$$\cos \omega(n)t, \ \sin \omega(n)t, \ \cos 2\omega(n)t, \ \sin 2\omega(n)t, \ \cdots, \cos \alpha\omega(n)t, \ \sin \alpha\omega(n)t, \cdots$$

or, more compactly,

$$e^{i\omega(n)t}, \ e^{2i\omega(n)t}, \ \cdots, e^{\alpha i\omega(n)t}, \ \cdots \qquad \text{where } e^{i\theta} = \cos\theta + i\sin\theta$$

where α is always an integer. Writing these frequencies as $\omega(n, \alpha)$ Heisenberg pointed out that this series is characterized by the rule

$$v(n, \alpha) + v(n, \beta) = v(n, \alpha + \beta).$$

All these frequencies contained the same value of n specifying one definite state of motion.

Quantum frequencies did not combine in this way. They always referred to *two* states of the system, not just one. The Bohr–Einstein frequency condition said that

$$v(n, n - \alpha) = \frac{1}{h}[W(n) - W(n - \alpha)],$$

where *both* labels are integers. These frequencies then obeyed the rule that

$$v(n, n - \alpha) + v(n - \alpha, n - \alpha - \beta) = v(n, n - \alpha - \beta)$$

or

$$v(n - \beta, n - \alpha - \beta) + v(n, n - \beta) = v(n, n - \alpha - \beta).$$

Here the second index of one frequency had to match the first index of the other, a requirement that replaced the classical requirement that the index n in all frequencies had to be the same. This way of adding frequencies embodied the combination principle.

Next was the question of what replaced the Fourier series of classical mechanics. The Fourier series was the sum of frequency factors $e^{i\omega(n)\alpha t}$ for various values of α and a constant value of n, each one multiplied by a coefficient $a_\alpha(n)$, the product being labeled by the two indices n and α, and the summation extending over all values of the integer α. The result was a numerical function of the time and of the index n specifying the state of the classical motion. Heisenberg suggested a quantum analog of this that consisted of multiplying each quantum frequency factor $e^{i\omega(n,\,n-\alpha)t}$ by a coefficient $a(n, n - \alpha)$, giving him an array of symbols depending on two indices, n and $n - \alpha$. Here the two indices were of the same kind, each labeling a quantum state of the system, so there was no obvious way to combine them into a numerical function. He kept them as simply an array and sought a way to use it.

Now he could approach the question he had posed earlier: If the array $a(n, n - \alpha)\, e^{i\omega(n,\,n-\alpha)t}$ represented $x(t)$, what array represented $x(t)^2$? He called it $b(n, n - \beta)\, e^{i\omega(n,\,n-\beta)t}$. Clearly it should involve products of the elements of the array representing $x(t)$, and the quantum way of combining frequencies suggested that an appropriate form would be

$$b(n, n - \beta)\, e^{i\omega(n,\,n-\beta)t} = \sum_{\alpha=-\infty}^{\infty} a(n, n - \alpha)\, e^{i\omega(n,\,n-\alpha)t} a(n - \alpha, n - \beta)\, e^{i\omega(n-\alpha,\,n-\beta)t}.$$

Each term on the right side of this equation depended on the time in the same way as the left side did, and all the products of array elements consistent with this way of combining frequencies were included with equal weights.

The rule for multiplying an array by itself could be extended to multiplying two different arrays. This had the surprising result that the product of two arrays depended on the order of the two factors; the array representing xy was not necessarily the same as the one representing yx. The algebra of these arrays was *noncommutative*. This could lead to ambiguities; for example, in the formula for the radiation by a moving and accelerating charge, the product of the velocity and the acceleration occurred. In which order should the factors be written? Often there were reasons for choosing one order rather than the other, but at this stage no hard and fast rules could be given.

Having established a way to manipulate the arrays representing physical quantities, Heisenberg turned his attention to applying it to physics. The first thing to which he chose to apply it was a version of the dispersion relation that Kramers and he had developed earlier, a version that had also been arrived at independently by Thomas and by Kuhn[5]. This led to a *quantum condition* that read

$$h = 4\pi m \sum_{\alpha=0}^{\infty} \{|a(n, n + \alpha)|^2\omega(n, n + \alpha) - |a(n, n - \alpha)|^2\omega(n, n - \alpha)\}.$$

Like Kramer's dispersion relation, this contained terms connecting the state n with higher ones and others connecting it with lower terms, the two having the characteristic difference of sign. This equation ensured that Kramers's equation for the moment induced in an atom by a wave whose frequency was far above any of the transition frequencies would take on the classical form originally obtained by J. J. Thomson, a valuable check on consistency with the correspondence principle.

Next Heisenberg took as an example an anharmonic oscillator. A harmonic oscillator would not have sufficed because the classical equation of motion did not contain any products of physical quantities and it led to a Fourier series with just one term. The anharmonic oscillator was difficult to solve exactly, but it could be treated easily if it differed only slightly from a harmonic one so that perturbation theory could be used. The equation of motion to be solved was

$$\frac{d^2x}{dt^2} + \omega_0^2 x + \lambda x^2 = 0,$$

with λ a small number. This equation looked exactly like the classical equation of motion. Its quantum-theoretic aspect lay in the fact that x was an array, and the term λx^2 had to be evaluated by means of Heisenberg's new algebra rather than by the multiplication of simple numerical functions.

To apply perturbation theory, a solution of the equation was assumed to have the form of a power series in the small parameter λ. This was substituted into the equation of motion, and then the coefficients of various powers of λ were equated, yielding an infinite set of equations. To an accuracy proportional to λ, these could be solved subject to the preceding quantum condition to get

$$a^2(n, n-1) = \frac{(n + \text{const})h}{\pi m \omega_0}.$$

Now quantum systems had ground states n_0, the states of lowest energy. The above quantity had to vanish if $n = n_0$ because there could be no transition to a lower state $n_0 - 1$. On choosing the label for the ground state to be $n_0 = 0$, the constant had to vanish. Then having the value of $a(n, n-1)$ Heisenberg calculated the energy and obtained

$$W = \frac{(n + 1/2)h\omega_0}{2\pi}.$$

The extra $1/2$ in the numerator did not occur in the old way of quantizing the oscillator, but the steps between the energies of successive states was the same.

After some further examination of anharmonic oscillators and of rotators, Heisenberg concluded his paper with this hopeful but reserved statement:

> Whether a method to determine quantum-theoretical data using relations between observable quantities, such as that suggested here, can be regarded as satisfactory in principle, or whether this method after all represents far too rough an attack on the physical problem of constructing a theoretical quantum mechanics, obviously above all a very intricate problem, can be decided only by a more thorough mathematical investigation of the method which has been very superficially used here.

With this paper a *quantum mechanics* to replace classical mechanics had arrived. As presented in this first paper it was far from a neat, obviously self-consistent theory, but this deficiency was soon remedied. A detailed account of progress from here on is beyond the scope of this book, but very briefly, the sequence of developments was as follows.

September 27, 1925

In a paper received at the *Zeitschrift für Physik* Born and Jordan recognized Heisenberg's multiplication of arrays representing observables as matrix multiplication.[6] This incorporated Heisenberg's algebra of observables into an established field of mathematics and made the tools of linear algebra available to help in the work of reformulating and solving the quantum mechanical equations that had been proposed.

November 7, 1925

The *Proceedings of the Royal Society* received a paper from P. A. M. Dirac, in whose introduction he stated:

> In a recent paper Heisenberg puts forward a new theory, which suggests that is not the equations of classical mechanics that are in any way at fault, but that the mathematical operations by which physical results are deduced from them require modification. *All* the information supplied by the classical theory can thus be made use of in the new theory.[7]

He then went on to develop "quantum algebra" and to propose far-reaching analogies between the formulations of classical and quantum mechanics.

November 16, 1925

The *Zeitschrift für Physik* received a paper from Born, Heisenberg, and Jordan which they referred to as Part II of the previous papers they had written separately.[8] It extended the theory to systems of many degrees of freedom and established its connection with the mathematics of Hermitian forms, which made many mathematical tools available for solving physical problems.

January 17, 1926

The *Zeitschrift für Physik* received Pauli's paper "On the Hydrogen Spectrum from the Standpoint of the New Quantum Mechanics," whose abstract read:

> It is shown that the Balmer terms for an atom with a single electron arise correctly from the new quantum mechanics and that all the difficulties that arose in the previous theories from the need to exclude singular motions that occur especially in the case of crossed fields disappear in the new theory.[9]

This was the first important problem of more than one degree of freedom solved, and Heisenberg's goal of avoiding all the difficulties arising from attributing meaning to the classical motion of electrons in atoms was well on the way to being achieved. The method was restricted to hydrogen because it exploited the fact that the direction of the major axis of the classical orbit is constant. But this is not the case in the alkalies, for which the classical orbits are rosettes.

January 22, 1926

The *Proceedings of the Royal Society* received a second paper from Dirac, in which he extended his work on quantum algebra and formulated the problem of the hydrogen atom.[10] Although he did not get as far in solving it as Pauli had in his paper, he did make important contributions to the formulation of central-force problems.

We could say, then, that by January 1926 a *theory* of quantum mechanics existed that was self-contained and that had led to the actual solution of some simple but important problems that had not been solved in a satisfactory way by any earlier theory.

Notes

1. D. R. Hartree, *Philosophical Magazine* **46**, 1090 (1923).
2. H. A. Kramers and W. Heisenberg, *Zeitschrift für Physik* **31**, 681 (1925).
3. W. Heisenberg, *Zeitschrift für Physik* **35**, 59 (1925); an English translation is included in B. L. van der Waerden, trans. *Sources of Quantum Mechanics* (New York: Dover, 1967), p. 261. I have taken my quotations from this version.
4. Ibid., p. 263.
5. W. Thomas, *Zeitschrift für Physik* **33**, 408 (1925); W. Kuhn *Naturwissenschaften* **13**, 627 (1925).
6. M. Born and P. Jordan, *Zeitschrift für Physik* **34**, 558 (1925).
7. P. A. M. Dirac, *Proceedings of the Royal Society* A **109**, 642 (1925).
8. M. Born, W. Heisenberg, and P. Jordan, *Zeitschrift für Physik* **35**, 557 (1926).
9. W. Pauli, *Zeitschrift für Physik* **36**, 336 (1926).
10. P. A. M. Dirac, *Proceedings of the Royal Society* A **110** (561 (1926).

Amplification

Heisenberg was concerned with the mathematical representation of a physical variable $x(t)$ in both classical and quantum theories. In classical physics, dealing with a periodic system with one degree of freedom, one could write it as a Fourier series,

$$x(n, t) = \sum_{\alpha = -\infty}^{\infty} \mathscr{A}_\alpha(n) \exp[i\omega(n)^\alpha t].$$

The index α over which the sum was taken labeled the overtones of the fundamental frequency $\omega(n)$ of the nth state motion and was of a kind or "weight" distinct from the quantity n labeling the state of motion. At this point Heisenberg wrote:

> A similar combination of the corresponding quantum-theoretical quantities seems to be impossible in a unique manner and therefore not meaningful, in view of the equal weight of the variables n and $n - \alpha$. However, one may readily regard the ensemble of quantities $\mathscr{A}(n, n - \alpha) \exp[i\omega(n, n - \alpha i)$ as a representation of the quantity $x(t)$ and then attempt to answer the above question: how is the quantity $x(t)^2$ to be represented?

The classical Fourier series for x^2 could be written with coefficients $\mathscr{B}_\beta(n)$ replacing the coefficients $\mathscr{A}_\alpha(n)$, where

$$\mathscr{B}_\beta(n) \exp[i\omega(n)\beta t] = \sum_{\mu = -\infty}^{\infty} \mathscr{A}_\alpha(n)\mathscr{A}_{\beta - \alpha}(n) \exp[i\omega(n)(\alpha + \beta - \alpha)t].$$

Here the harmonics on the two sides of the equation were identified. Heisenberg needed the quantum analog of this. He suggested that the appropriate one was

$$\mathscr{B}(n, n - \beta) \exp[i\omega(n, n - \beta)t]$$
$$= \sum_{\alpha = -\infty}^{\infty} \mathscr{A}(n, n - \alpha)\mathscr{A}(n - \alpha, n - \alpha - \beta) \exp[i\omega(n \, n - \beta)t]$$

because this made the frequencies combine according to the Ritz combination principle rather than by identifying harmonics. This way of treating products had occurred in the dispersion relation paper by Kramers and Heisenberg without being assigned broad significance. The ensemble of $\mathscr{B}$'s was to represent x^2 in the same way that the ensemble of $\mathscr{A}$'s represented x.

It was this rule for finding the array representing the product of other arrays that was recognized by Born and Jordan as being the same as the rule for multiplying two matrices together. If two matrices A and B consisting of elements A_{ij} and B_{ij} respectively, are multiplied together, their product C has elements given by

$$C_{ij} = \sum_k A_{ik}B_{kj}$$

which is of the same form as Heisenberg's rule, the index summed over being the second index of the first factor and the first index of the second factor. Heisenberg's arrays representing physical observables in quantum mechanics were well-known mathematical structures, even if unfamiliar to most physicists, for which the rules of manipulation were firmly established. This made rapid application of the theory possible.

Schrödinger's Wave Mechanics

During the first six months of 1926 Erwin Schrödinger presented the world with a series of four "contributions" entitled "Quantization as an Eigenvalue Problem" that presented an approach to quantum theory that looked entirely different from Heisenberg's. In a paper interpolated between the second and third contributions he established the connection between his theory and Heisenberg's even though they had such distinct forms. In the introduction to this paper Schrödinger contrasted the two approaches:

> Above all, the departure from classical mechanics appears to develop in diametrically opposed directions in the two theories. Heisenberg replaced the classical continuous variables by a system of discrete numerical quantities (matrices) depending on a pair of integer indices that were determined by *algebraic* equations. The authors themselves characterized the theory as a "true discontinuum theory." Wave mechanics, on the other hand, signifies a precisely reverse step from classical mechanics to *continuum theory*. In place of a picture in which a finite number of variables are determined by a finite number of total differential equations, there appears a continuous field-like picture in configuration space governed by a single . . . partial differential equation.[2]

The first contribution announced the intention

> . . . to show in the simplest case of the (non-relativistic and unperturbed) hydrogen atom that the usual prescription for quantization can be replaced with another requirement in which no word of "integers" any longer appears. Rather the integers appear in the same natural way as the inter number of nodes in a vibrating string.

Schrödinger then proceeded to arrive at a wave equation using standard methods starting from a variational principle. This principle was constructed from the classical expression for the energy as a function of the coordinates and momenta in a completely novel way that he said was suggested by de Broglie's association of a wave with a particle. In effect Schrödinger was developing a wave equation for de Broglie's waves. After finding the equation he looked for solutions that represented standing waves rather than the traveling waves that de Broglie had considered. These waves he required to be finite at the center of the atom and to vanish or remain finite at infinitely large distances from it. There were many solutions satisfying these conditions. There was a discrete set that vanished at infinity for which the energy of the atom was negative; these described bound states of the electron and had as energies exactly the Balmer energies. For each energy there were just the correct number of independent solutions to account for the fine structure and the splittings of lines in electric fields, but not for the anomalous Zeeman effect. Then there was a continuous set for which the energy was positive; these corresponded to unbound electrons, to the classical hyperbolic orbits of scattering processes rather than the elliptical orbits of the bound states.

All of these solutions were counterparts of states that Pauli found in his solution of the Heisenberg equations for the hydrogen atom.

The Schrödinger approach to quantum mechanics had the great advantage of using more familiar mathematics and presenting a more visualizable solution, a wave function in configuration space, although the physical interpretation of this wave was obscure, especially for atoms with more than one electron for which the configuration space was of six or more dimensions. These difficulties of interpretation, however, were distinct from the difficulties of the pre–Heisenberg–Schrödinger times when even the formulation of quantum conditions demanded a mass of contradictory rules. A welcome feature of Schrödinger's wave function for the ground state of the hydrogen atom was that it was spherically symmetrical; there was no distinguished orbital plane, as in the previous semiclassical pictures.

Quantum mechanics now had two distinct but equivalent formulations as an at least potentially self-consistent theory. The quantum era had arrived, but it did not bring an end to controversy. The interpretation of the new quantum kinematics was, and still is, a source of both conceptual discussion and experimental exploration of its consequences in places where it contradicts deep-rooted intuitions of physicists and others, especially for questions of "physical reality" and causality. So far, in 1992, all the experimental tests have decided in favor of the quantum kinematics. More than that cannot be said.

Notes

1. The first contribution appeared in the *Annalen der Physik* **79**, 361 (1926); a translation of the series by J. F. Shearer and W. M. Deans was published by Blackie & Son, London, in 1928.

2. E. Schrödinger, *Annalen der Physik* **79**, 734 (1926).

Epilogue

The thirty years from 1895 to 1925 saw as big a change in our view of the world as had occurred since the time of Isaac Newton when classical mechanical acquired the role of organizer of nature. In George Gamow's phrase, it was "thirty years that shook physics."[1] Not that all was serene during the intervening centuries. There was many new phenomena discovered and many new ideas formulated that caused what at the time seemed major upsets and changes of the view of phenomena but that did not change what might be called the *kinematics* of physics very much. The wave theory of light displaced the corpuscular theory and brought new emphasis to the mysterious ether that filled the void that, in the view of many, had to be filled at any intellectual cost. The mechanical theory of heat displaced the caloric theory and brought with it new reality for the atoms and molecules proposed and opposed by many chemists and physicists. The electromagnetic field appeared and threatened to displace action at a distance everywhere, but it succeeded only for electromagnetic forces. Gravity remained safely Newtonian even while Coulomb's law became part of a local field in which disturbances propagated from point to point at a finite speed. At the end of this period, however, material bodies still had definite positions and definite velocities; their energies and angular momenta could take on continuous ranges of values; and they remained subject to Newton's laws of motion. Time flowed equably at the same rate for everyone everywhere. Euclid's geometry was the geometry of space. The kinematics, the means of describing phenomena, was that of Aristotle, even though the physics, the description of phenomena, was not.

During the time span covered by this book, the kinematics underwent two major changes. The first was that introduced by Einstein when he made space and time into related aspects of a larger concept, space-time. Lorentz had not contemplated any such change when he introduced his local time, and neither had FitzGerald or Lorentz when they introduced their contractions in the direction of motion. Similarly, Planck had not contemplated any change of kinematics when he had to introduce his energy quantum. Yet eventually the one led to the combination of space and time into a single larger concept of space-time with unsettling features such as the FitzGerald–Lorentz contraction and the nonexistence of an absolute definition of simultaneity, and the other led to a new way of manipulating the symbols representing the positions of bodies.

The change in kinematics introduced by Einstein in his 1905 paper on special relativity was sudden and drastic. It combined two previously utterly distinct concepts into a single larger one, but it did not alter the way in which the positions of bodies and the times of events were specified *in a given reference frame*. It did elevate c, the speed of light, to an exalted position in physics; it was no longer simply the greatest known speed but was the greatest attainable speed. Einstein did not spark a struggle to comprehend the meaning of the change he had introduced. Most of the results obtainable by his theory had been obtainable by the previous theories at the price of some assumptions, but not outlandish ones. By some it was rejected and by more it was accepted, however reluctantly, once it was understood, but its significance was not totally obscure to either group.

The change in kinematics introduced by Heisenberg in 1925 was, of course, also sudden and drastic. Before it appeared, the kinematics of a particle referred to a given reference frame were as always, though afterward they were very different. The background of this change and the obscurity of its meaning were, however, not at all like those of Einstein's change. The results to be obtained by a consistent use of Heisenberg's kinematics not only were in accord with experiment, but they also were previously obtainable only by using *ad hoc* and inconsistent sets of rules that gave no picture of the underlying processes. Th principal topic of this book has been the slow, difficult, imaginative, and frustrating struggle to achieve this new kinematics. One branch of the way began with a purely experimental result, the measurement of a number, the electron charge e. This was no ordinary number once J. J. Thomson had identified the electron as a universal constituent of atoms. Another branch began when Planck arrived at a formula that described with uncanny accuracy the spectrum of blackbody radiation. His attempt to relate this formula to the established concepts of physics led him to introduce energy quanta of a size proportional to the frequency in the processes of energy gain or loss by harmonic oscillators. His magical formula then contained two fixed numbers with physical meanings: One was the mysterious proportionality constant h giving the size of his energy quanta; the other was the constant k, Boltzmann's constant, giving the entropy of a state when the number of microscopic ways of achieving that state was specified. Knowledge of the latter from the measurement of the blackbody radiation spectrum permitted the evaluation of Avogadro's number and then of the electron charge. At the early date of 1905, then, all four of these constants had acquired the status of important players in our drama, although the starring role of h was not yet evident. There would be many failing attempts to write h out of the script.

In the end it was the finiteness of h led to the new quantum kinematics. The attempts to make the behavior of electrons in atoms fit into Aristotle's or even Einstein's kinematics always came to grief. As Heisenberg observed, experiment seemed unable to locate an electron at a point in an atom. There were irreducible elements in phase space whose size was determined by h, which meant that the phase point specifying the state of motion of an electron could not be specified more precisely than that it lay in one of these elements. This made meaningless the application of Newton's equation of motion in the accepted way when the motion was on a scale such that this imprecision had observable consequences for the subsequent motion. The wrenching feature of the quantum kinematics was that it altered the most basic relationship between numbers and physical variables, a relationship that was so ingrained that it was almost inconceivable that it should be questioned, to say nothing of its being changed. The consequences of this alteration are still unsettling, despite the agreement between these consequences and the results of the most detailed and subtle experiments designed to distinguish between the old and the new ways of relating the variables describing a system to the numbers giving the results of experiments on that system.

Note

1. George Gamow, *Thirty Years That Shook Physics* (Garden City, NY: Doubleday, 1966).

Suggested Readings

BIOGRAPHICAL

Niels Bohr

Niels Bohr's Times. By A. Pais. New York: Oxford University Press, 1991.

Niels Bohr, the Man, His Science, and the World They Changed. By Ruth E. Moore. New York: Knopf, 1966.

Ludwig Boltzmann

Ludwig Boltzmann: Man, Physicist, Philosopher. Ed. Engelbert Broda. Woodbridge, CT: Oxbow Press, 1983.

Marie and Pierre Curie

Pierre Curie, with *Autobiograhical Notes.* By Marie Curie. New York: Dover, 1963.

Madame Curie. By Eve Curie. Paris: Gallimard, 1938; New York: Doubleday, 1938.

Paul Ehrenfest

Paul Ehrenfest. By M. J. Klein. Amsterdam: North–Holland, 1970.

Albert Einstein

Subtle Is the Lord. By A. Pais. New York: Oxford University Press, 1982.

Albert Einstein: Philosopher-Scientist. Ed. P. A. Schilpp. New York: Library of Living Philosphers, Tudor Publishing, 1951.

Albert Einwtein, Creator and Rebel. By B. Hoffmann, with H. Dukas. New York: Viking, 1972.

H. A. Lorentz

H. A. Lorentz: Impressions of His Life and Work. Ed. G. L. Haas-Lorentz. Amsterdam: North–Holland, 1957.

R. A. Millikan

Autobiography. New York: Prentice-Hall, 1950.

Max Planck

Scientific Autobiography and Others Papers. New York: Philosophical Library, 1949; repr. Westport, CT: Greenwood Press, 1968.

Lord Rutherford

Rutherford and the Nature of the Atom. By E. N. Andrade, Garden City, NY: Doubleday, 1964.
Lord Rutherford. By Norman Feather. London: Priory Press, 1973.
Rutherford. By David Wilson. Cambridge, MA: MIT Press, 1983.
Rutherford, Recollections of the Cambridge Days. By Oliphant. London, Elsevier, 1972.

Frederick Soddy

The Self-splitting Atom. By Thaddeus J. Trenn. London: Taylor and Francis, 1977.

J. J. Thomson

Recollections and Reflections. New York: Macmillan, 1937.

ORIGINAL PAPERS

Various collections of original papers contain many of the most crucial, but not necessarily the clearest, contributions. These collections are in many cases more accessible than the journals in which the papers appeared, and in many cases the papers have been translated into English. I list a few such that I have found particularly useful.

Classical Scientific Papers: Physics. Introduction by Stephen Wright. New York: American Elsevier, 1965. Referred to as *Classical.*
Sources of Quantum Mechanics. By B. L. van der Waerden. New York: Dover, 1967. Referred to as *Sources.*
The Question of the Aomt. Ed. Mary Jo Nye. San Francisco: Tomash, 1984. Referred to as *Question.*

Investigations on the Theory of the Brownian Movement, Papers of A. Einstein. Ed. R. Fürth. Trans. A. D. Cowper. New York: Dover, 1956. Referred to as *Brownian*.

The Principle of Relativity, Papers of Lorentz, Einstein, Weyl, and Minkowski. New York: Dover, 1952. Referred to as *Relativity*.

Papers of some accessibility to the nonspecialist reader include the following, listed by subject:

Brownian motion

Einstein's original 1905 paper [*Annalen d. Physik* 17, 549 (1905)], Paper I in *Brownian*, can be read with profit even if the details as included in the mathematics are skipped. Einstein's paper *The Elementary Theory of the Brownian Motion* [*Zeitschrift für Elektrochemie* 14, 235 (1908)] is simpler. It is included in *Brownian*.

A large part of Perrin's review paper [*Annales de chemie et de physique*, September 1909] is accessible. It contains descriptions of the experimental work that established the validity of the theory. An English translation appears in *Question*.

Radioactivity as transmutation of chemical elements

The paper by Rutherford and Soddy [*Philosophical Magazine* 4, 370 (1902)], Paper I in *Classical*, is truly a classic and repays reading from beginning to end.

First quantization of the hydrogen atom

The first paper of the trilogy [*Philosophical Magazine* 26, 1 (1913); *Collected Works*, vol. 2, p. 159] that Bohr published in 1913 it difficult, not because of the technicality of the paper, but because it broke totally new ground and the ideas presented had not yet become clear in Bohr's mind. Bohr was one to see problems as parts of large conceptual structures, and so he could not write papers that were narrowly focused on a few particulars. Rutherford, Bohr's mentor in the early days, always thought his papers were too long and wordy.

Additional quantization rules

These papers tend to be highly technical. One of the least difficult is Sommerfeld's first paper [*Annalen der Physik* 51, 1 (1916)], as the subject had not yet developed enough to permit the exploitation of the technical methods used in celestial mechanics. This situation did not last long. I know of no English translation of this paper.

Quantum theory of radiation

The "one equation paper" by Bohr–Kramers–Slater [*Philosophical Magazine* 47. 785 (1924)] is a wonderful illustration of how difficult the formulation of a new theoretical picture is and how ambiguous the choices are that must be made among things, some of which must be given up. Included in *Sources*.

Relativity

Before 1905 Lorentz had arrived at many of the conclusions that Einstein later codified in his theory of relativity. A wonderful example of this is his introduction of the Lorentz contraction in his paper *Versuch einer Theorie der elektrischen und optischen Erscheinungen in bewegten Körpern*, Leiden, 1895. An English translation of the relevant sections appears in *Relativity*.

The first paper of Einstein's papers on special relativity [*Annalen d. Physik* 17, 891 (1905)], in which he discusses the consequences of his postulate that the speed of light is independent of the motion of its source in conjunction with the postulate that the laws of electrodynamics and optics are valid in all frames in which the equations of mechanics hold good, is quite readable. Later sections use Maxwell's equations explicitly and are therefore much less transparent to readers not familiar with them. An English translation appears in *Relativity*.

Minkowski's lecture (unpublished) on the geometrization of relativity is easier to read than are most papers on this subject. The ideas are straightforward but demand close attention if the argument is to be understood. An English translation is included in *Relativity*.

There is a vast secondary literature on all the optics I covered in the book, much of it broader in scope and less detailed in its treatment of the development of concepts and some on special subjects that is much more detailed. A few noteworthy books are listed here.

F. Cajori. *A History of Physics*. New York: Dover, 1962.

I. B. Cohen. *Revolution in Science*. Cambridge, MA: Harvard University Press, 1985.

George Gamow. *Thirty Years That Shook Physics*. Garden City, NY: Doubleday: 1966.

Alex Keller. *The Infancy of Atomic Physics*. Oxford: Clarendon Press, 1983.

Thomas Kuhn. *Black-body Theory and the Quantum Discontinuity*. New York: Oxford University Press, 1978.

A. Pais. *Inward Bound*. New York: Oxford University Press, 1986.

Emilio Segrè. *From X-Rays to Quarks*. San Franciso: Freeman, 1980.

E. T. Whitaker. *A History of the Theories of Aether & Electricity*. London: Thomas Nelson & Sons, 1951, 1953; New York: Dover, 1989.

Index